Atlas of Structural Geological Interpretation from Seismic Images

Atlas of Structural Geological Interpretation from Seismic Images

Edited by

Achyuta Ayan Misra
Reliance Industries Ltd
Navi Mumbai
India
Achyuta.Misra@ril.com; achyutaayan@gmail.com

Soumyajit Mukherjee
Indian Institute of Technology Bombay
Mumbai, Maharashtra
India
smukherjee@iitb.ac.in

WILEY Blackwell

This edition first published 2018

Registered Office(s)
John Wiley & Sons, Inc., 111 River Street, Hoboken, NJ 07030, USA
John Wiley & Sons Ltd, The Atrium, Southern Gate, Chichester, West Sussex, PO19 8SQ, UK

Editorial Office
9600 Garsington Road, Oxford, OX4 2DQ, UK

For details of our global editorial offices, customer services, and more information about Wiley products visit us at www.wiley.com.

Wiley also publishes its books in a variety of electronic formats and by print-on-demand. Some content that appears in standard print versions of this book may not be available in other formats.

Library of Congress Cataloging-in-Publication Data

Names: Misra, A. A. (Achyuta Ayan), editor. | Mukherjee, Soumyajit, editor.
Title: Atlas of structural geological interpretation from seismic images / edited by Achyuta Ayan Misra and Soumyajit Mukherjee.
Description: Hoboken, NJ : John Wiley & Sons, 2018. | Includes bibliographical references and index. |
Identifiers: LCCN 2017049885 (print) | LCCN 2017054608 (ebook) | ISBN 9781119158349 (pdf) | ISBN 9781119158356 (epub) | ISBN 9781119158325 (cloth)
Subjects: LCSH: Seismic tomography–Atlases.
Classification: LCC QE538.5 (ebook) | LCC QE538.5 .A85 2018 (print) | DDC 551.22–dc23
LC record available at https://lccn.loc.gov/2017049885

Cover Design: Wiley
Cover Image: Devonian, Permo-Triassic and Jurassic extensional faults at the transition between the Viking Graben (right hand side) and the Crawford-Skipper Basin on the East Shetland Platform (left hand side), UK North Sea.
Courtesy of PGS (3D GeoStreamero survey Q16-2013).

Set in 10/12pt Warnock by SPi Global, Pondicherry, India

10 9 8 7 6 5 4 3 2 1

In memory of my grandmother, Shankari Chakraborty, for her unconditional love and inspiration: Achyuta Ayan Misra

In loving memory of my grandmother, Namita Mukherjee, who left the world on 12 October 2016: Soumyajit Mukherjee

Contents

Contributors

Victor Alania
I. Javakhishvili State University
Institute of Geophysics
Tbilisi
GEORGIA

Achyuta Ayan Misra
Exploration Division
Reliance Industries Ltd
Mumbai
INDIA

A. Chabukiani
Institute of Geophysics
I. Javakhishvili State University
Tbilisi
GEORGIA
and
State Agency of Oil and Gas
Ministry of Energetics
Tbilisi
GEORGIA

Swagato Dasgupta
Exploration Division
Reliance Industries Ltd
Mumbai
INDIA

M. Diraison
Institut de Physique du Globe de Strasbourg (IPGS)
UMR 7516-CNRS
Université de Strasbourg/EOST
Strasbourg
FRANCE

O. Enukidze
Institute of Geophysics
I. Javakhishvili State University
Tbilisi
GEORGIA

Y. Géraud
Ecole Nationale Supérieure de Géologie
Université de Lorraine
UMR 7359-GéoRessources
Vandoeuvre-lès-Nancy cedex
FRANCE

K. Gogrichiani
State Agency of Oil and Gas
Ministry of Energetics
Tbilisi
GEORGIA

Pekka Heikkinen
Institute of Seismology
Department of Geoscience and Geography
University of Helsinki
Helsinki
FINLAND

Annakaisa Korja
Institute of Seismology
Department of Geoscience and Geography
University of Helsinki
Helsinki
FINLAND

Hemin A. Koyi
Department of Earth Sciences
Uppsala University
Uppsala
SWEDEN

Charlotte M. Krawczyk
Helmholtz Centre Potsdam
GFZ German Research Centre for Geosciences
GERMANY
and
TU Berlin
GERMANY

Marco Ligi
Institute of Marine Sciences – National Research Council (ISMAR-CNR)
Bologna
ITALY

Abhimanyu Maitra
Drilling and Completions
Reliance Industries Ltd
Mumbai
INDIA

Z. Mikeladze
State Agency of Oil and Gas
Ministry of Energetics
Tbilisi
GEORGIA

Neil C. Mitchell
School of Earth, Atmospheric and Environmental Sciences
The University of Manchester
Manchester
UK

Isabelle Moretti
GDF SUEZ
Courbevoie
FRANCE

Soumyajit Mukherjee
Department of Earth Sciences
Indian Institute of Technology Bombay
Mumbai
INDIA

Charles Naville
IFP
Division Géologie - Géochimie
Rueil-Malmaison
FRANCE

Michal Nemcok
EGI at University of Utah
Salt Lake City
UT
USA
and
EGI Laboratory at SAV
Bratislava
SLOVAKIA

Kaisa Nikkilä
Institute of Seismology
Department of Geoscience and Geography
University of Helsinki
Helsinki
FINLAND

Paolo Pace
Dipartimento di Ingnegneria e Geologia
Università degli Studi 'G. d'Annunzio' di Chieti-Pescara
Chieti
ITALY

present address
GEPlan Consulting, Petroleum Geosciences
Ferrara (FE)
ITALY

Stefano Patruno
Petroleum Geo-Services (PGS)
Weybridge
UNITED KINGDOM

Joachim Place
Formerly at Department of Earth Sciences
Uppsala University
Uppsala
SWEDEN

U. Polom
Leibniz Institute for Applied Geophysics
Hannover
GERMANY

William Reid
Petroleum Geo-Services (PGS)
Weybridge
UNITED KINGDOM

Neeraj Sinha
Exploration
Reliance Industries Ltd
Ghansoli
Mumbai
INDIA

G. Tatishvili
State Agency of Oil and Gas
Ministry of Energetics
Tbilisi
GEORGIA

Timo Tiira
Institute of Seismology
Department of Geoscience and Geography
University of Helsinki
Helsinki
FINLAND

Zhenyun Wu
Earthquake Administration of Jiangsu Province
Nanjing
Jiangsu
CHINA

Ashok Yadav
Exploration
Reliance Industries Ltd
Mumbai
INDIA

Hongwei Yin
School of Earth Science and Engineering
Nanjing University
Nanjing
Jiangsu
CHINA

Junzhang Zheng
PetroChina Research Institute of Petroleum
and Development
Beijing
CHINA

Preface

Analyses of geological structures from field exposure of rocks have been one of the important and intriguing disciplines in (applied) Earth Sciences. Since the 1990s, especially after the boom of three-dimensional seismic technologies, reflection seismic data in two and three dimensions have become the modern 'field' for geologists. A large number of exploration geoscientists analyze reflection seismic data to interpret sedimentary sequences and deformation structures to provide prospective leads for exploration in their respective industries. Today, seismic data are used extensively in industries such as hydrocarbon and mineral exploration, carbon dioxide sequestration, sea floor hazard surveys for cables and pipelines and so on. This book primarily deals with reflection seismic data in the hydrocarbon industry. The last atlas on interpretation of seismic images for structural information was in 2006 (Shaw *et al.*, 2006). That atlas presents seismic images in black and white print and covers compression-related structures. In keeping up with the growing usage of reflection seismic in the industry, interpretations of such seismic images have been included in PhD coursework, postgraduate and graduate level courses in several universities. There is a scarcity of text books, especially with an Indian background, in this field presumably due to the lack of an interface between universities and industries. We hope this book fills that requirement.

We have collated here seismic examples from North and South America, Africa, Europe, Asia and Australia. We received contributions from fifteen authors from various countries, including the United States of America, the United Kingdom, Germany, Sweden, Slovakia, India, Italy, China, Georgia and Finland.

Whom is this Atlas for?

This book is primarily designed for graduate students in Earth Sciences, researchers and new entrants in industry who are interested in seismic interpretation. The key geophysical- and geological concepts are explained in the first two chapters. Structural geological examples with illustrations and explanations that will guide students and industry professionals are then provided.

What this Atlas is for?

This atlas is intended to be a compilation of examples of different types of geological structure that can act as a quick guide to find analogous structures. Seismic expression of different geological structures: faults and folds have been covered extensively.

What this Atlas is *not* for?

This book is designed to provide examples for a few deformation structures and does *not* deal with the uncertainties or several alternate seismic interpretations. Readers can refer to the online Virtual Seismic Atlas project (www.seismicatlas.org) to see different interpretations of the same data. To be able to make multiple interpretations or to logically comprehend those made by others, a sound knowledge of seismic processes and interpretation is necessary. We hope that readers will pick up that from this book. A plausible seismic interpretation is normally reached in the industry by having more than one team interpreting the same data and then converging to a single logical explanation after integrating other geoscientific data such as from wells, outcrops, reservoirs and so on to reduce uncertainty. We have also included an uninterpreted seismic section for every interpreted section, so that readers may feel free to interpret the data as per their conceptualization.*

Seismic interpretation is evolving and is also subjective. Since there is no final word, different interpretations and suggestions from readers will be thankfully accepted.

Refer this book as:
Misra, A.A. and Mukherjee, S. (Eds), 2018. Atlas of Structural Geological Interpretation from Seismic Images. Wiley Blackwell, Oxford, pp. xxx. ISBN: 9781119158325

Reference

Shaw, J.H., Christopher, C.C. and Suppe, J. 2006. *Seismic Interpretation of Contractional Fault-Related Folds: An AAPG Seismic Atlas.* AAPG Studies in Geology #53, American Association of Petroleum Geologists, Tulsa, OK.

Refer individual chapters as:
Place, J., Diraison, M., Géraud, Y. and Koyi, H.A. 2018. Horst Inversion Within a Décollement Zone During Extension Upper Rhine Graben, France. In: Misra, A.A., Mukherjee, S. (Eds), 2018. Atlas of Structural Geological Interpretation from Seismic Images. Wiley Blackwell, Oxford, Chapter 7, 44–46. ISBN: 9781119158325

Achyuta Ayan Misra, Soumyajit Mukherjee
Achyuta.Misra@ril.com, achyutaayan@gmail.com
soumyajitm@gmail.com

*Readers are welcome to interpret the seismic images presented in this book in alternative ways and inform the contributing authors and the first co-editor, Dr Achyuta Ayan Misra, by e-mail.

Acknowledgements

The co-editors are grateful to the external reviewers of the book proposal: especially for reminding caution be adopted in producing such a book. A Research Sabbatical provided by IIT Bombay to Soumyajit Mukherjee (SM) for the year 2017 helped him to co-edit. The Wiley Blackwell team's (Ian Francis, Delia Sandford, Andrew Harrison, Sonali Melwani, Shummy Metilda and Kelvin Matthews) support has been fantastic. Achyuta Ayan Misra (AAM) thanks all the present and the past colleagues at Reliance Industries Ltd (RIL), especially in the Regional Geology Group, for numerous discussions, which improved his seismic interpretations. AAM owes special thanks to Neeraj Sinha (Head, Exploration), Suraj Sharma (Head, Geology) and Pramod Srivastava (Head, Geophysics). Discussions with Mainak Choudhuri, Sudipta Sinha, Sandipan Saha, Swagato Dasgupta, Krishna Murari, Abhimanyu Maitra, Arijit Ghosh, and Biplab Mukherjee (RIL), Debyoti Guha (Statoil), Michal Nemčok (EGI Utah) and many others were particularly helpful. Mr Gautam Sen, presently an independent consultant in oil and gas exploration, is thanked for providing internal review on the Preface, Chapters 1, 2 and 7. Both AAM and SM thank RIL for providing the seismic data, without which this atlas would be very difficult to prepare. They also thank the contributors and reviewers involved in this atlas. SM is grateful to IIT Bombay for providing a CPDA grant. AAM thanks his wife, Chaitali and daughters Ritocheta and Rishita for support, and sincerely thanks his parents for sharing responsibilities while he compiled this atlas.

1

Introduction to Seismic Data

Achyuta Ayan Misra and Ashok Yadav*

Reliance Industries Ltd, Mumbai, India
** Achyuta.Misra@ril.com; achyutaayan@gmail.com*

1.1 Seismic Reflection Method

Whenever a reflection seismic section is mentioned, something similar to Figure 1.1 comes to mind. The process leading to the generation of such a section is briefly discussed in this chapter. There is a large volume of literature detailing all the processes and their variations (e.g., Sheriff and Geldart, 1995; Yilmaz, 2001; Liner, 2004; Ashcroft, 2011; Herron and Latimer, 2011; Onajite, 2014). Only a brief account is given here to build the platform for the following chapters.

Seismic waves propagate through the Earth at velocities that depend on the acoustic impedance and density of the medium through which they travel. The acoustic impedance, Z, is expressed by (Liner, 2004):

$$Z = V\rho \tag{1.1}$$

where V is the seismic wave velocity and ρ is the rock density. If the rock varies in density in several directions, one can work with the "effective density" deduced in Mukherjee (2017, 2018, in press).

When a seismic wave propagating through the Earth encounters a boundary between two materials of different acoustic impedances, a part of the energy reflects off the interface while the remainder refracts through it. Seismic reflection prospecting involves generating seismic waves at the surface, which propagate into the subsurface, and capture the reflected wavefronts from the different interfaces while propagating. At each layer most of the energy is transmitted or refracted and a part reflects back (Sheriff and Geldart, 1995; Yilmaz, 2001; Liner, 2004; Ashcroft, 2011; Herron and Latimer, 2011; Onajite, 2014).

To generate the disturbance, a 'shot' or a vibration is made on the sea surface or on Earth's surface in onland. As the wave propagates into subsurface, each layer reflects the wave at multiple incidence angles and these reflected waves are measured at the surface by receivers, which are hydrophones on water and geophones on land (Figure 1.2). The distance between the source and the receiver is termed the 'offset'. The data from receivers near the source are called 'near offset' and those far away as 'far offset'. The near receivers receive the reflected signal quicker than those further away from the source, so the response of the same boundary will appear progressively later (Figure 1.3).

There are two types of seismic waves: (i) P-waves (longitudinal /compressional /body waves), where the particle motion is parallel to the direction of wave propagation, and (ii) S-waves (shear /transverse waves), where particles move perpendicular to the wave propagation direction. P-waves convert into S-waves and vice versa when they transmit or reflect across a boundary, where there is a phase change i.e. solid to liquid/gas or liquid/gas to solid. Pore spaces have liquid/gas and thus this conversion is very common. Both P and S waves follow Snell's law of reflection and refraction (Yilmaz, 2001). The angle of incidence equals the angle of reflection; the incident ray, the reflected ray, and the normal to the plane of incidence are co-planar (Figure 1.4). The refracted seismic waves also follow Snell's law, which states:

$$\frac{Sin\,\theta_1}{V_{PR}} = \frac{Sin\,\theta_2}{V_{PT}} = \frac{Sin\,\phi_1}{V_{SR}} = \frac{Sin\,\phi_2}{V_{ST}} \tag{1.2}$$

where θ_1 is the angle of incidence, V_{PR} the velocity of reflected P-wave, θ_2 the angle of transmitted P-wave, V_{PT} the velocity of transmitted P-wave, φ_1 the angle of the reflected S-wave, V_{SR} the velocity of the reflected S-(converted from P-) wave, φ_2 the angle of the transmitted S-wave and V_{ST} the velocity of the transmitted S-wave.

1.2 Seismic Data Acquisition

Earth's interior can be imaged by reflection seismic data much like remote sensing satellites image the Earth surface. Rock layers, for example sands and shales, differ

Atlas of Structural Geological Interpretation from Seismic Images, First Edition. Edited by Achyuta Ayan Misra and Soumyajit Mukherjee.

Figure 1.1 A seismic section showing reflections from sedimentary boundaries. Seismic data courtesy Reliance Industries Ltd. Reproduced with permission from the Directorate General of Hydrocarbons (DGH), India.

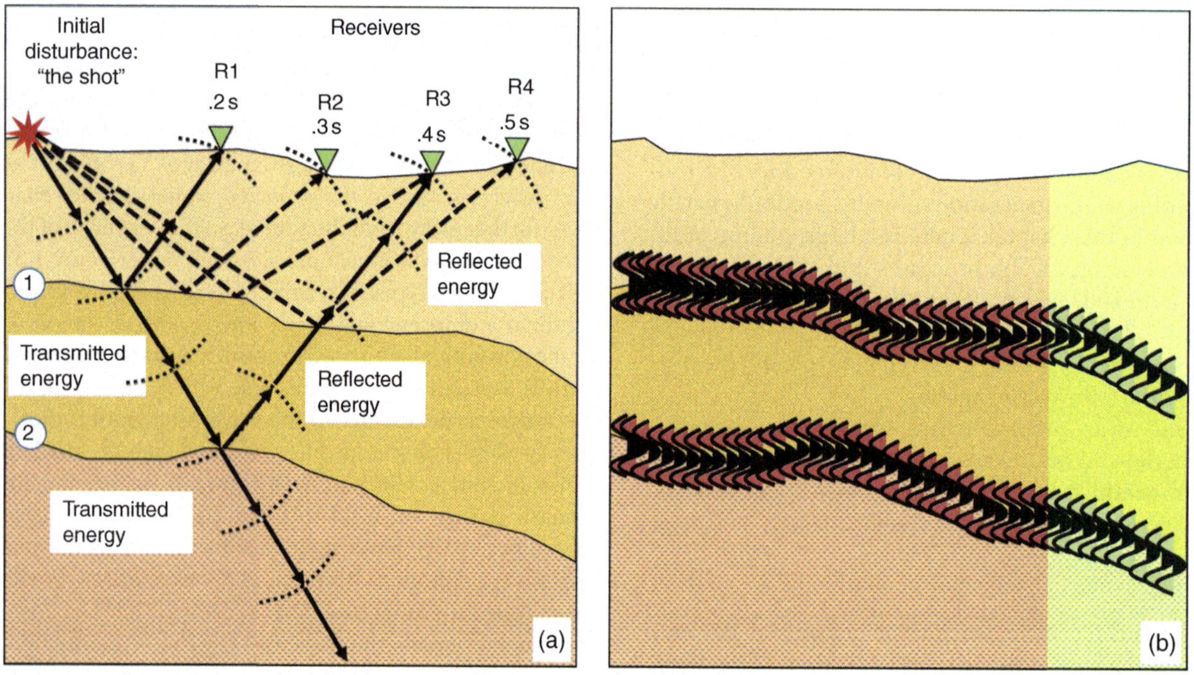

Figure 1.2 (a) Schematic diagram showing layout of transmitted and reflected energy from the shot point and to the individual receivers. At every boundary, a part of the energy is transmitted and another part is reflected. The latter reach the receivers at the surface. Note there are two reflectors, numbered (1) and (2). R_n: receiver number; (b) shows the corresponding simplified seismic wriggle traces.

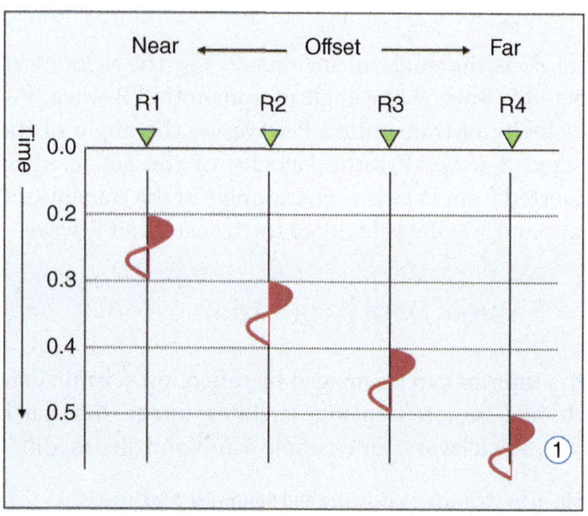

Figure 1.3 Schematic seismic response of reflector (1) on the receivers as shown in Figure 1.1.

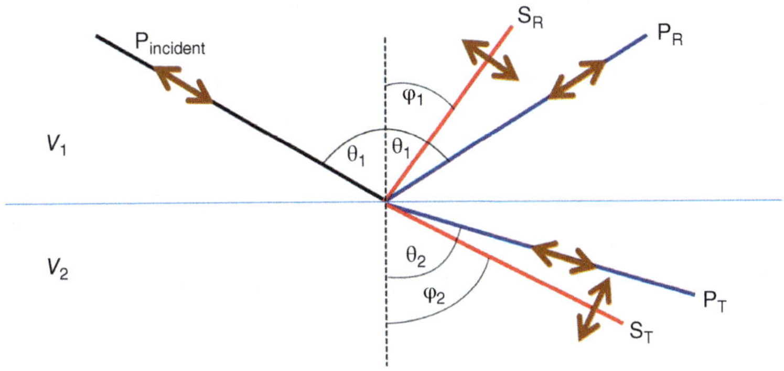

Figure 1.4 Schematic diagram showing mode conversion of incident P wave in P- and S-waves at a boundary of two lithologies with different velocities V_1 and V_2. The refracted (transmitted) waves follow Snell's law. S_R: Reflected mode converted S-wave; P_R: reflected P-wave; S_T: Transmitted mode converted S-wave; P_T: transmitted P-wave.

in density. The acoustic impedance (Equation 1.1) to seismic wave velocity passing through the layers thus differs and it is reflections of the wave that are imaged. These layers appear as reflections in a seismic record and are *interpreted* to reconstruct the geological architecture. This reconstruction is challenging and takes time. To understand the data properly, the data acquiring process has to to be understood.

Seismic data acquisition parameters are guided as per requirements. For example, if a large NE trending anticline of 100 km^2 area at 3000 m below sea level is to be imaged, the spacing between individual lines in a 2D seismic survey should be 10–20 km. On the other hand, when a 50–100 m long reservoir fault demands mapping, a high resolution 3D volume will be acquired (see Section 1.3.4 for a discussion on resolution). The trend of the structure will determine the orientation (azimuth) of the receiver lines, for example the NE trending structure will need NW oriented receiver lines. Again, for crustal architecture studies (Misra *et al.*, 2015, 2016) a larger record length (Box 1.1), a high-energy source, a larger offset etc. yield a good image.

Three primary environments can be envisaged for seismic data acquisition: (i) land, (ii) marine, and (iii) transition zones.

1.2.1 Land Acquisition

Seismic land acquisition involves acquiring seismic data in any part of the Earth exposed aerially: deserts, forests, swamps, mountains, inhabited areas, tundra, permafrost etc. Acquiring data may involve logistic problems and labour. 2D data are commonly acquired in logistically difficult areas. 3D seismic volumes are normally acquired before firming up *a certain play or some particular prospect(s)* prior to drilling in hydrocarbon exploration.

The energy sources commonly used on land for acquisition of seismic data are (Dobrin and Savit, 1988; Onajite, 2014; Schlumberger, 2016):

- Explosives: the most commonly used explosives are dynamite and primacord. The source can be placed below the weathered rock layer. This reduces the noise of the topmost layer. However, it often requires drilling of expensive 3–100 m deep shot holes and is difficult to handle. It can also have a significant cultural and environmental impact.
- Vibrators: one of the most commonly used sources presently. They have low cultural and environmental influences, high signal-to-noise ratio and it does not involve drilling holes. However, the topmost soil layer can cause noise and, thus, recording and processing can complicate.
- Gas or air guns: these can be used in rough terrains. Shot holes are not required and the guns are not expensive. However, energy penetration is limited and noise is high.
- Weight drop: these are also relatively inexpensive and holes need not be drilled. Unfortunately, different sources in an array cannot be synchronized and noise is high.

The receivers, i.e. geophones are the mechanical/analogue devices that convert mechanical ground movement into electrical energy. Single geophones are often connected together to form geophone arrays, which maximize signal-to-noise ratio. Recording stations gather the response from the receivers and perform basic processing to render the data readable. The recording process further controls parameters such as record length, sample rate etc.

1.2.2 Marine Acquisition

This involves acquiring seismic data in shallow (~10–40 m) to ultra-deep (>1500 m) water by specialized vessels with arrays of hydrophones (receivers) towed by one or more cables, called streamers. A seismic source comprising an air gun or an array of air guns is also towed together with the geophones.

Two types of energy sources are used in marine seismic surveys (Games and Wakefield, 2014; Onajite, 2014; Schlumberger, 2016):

Box 1.1 Important terms in seismic acquisition and processing.

1) **Acoustic impedance (AI):** a product of density and velocity. The greater the AI, the stronger the reflection. Acoustic impedance, $Z = V\rho$.
2) **Amplitude:** reflection strength.
3) **Fold:** not the fold in structural geology! For the one in structural geology, see Chapter 4. The geophysical fold is the numbers of traces in a CMP gather, so if data has 10 traces, the "foldage" of the data is 10 and when it has 60 traces, the foldage is 60. The more the foldage of the data, the clearer will the imaging, due to a higher signal-to-noise ratio.
4) **Migration:** restoring a dipping reflector to its correct subsurface position.
5) **Minimum phase wavelet:** front loaded energy, that is at time zero minimum energy and elsewhere maximum.
6) **Offset**: Distance between the source and the receiver.
7) **Receiver**: **geophone** (on land) or **hydrophone** (offshore); devices to measure ground movement or sound from the shot.
8) **Record length**: predetermined time after which the receiver measures the reflected seismic waves. A larger record length indicates greater depth of imaging.
9) **Reflection coefficient:** type and size of acoustic impedance change.
10) **Reflections**: acoustic waves reflected from an interface of contrasting lithologies. If the boundaries are gradational, the reflections may be chaotic or comprise a number of reflections, depending on the frequency of the data.
11) **SEGY**: or SEG-Y, is a popular format for storing seismic data. Controlled by the Society of Exploration Geophysicists. Associated file extensions: .segy and .sgy.
12) **Seismic reflector:** boundary across which the competence changes. Also called **acoustic-impedance boundary.**
13) **Shot**: initial disturbance/sounding/explosion; generated by explosives or seismic vibrators on land and by pressurized-air guns offshore.
14) **Shot point**: geographic location of the shot, measured precisely by positioning systems.
15) **Source pulse or wavelet**: Resulting sound wave from the shot. This wavelet is commonly used for zero-phasing the seismic data (see "Zero-phase wavelet").
16) **Streamer feathering**: only in marine acquisition; the deviation of the streamer array away from the linear towing direction. This may happen due to water currents or change in towing direction.
17) **Streamer length**: length of the cable on which the hydrophones are attached. The length of the streamers depends on the depth of the objective. Typically, the streamer lengths equal the target depth.
18) **Trace:** stream of reflections recorded by geophone.
19) **Two-way time:** time from shot to recording. It is the time required for the wave to reach the reflector from the source and the reflected wave to reach the receiver.
20) **Zero-phase wavelet:** the wavelet has maximum energy at time zero.

- Water guns: these use a piston that pushes water very fast to generate a bubble, which collapses to produce a large vibration/seismic energy. Water guns require specialized processing due to pressure variations that occur before the main pressure pulse.
- Air guns: these are the most commonly used sources presently. Air is kept at very high pressure within chambers. The compressed air is released causing a large vibration and forming an air bubble. The bubble continues to oscillate and generates a series of smaller decaying vibrations. A number of air guns are used in an array to minimize the size of the bubble and to make the shape of the bubble nonspherical. The oscillating air bubble minimizes the decaying vibration.

Receivers are (array of) hydrophones attached inside a jacketed streamer filled with a liquid lighter than water. A combination of weights balances the streamer at a predetermined depth below the water surface. The length of a streamer, depth below the water surface and spacing between two consecutive streamers are critical as it constrains the depth of the target/objective that can be imaged. The streamer length is usually the same as the depth of the objective.

Many times ocean bottom cable surveys are used as a cost-effective method to obtain repeated (of the order of few months) or 4D seismic data over producing fields. 4D surveys observe the depletion patterns of reservoirs and identify target areas. Permanently deployed cables used to get numerous 4D seismic data are known as "life of field seismic" (Barley and Summers, 2007).

1.2.3 Transition Zones

There are varied types of landforms like shores/beaches, deltas and estuaries, reefs, river mouths, swamps, marshlands etc. In such zones the water depth prevents seismic land acquisition techniques. Conventional seismic acquisition vessels cannot work. A mix of both land and marine techniques are used; for the latter, smaller boats than those used in open oceans are employed.

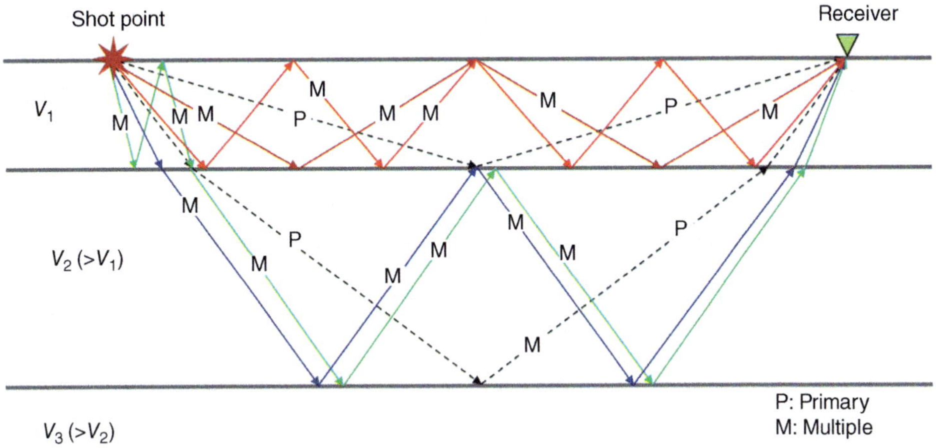

Figure 1.5 Formation of multiples in seismic acquisition for one shot point and receiver set. Originated receiver set. Dotted lines: primaries; solid lines: multiples.

1.3 *Processing* Seismic *Data*

The acquired seismic data have to be processed because they contain noise related to acquisition problems. Moreover, the geophysical data have to be modified to represent the geology of the subsurface. Seismic data processing is neither a one-step nor an entirely computer-based automatic process. Rather, it requires a good understanding of the data, especially their limitations, and exploration or production objectives. A good resource for studying processing is Yilmaz (2001). Only key methods/features in seismic data processing, namely (i) noise, (ii) gathers, (iii) migration and stacking and (iv) resolution, are mentioned here so that the geologist understand the data before interpretation.

1.3.1 Noise

The signal part of the data is termed as *primaries*. The most important processing workflow deals with the removal of seismic noise, because noise is expected in all the seismic data. Any amplitude or reflection event not generated from a geological boundary is termed *noise* in the seismic data. Noise is inevitable in seismic data, as numerous sources of noise exist even in the most optimal acquisition conditions. The most basic noise may result from faulty equipment; human error may turn out to be the most erratic one. The wind blowing, flowing water during marine acquisition, cable vibration, lightning or even a dog bark can induce noise! Seismic noise may also arise when the lithological parameters, such as density and velocity, are presumed incorrectly (Tarantola, 1986). Sometimes, a high amplitude reflection – commonly the sea bed – may appear repeatedly in the seismic record below it; these are referred to as *multiples* – reflected energies from other interfaces, as in

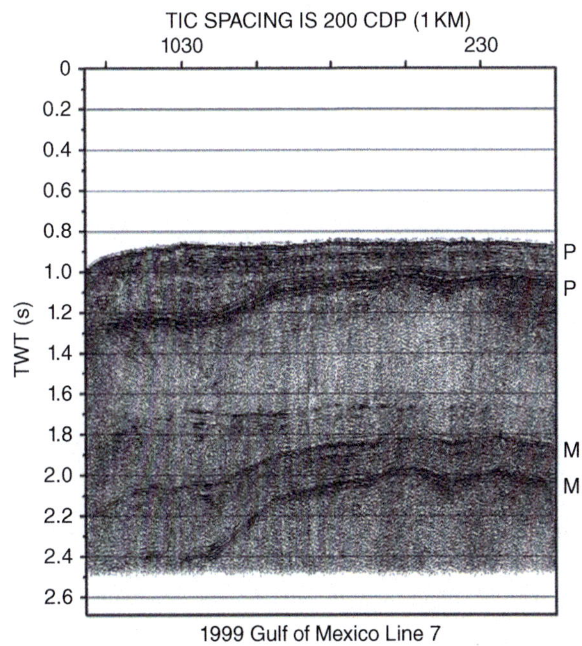

Figure 1.6 Multiples in a seismic record. P: primaries; M: multiples. Note that the multiples appear at approximately twice the depth of the primaries. Seismic section courtesy US Geological Survey, Department of the Interior/USGS

Figure 1.5. Multiples are very common in seismic data and need to be identified for a perfect interpretation. Usually multiples are removed in the processing workflow; though some may persist. Multiples mimic their primaries and appear at twice (or in multiples of two) the depth of the primary (Figure 1.6).

Noise in the seismic data can be classified based on several parameters. We present two common types of noise found in seismic data; they are based on genesis and nature (Figures 1.7 and 1.8). Classifying noise helps

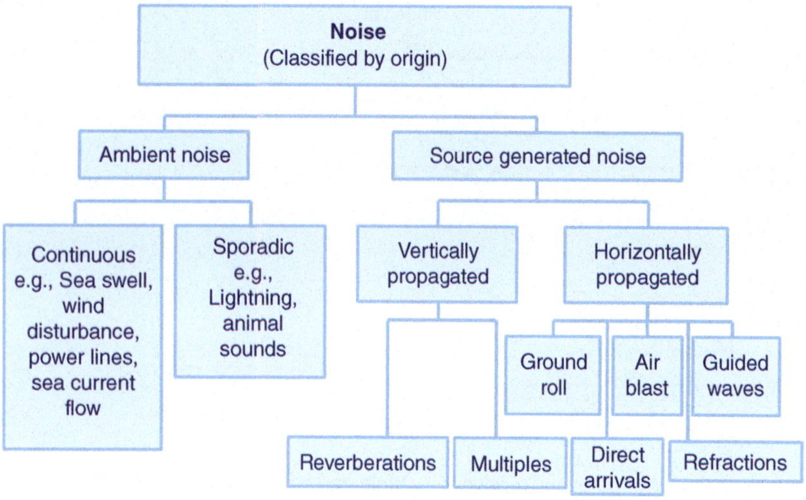

Figure 1.7 Noise in seismic data classified by origin. Adapted from Prof C.H. Mehta's lecture material.

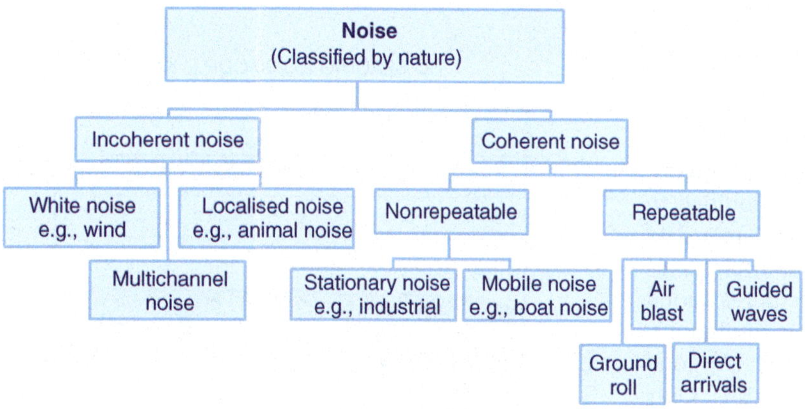

Figure 1.8 Noise in seismic data classified by nature. Adapted from Prof C.H. Mehta's lecture material.

to understand the form of noise and whether it can be removed easily or whether additional processes will be required. For example, sporadic and random noise is difficult to identify and remove. Yilmaz (2001) describes noise removal comprehensively.

As will be discussed in Section 1.4, certain seismic attributes (coherency, dip, azimuth etc.) can be used to quickly interpret/detect complex structures, for example faults and fractures. It is important to understand the effect of noise in the seismic data, which may manifest itself as geological structures.

In spite of the best efforts in seismic data acquisition and processing, these data contain both signal and noise. Unless the seismic interpreter recognizes the characteristics of seismic noise, there is a real danger that noise in the seismic data may be misinterpreted as a structural feature. Knowledge about the source and nature of nongeological features that appear on attribute maps is, therefore, crucial (Hesthammer and Fossen, 1997).

Thus, it is possible that noise can masquerade as interesting structural features; only an experienced structural interpreter can differentiate between these artefacts and a real feature. Certain tools can be useful to understand the origin and effect of noise on seismic data; one such tool is seismic forward modelling. A conceptual geological model can be converted to an acoustic model and can be forward modelled to get synthetic seismic data, interpretation of this synthetic seismic data in conjunction with the real seismic data can discriminate the nongeological features present in the real seismic data.

The other tool that can be used is a measurement of the quality of seismic data. Signal-to-noise ratio (SNR) is one quantity that can provide a measurement of the quality of the seismic data; it does not directly help in structural interpretation but can be quite useful in laying down the expectation from structural interpretation. The interpreter will have an idea of the level of noise present in the data and will, thus, be warned against overinterpretation of the structural features.

A simple definition of the SNR was introduced by Yang and Chengchu (1997). SNR is a ratio of the power of signal to the power of noise, represented by the equation:

$$\mathrm{SNR} = \left(\mathrm{Sa} + \mathrm{n} - \mathrm{Sn}\right) / \mathrm{Sn} \tag{1.3}$$

where SNR is the signal-to-noise ratio, 'Sa + n' the seismic data with noise, 'Sn' the noise in the seismic data, 'a' the seismic signal and 'n' the seismic noise.

1.3.2 Gathers

Gathers are the raw data form and are very important for the processing geophysicists. An interpreter must also have some knowledge about gathers to make a sound interpretation. Gathers are used to visualize the seismic data for errors, noise and geological signatures. These are displays of seismic trace with a common acquisition parameter. The most common types are:

- Common shot gather: the common acquisition parameter is the shot point in this type of a seismic gather. However, different receiver locations are considered. It is the result of a single shot and contains information about signal and noise produced from one specific shot (Figure 1.9a, 1.9b).
- Common receiver gather: as the name suggests, the common acquisition parameter is the receiver. These gathers contain information for all shot points for a particular receiver. It contains signal and noise associated with a particular receiver (Figure 1.9c, 1.9d).
- Common mid-point (CMP) gather: in this sort of gather, the incidence point of the seismic wave is same for all the traces. So, data from several shot gathers are combined to create a CMP gather. These gathers are the starting point for most of the steps in a processing workflow. A 'fold' in seismic data processing is the number of traces in a CMP gather (Figure 1.9e, 1.9f).
- Common offset gather: the common parameter here is that the offset is always kept equal for all the traces in the gather. It is effectively a single fold section for a specific offset. It is useful in some processing workflows, for example migration (Figure 1.9g, 1.9h).
- Common depth point (CDP) gather: the constant parameter is the depth of the reflector. In case of a horizontal bed, CMP = CDP. However, for a dipping bed they are unequal. Here the traces are reflections arriving from a particular depth. However the offsets for each shot point receiver pair differ (Figure 1.9i, 1.9j).

1.3.3 Migration and Stacking

After removing certain types of noise from gathers other steps in data processing follow, two of which are migration and stacking. Migration and stacking are important steps because they lead to the geophysical data representing the geology of the subsurface. In the case of horizontal layers in the subsurface, the boundaries are imaged at their correct positions in a zero-offset section. However, dipping layers are not imaged at their correct position in a zero-offset seismic profile, because the reflected rays sway laterally (Figure 1.10) when they travel from the reflector to the receiver and their response is perceived at an apparent reflection point. This apparent reflection point is not located on the reflector itself but shifts laterally and above it. The shift increases with depth, dip of the reflector and velocity. Migration aims to position the reflection correctly.

Thus, in the case of complex geometries such as antiforms, synforms, salt bodies (e.g., Mukherjee *et al.*, 2010; Mukherjee, 2011) and etc. present in the subsurface, the image becomes unusable due to the above mentioned phenomena. In such situations, migration must inevitably be carried out to position the boundaries perfectly in the subsurface (Figure 1.11). Migration has additional benefits:

- its mimics the wave propagation more accurately, thus producing realistic images of the subsurface;
- prestack migration generates more precise velocities;
- prestack migration is essential for amplitude versus offset (AVO) analysis (Aki and Richards, 1980; Ostrander, 1984; Shuey, 1985);
- migration also enhance spatial resolution.

Migration can be carried out either in the time domain or the depth domain. In the time domain, a root mean square velocity for the entire section is used. The depth domain, on the other hand, uses interval velocities for each layer. Time-migrated data have a smooth velocity model that does not allow much lateral variation in velocity, whereas the depth-migrated data employ a complex velocity model with strong lateral variation. Therefore, time-migrated data are less sensitive to erroneous velocity than the depth-migrated data. However, when the subsurface has strong lateral variation in velocity, time-migrated data cannot position correctly the reflectors in the subsurface.

The traces in a gather (Section 1.3.2), most commonly the CMP gather, are added together or '*stacked*' after migration. This generates an approximate zero-offset trace for the gather. This process sums up the response for all the traces in the gather, resulting in nullifying some noise and amplifying signals.

1.3.4 Resolution of Seismic Data

Features identifiable on the seismic data are controlled by its resolution, which provides the ability to distinguish two separate features/events. Thus, resolution of the seismic data limits observation or interpretation of small-scale structures. Resolution in seismic data is considered in both the vertical and horizontal perspectives. In the vertical scale, wavelength (λ) is the *measure* of the resolution. If two subhorizontal events or reflections are to be distinguished, they must differ by at least λ/4

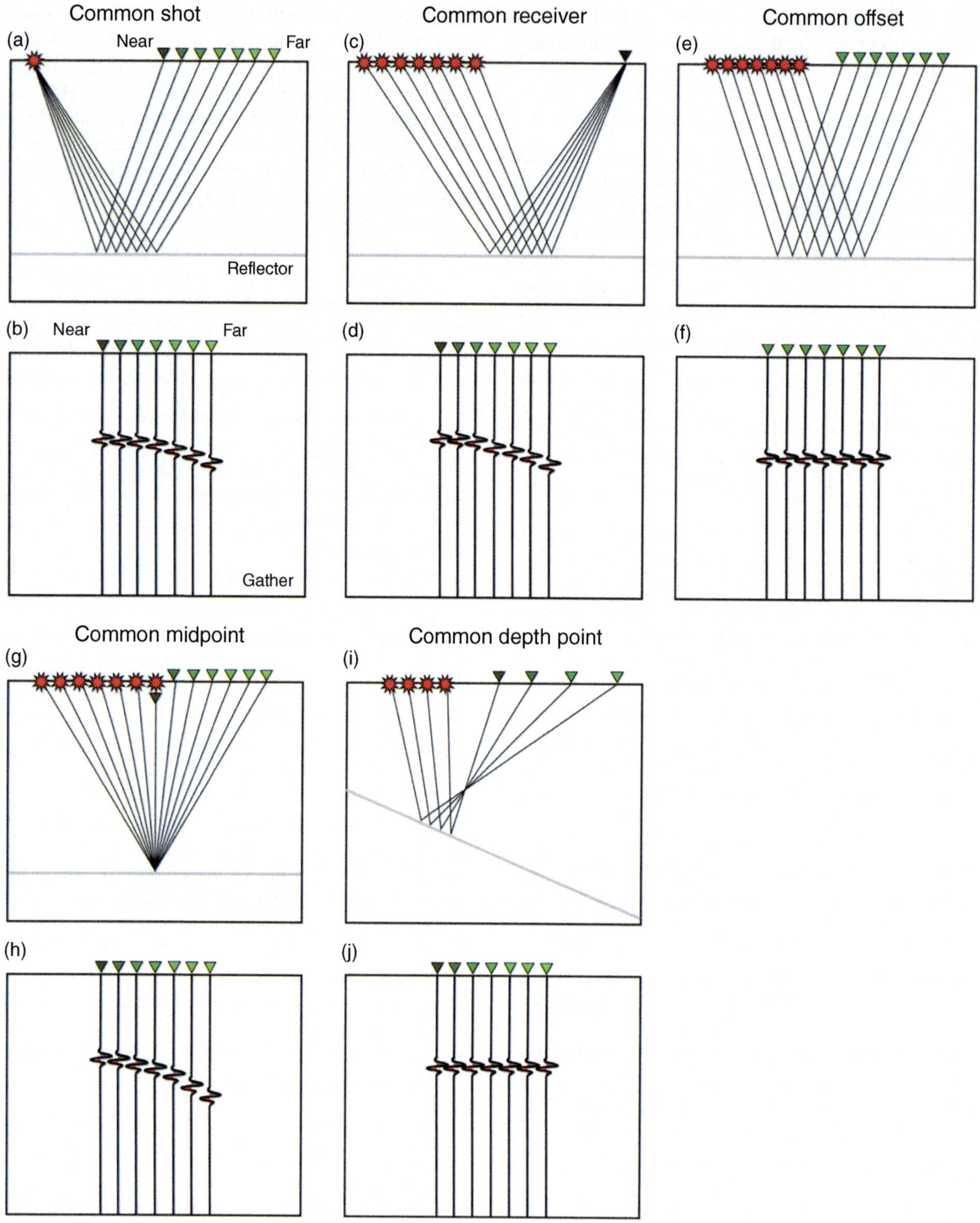

Figure 1.9 Top row: schematic seismic ray path diagram; bottom row: gather for the set-up above. (a, b) Common shot gather; (c, d) common receiver gather; (e, f) common offset gather. Green triangles, dark to light: Receivers, near to far. Red stars: shot; Green triangles: receivers. (g, h) Common midpoint gather; (i, j) common depth point gather. Originated point point gather. Red stars: shot; Green triangles: receivers.

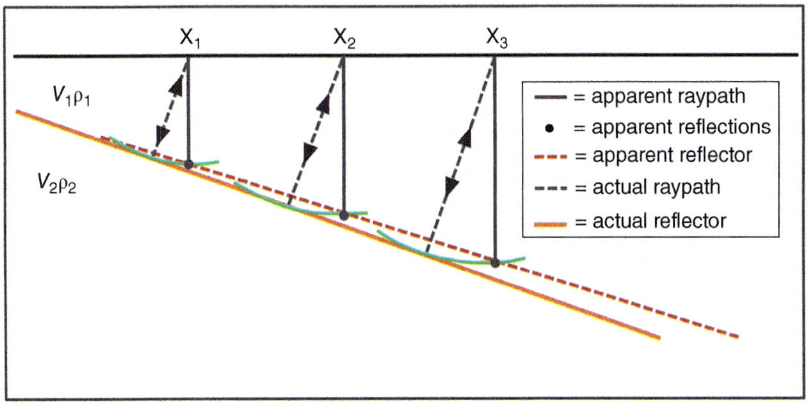

Figure 1.10 Seismic ray path schematic diagram for a dipping reflector. The sway of the seismic wave is shown in green.

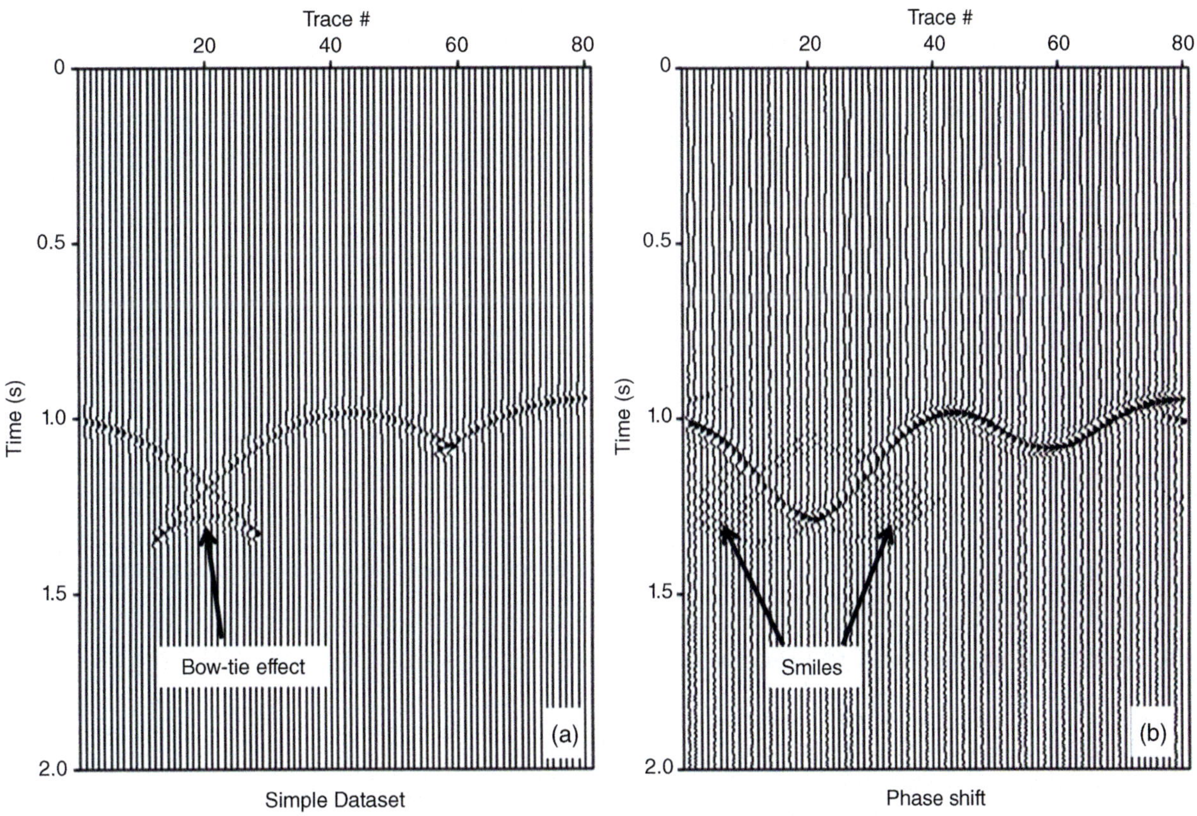

Figure 1.11 (a) Unmigrated simple seismic image; (b) migrated image of the same. Notice the appearance of the syncline in (b), which was impossible to detect in (a). The typical signature of the syncline in (a) is called a 'bow-tie' effect. There is some migration induced noise in (b), which is known as 'smiles'. Smiles may occur due to velocity mismatch between the real world lithology and the model. Output from Seismic Unix © Stockwell and Cohen, 2008.

(Sheriff, 1985). In seismic data, velocity increases and frequency drops with depth and, thus, resolution falls. Consider a shallow event with the following parameters:

Velocity (v): 2000 m s^{-1}
Frequency (f): 50 Hz
Therefore, wavelength (λ) = v/f = 2000/50 = 40 m
Thus, the resolution is ~ $\lambda/4$ = 10 m.

Now, consider a deep event with the following parameters:

Velocity (v): 3000 m s^{-1}
Frequency (f): 20 Hz
Therefore, Wavelength (λ) = v/f = 3000/20 = 150 m
Thus, the resolution is ~ $\lambda/4$ = 37.5 m.

So, what can be identified on a seismic image? Figure 1.12 shows an approximately 10-m high outcrop with

Figure 1.12 Deccan trap outcrop in Kharghar Hills, Mumbai/Bombay, compared to a wavelet of the most dominant frequency, out of the large number of frequencies it can represent, is shown here. This lithology is basalt, which has a velocity ranging between 5000 and 6000 ms^{-1}. So even at very shallow levels the best possible resolution will be 25–30 m. Note the intense small scale deformation accompanied with large-scale fault planes (F). Courtesy of Achyuta Ayan Misra.

Figure 1.13 An approximately 200-m high outcrop in the Spiti Valley, Himachal Pradesh, India, of Mesozoic limestones (light beige to brownish) and shales (dark grey). A possible seismic trace of the most dominant frequency, out of the large number of frequencies it can represent, is placed alongside. This trace does not indicate the typical seismic response for the lithologies. Note the fold in the lower right corner of the photograph. This fold will barely be resolved in a seismic data. The bed-perpendicular fractures in the limestone, also, will not be resolvable. However, the horizontal layers of rock may be resolved. Houses (white) near the foothill as markers. Courtesy of Achyuta Ayan Misra.

observable deformation on an outcrop scale. However, comparing that with a seismic wavelet, it can be noted that the deformation at the outcrop scale is far below the seismic resolution. Thus, it is nearly impossible to 'see' such features using seismic data. Figure 1.13 shows another outcrop, approximately 200 m high. Here, the beds may be resolved but the deformation will not. Looking at global landmarks of mentionable height, only the tallest of them can be successfully resolved in seismic data (Figure 1.14). The 53-m high leaning tower at Pisa would barely be resolved; the same is true for the Taj Mahal (Agra, India) or the Statue of Liberty (New York, USA). The other towers in Figure 1.14 will be successfully resolved.

Resolution considered in the horizontal sense, known as lateral resolution, is also important during seismic interpretation. The seismic wave propagates through the subsurface in the form of a spherical wavefront as shown in Figure 1.15. As the wavefront propagates away from the source, it spreads out to a larger area. Reflections from a reflector or surface are not from a point but from a region depending upon the dimensions of this wavefront. Thus, signals received from this region (AA′) at the same interval of time are indistinguishable. This region is fairly large, a roughly circular area of the reflecting surface, and is called the first Fresnel zone. The radius of this zone is often taken as the horizontal resolution for unmigrated seismic data. 3D migration tends to collapse this Fresnel zone and, thereby, increase the horizontal resolution of the seismic data.

The horizontal resolution is given by the width of this Fresnel zone:

$$r = \frac{\upsilon}{2}\sqrt{\frac{t_0}{f}}. \tag{1.4}$$

where r is the radius of the Fresnel zone/Horizontal resolution, v the velocity, t_0 is the two-way travel time and f the frequency.

Newer techniques such as shear (S-) wave reflection seismic profiles can provide very high resolution. See Chapters 3.14, 3.15, 5.3 and 5.4, where S-wave reflection profiles are shown to give resolutions as small as 0.5–1 m. Such techniques require further studies to be used widely.

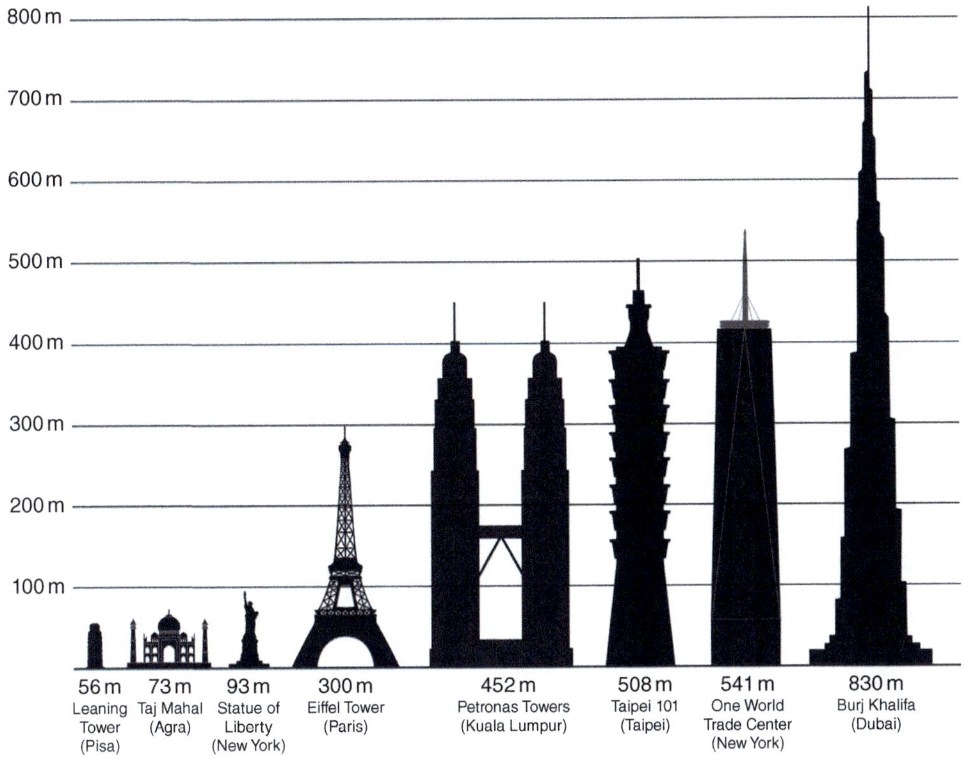

Figure 1.14 Comparison of some tall global structures. Those more than about 100 m high can be resolved with ease in seismic data.

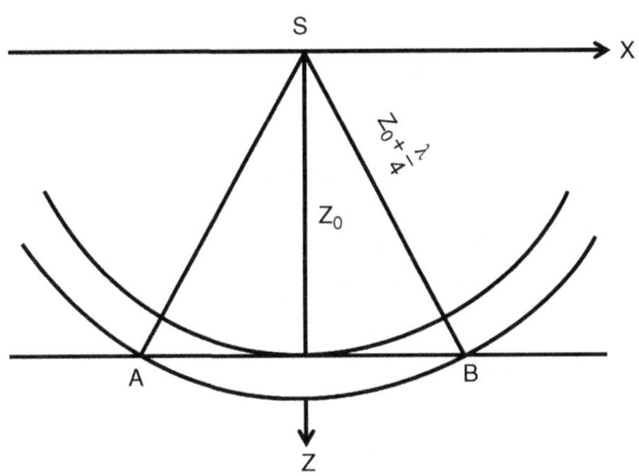

Figure 1.15 Definition of the Fresnel zone AB. S: source; X, Z: horizontal and vertical direction; Z_0: vertical depth to the reflector and Lamba (λ) is the wavelength of the seismic wave.

1.4 Seismic Attributes

Visualization and identification of many structures are sometimes easily achieved using seismic attributes. Seismic attributes are measured, computed or implied derivatives from seismic data. There may be various inputs for computing a seismic attribute, such as:

- Single trace: attributes derived from a single migrated and stacked trace are called trace attributes (Taner and Sheriff, 1977; Taner *et al.*, 1979).
- Set of pre-stack traces: attributes derived from these are known as multitrace attributes (Taner and Sheriff, 1977; Taner *et al.*, 1979).
- Surface: a surface for input is usually an interpreted horizon gridded and interpolated to a surface. Attributes derived from surfaces are called surface attributes.
- Volume: a small part of a 3D seismic volume or an entire volume is sometimes used as an input to derive seismic attributes. Attributes thus derived are called volume-based attributes.

There are numerous seismic attributes and they are used for a number of requirements. There are some, for example, that help in highlighting specific features, *be it geomorphological or related to structural deformation,* in the subsurface. The selection of the attributes depends on the feature of interest that is to be analysed. The most common seismic attributes used to interpret deformation are mentioned here. There are several attributes for interpreting faults in seismic data (Chen and Sidney, 1997). Seismic attributes are employed primarily for interpreting deformation *along a* surface and *in* volume.

1.4.1 Coherency

This seismic attribute (Figure 1.16) measures the similarity between neighbouring seismic traces. It can be calculated along a horizon or within a volume. It is most commonly used to identify faults, channels or other

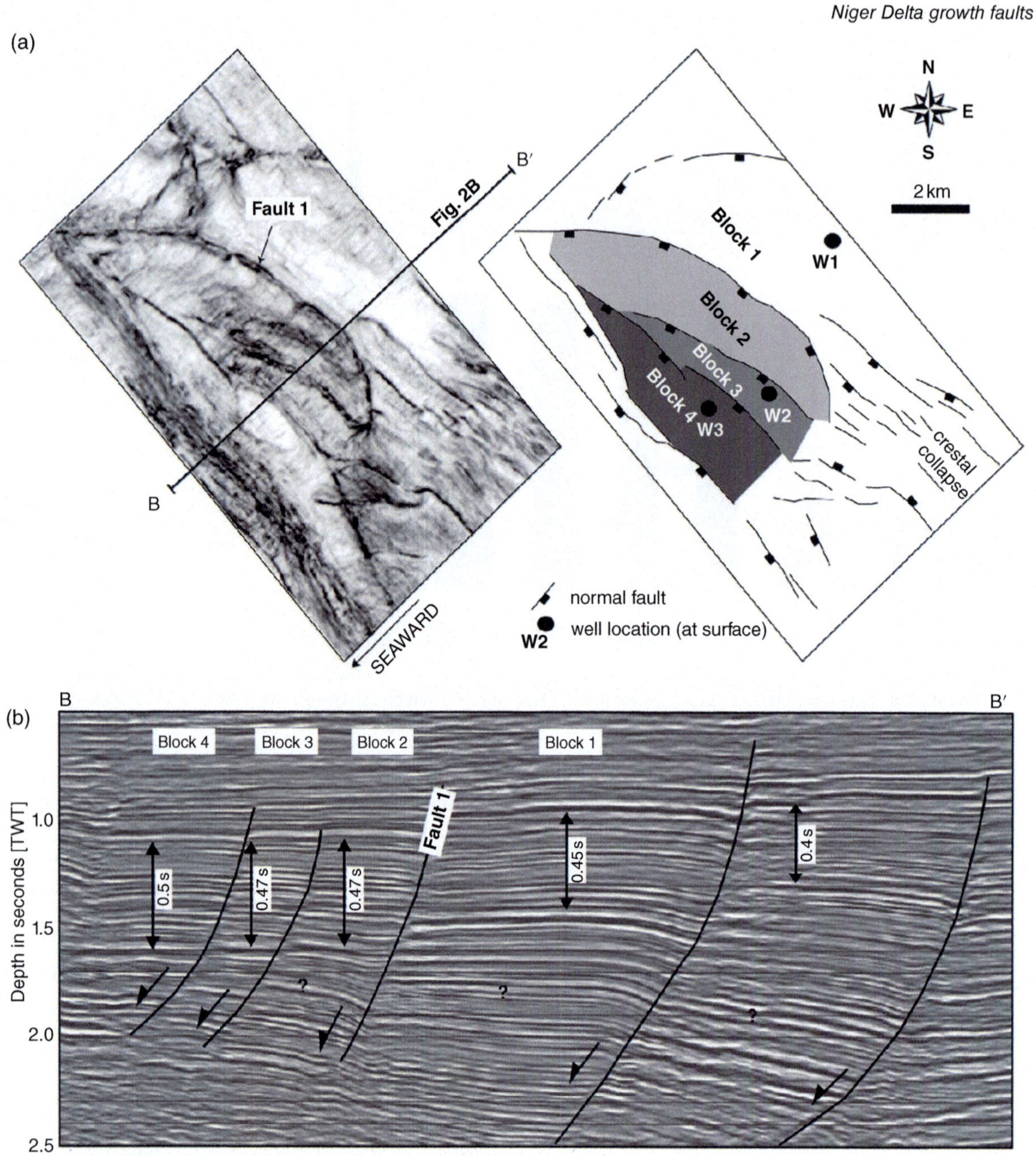

Figure 1.16 Example of coherence from the Niger Delta. (a) Coherency slice at 1.5 s TWT and the interpreted fault map; (b) reflection seismic section showing the faults. Reproduced with permission from Back *et al.* (2006).

discontinuities (Marfurt *et al.*, 1998). Coherency is quite useful, as it quickly evaluates large data sets when analysing deformation and quantifying faults or fractures.

1.4.2 Dip and Azimuth

Dip is an attribute that computes, for each trace, the best fit line with 2D data or plane with 3D data between neighbouring traces. Dip is computed on an interpreted horizon and the magnitude of dip is that of the said plane in degrees. This indicates paleogeography. However, when the changes in dip are sharp, faults may be indicated.

Azimuth is another attribute that is employed together with dip. Azimuth calculates the direction of maximum dip of the plane from the dip attribute. Azimuth is measured in degrees, clockwise from north. Thus, dip and azimuth attributes together compute dip and dip direction of an interpreted horizon. An example is shown in Figure 1.17.

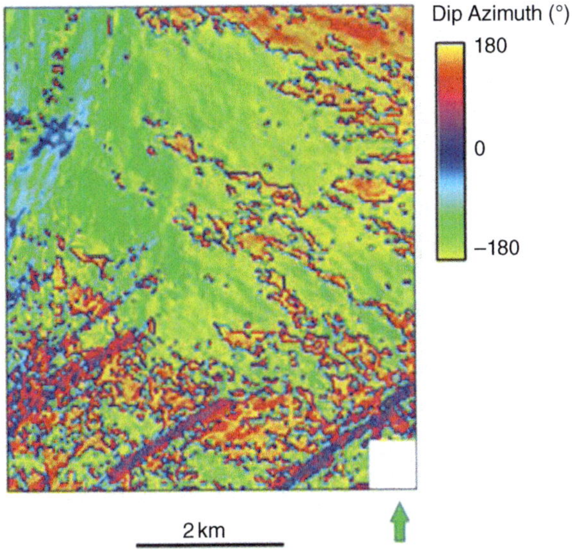

Figure 1.17 Volumetric dip–azimuth along a picked horizon. Note the NE striking and N-dipping faults in the lower part of the image. Reproduced from Paes and Marfurt (2016). AAPG © 2016 whose permission is required for further use.

1.4.3 Curvature

In general, curvature is a measure of the tightness of a curve at particular points. Thus, a larger curvature means the area is more deformed than an area with a lower curvature (Figures 1.18 and 1.19). For a 2D curve, curvature is the reciprocal of the radius of curvature (Chopra and Marfurt, 2007). The most important curvature attributes are the maximum (k_{max}) and minimum curvatures (k_{min}). Together, they are called *principal curvatures*. They are combined to derive other curvature attributes such as:

- Mean curvature $k_{mean} = 0.5 \times (k_{max} + k_{min})$ (1.5) – indicates the overall deformation of the area of interest.
- Gaussian curvature ($k_g = k_{max} \times k_{min}$) (1.6) – a good indicator of whether a surface is warped.
- Curvedness $r = (k^2_{max} + k^2_{min})^{1/2}$ (1.7) – a good measure of folding intensity.
- Shape index, $s = (2/\pi) \tan^{-1} [(k_{max} + k_{min})/(k_{max} - k_{min})]$ (1.8) – classifies the shape of the deformed geomorphology of the area: dome, ridge, saddle, valley and so on.

Figure 1.18 Sketch to show the idea of curvature along a 2D line. Anticlines have a positive curvature and synclines have a negative one. The anticline on the left has a higher curvature than the one on the right. That is because the one on the left is more tightly folded than the right.

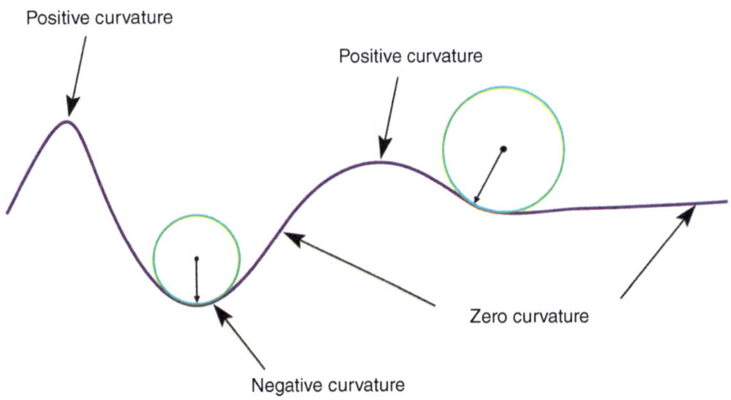

Figure 1.19 Calculated maximum curvatures for a domal surface. Note the linear faults that can be easily interpreted on this image. Reproduced from Bergbauer *et al.* (2003). AAPG © 2003 whose permission is required for further use.

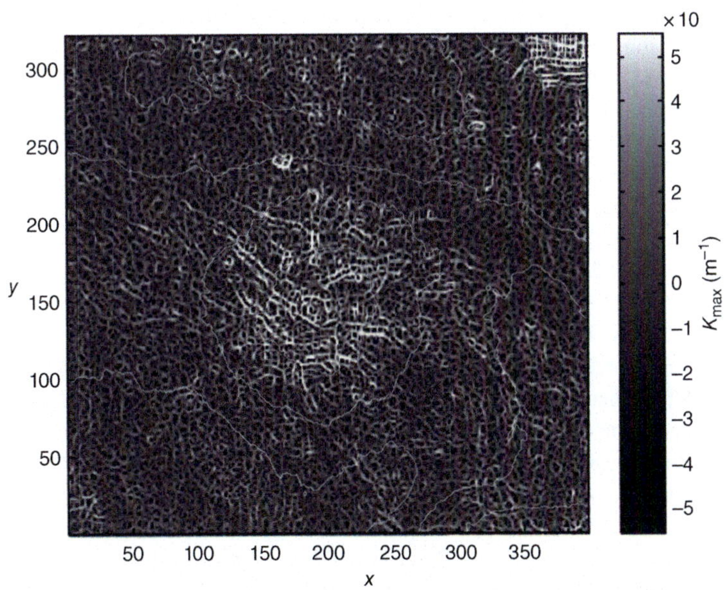

References

Aki, K. and Richards, P.G. 1980. *Quantitative Seismology: Theory and Methods*, vol. 1. W.H. Freeman and Co, San Francisco, CA.

Ashcroft, W. 2011. *A Petroleum Geologist's Guide to Seismic Reflection*. John Wiley & Sons Ltd, Chichester.

Back, S., Höcker, C., Brundiers, M.B. and Kukla, P.A. 2006. Three-dimensional-seismic coherency signature of Niger Delta growth faults: integrating sedimentology and tectonics. *Basin Research*, **18**, 323–337.

Barley, B. and Summers, T. 2007. Multi-azimuth and wide-azimuth seismic: Shallow to deep water, exploration to production. *The Leading Edge*, **26**, 456–457.

Bergbauer, S., Mukerji, T. and Hennings, P. 2003. Improving curvature analyses of deformed horizons using scale-dependent filtering techniques. *AAPG Bulletin*, **87**, 1255–1272.

Chen, Q. and Sidney, S. 1997. Seismic attribute technology for reservoir forecasting and monitoring. *The Leading Edge*, **16**, 445–448.

Chopra, S. and Marfurt, K. 2007. Seismic curvature attributes for mapping faults/fractures, and other stratigraphic features. *CSEG Recorder*, **32**, 37–41.

Dobrin, M.B. and Savit, C.H. 1988. *Introduction to Geophysical Prospecting*. McGraw-Hill Inc, Singapore.

Games, K.P. and Wakefield, N.D. 2014. The successful design, development and acquisition of a UHRS 3D seismic dataset. In: *Near Surface Geoscience 2014 – First Applied Shallow Marine Geophysics Conference*, 14–18 September, Athens, Greece, EAGE. doi: 10.3997/2214-4609.20142132.

Herron, D.A. and Latimer, R.B. 2011. *First Steps in Seismic Interpretation*. Society of Exploration Geophysicists, Tulsa, OK.

Hesthammer, J. and Fossen, H. 1997. The influence of seismic noise in structural interpretation of seismic attribute maps. *First Break*, **15**, 209–219.

Liner, C.L. 2004. *Elements of 3D Seismology*, 2nd edn. PennWell Books, Tulsa, OK.

Marfurt, K.J., Kirlin, R.L., Farmer, S.L. and Bahorich, M.S. 1998. 3-D seismic attributes using a semblance-based coherency algorithm. *Geophysics*, **63**, 1150–1165.

Misra, A.A., Sinha, N. and Mukherjee, S. 2015. Repeat ridge jumps and microcontinent separation: insights from NE Arabian Sea. *Marine and Petroleum Geology*, **59**, 406–428.

Misra, A.A., Banerjee, S., Kundu, N. and Mukherjee, B. 2016. Subsidence around oceanic ridges along passive margins: NE Arabian Sea. In: Mukherjee S., Misra A.A., Calvès, G. and Nemčok, M. (eds) *Tectonics of the Deccan Large Igneous Province*. Special Publications vol. 445, The Geological Society, London, pp. 119–149. doi: 10.1144/SP445.10.

Mukherjee, S. 2011. Estimating the Viscosity of Rock Bodies – a Comparison Between the Hormuz – and the Namakdan Salt Domes in the Persian Gulf, and the Tso Morari Gneiss Dome in the Himalaya. *The Indian Journal of Geophysical Union*, **15**, 161–170.

Mukherjee, S. 2017. Airy's isostatic model: a proposal for a realistic case. Arabian *Journal of Geosciences*, **10**, 268.

Mukherjee, S. 2018. Locating center of pressure in 2D geological situations. *The Journal of Indian Geophysical Union*, **22**, 49–51.

Mukherjee, S. (in press) Locating center of gravity in geological contexts. *International Journal of Earth Sciences*. doi: 10.1007/s00531-017-1560-z.

Mukherjee, S., Talbot, C.J. and Koyi, H.A. Viscosity estimates of salt in the Hormuz and Namakdan salt diapirs, Persian Gulf. *Geological Magazine*, **147**, 497–507.

Onajite, E. 2014. *Seismic Data Analysis Techniques in Hydrocarbon Exploration*. Elsevier, Waltham, MA.

Ostrander, W.J. 1984. Plane wave reflection coefficients for gas sands at non normal angles of incidence. *Geophysics*, **49**, 1637–1648.

Paez, L.I. and Marfurt, K.J. 2016. Using volumetric calculators for better accuracy. Search and Discovery Article #41757, AAPG. http://www.searchanddiscovery.com/pdfz/documents/2016/41757paez/ndx_paez.pdf.html (last accessed 5 September 2017).

Schlumberger 2016. *The Oilfield Glossary: Where the Oil Field Meets the Dictionary*. Schlumberger. http://www.glossary.oilfield.slb.com/en.aspx (last accessed 5 September 2017).

Sheriff, R.E. 1985. Aspects of seismic resolution. In: Berg, O.R. and Woolverton, D.G. (eds) *Seismic Stratigraphy II: An Integrated Approach to Hydrocarbon Exploration*. AAPG Memoir vol. 39, American Association of Petroleum Geologists, pp. 1–10.

Sheriff, R.E. and Geldart, L.P. 1995. *Exploration Seismology*. Cambridge University Press, Cambridge.

Shuey, R.T. 1985. A simplification of the Zoeppritz equations. *Geophysics*, **50**, 609–614.

Stockwell, Jr. J.W. and Cohen, J.K. 2008. The New SU User's Manual. Centre for Wave Phenomena, Colorado School of Mines, Golden, CO. ftp://ftp.cwp.mines.edu/pub/cwpcodes/sumanual_600dpi_a4.pdf (last accessed 5 September 2017).

Taner, M.T. and Sheriff, R.E. 1977. Application of Amplitude, Frequency, and Other Attributes to Stratigraphic and Hydrocarbon Determination. In: Payton, C.E. (ed) *Seismic Stratigraphy: Applications to Hydrocarbon Exploration*. AAPG Memoir vol. 26, American Association of Petroleum Geologists, pp. 301–328.

Taner, M.T., Koehler, F. and Sheriff, R.E. 1979. Complex Seismic Trace Analysis. *Geophysics*, **44**, 1041–1063.

Tarantola, A. 1986. A strategy for nonlinear elastic inversion of seismic reflection data. *Geophysics*, **51**, 1893–1903.

Yang, L. and Chengchu, L. 1997. Some methods for estimating the signal/noise ratio of seismic data. *Oil Geophysical Prospecting*, **2**, 12.

Yilmaz, O. 2001. *Seismic Data Analysis: Processing, Inversion and Interpretation of Seismic Data*. Society of Exploration Geophysicists, Tulsa, OK.

2

Seismic Structural Analysis

Achyuta Ayan Misra[1]* *and Soumyajit Mukherjee*[2]

[1] *Reliance Industries Ltd, Mumbai, India*
[2] *Indian Institute of Technology Bombay, Mumbai, India*
* *Achyuta.Misra@ril.com; achyutaayan@gmail.com*

2.1 Introduction

Structural analysis is the study of the processes that create basins/depressions on the Earth's surface in which sediments accumulate (Tearpock and Bischke, 2002 and Rowland *et al.*, 2013). It also includes analysing the processes that may have deformed the sedimentary fill occupying the basins. The scale of analysis typically ranges from the regional scale, for example plate tectonics, to the microscopic scale, for example deformation microstructures, and even smaller if data permit (Passchier and Trouw, 2005; Mukherjee, 2011, 2013, 2014a, 2015, 2017). The basis of seismic structural analysis is structural interpretation of seismic data. Thus, seismic structural analysis is limited, covering the regional scale down to tens of metres, which is the maximum possible resolution of seismic data (Chapter 1.3.4). Seismic data comprises three parameters (Liner, 2004) (i) seismic energy travel time, (ii) amplitude of the seismic wave and (iii) information about the seismic waveform. For seismic structural analysis only the first of these parameters is used whereas for stratigraphic analysis all the parameters must be used. The various advantages and limitations of using seismic data for structural analysis are presented in Section 2.2.

The main aim of the interpreter or the structural geologist is (i) to make structural geological observations of the 'present-day form' and represent them; and (ii) to interpret the processes that led to the 'present-day form'. The present-day form may not be the final form, for example several regions on the Earth reactivate/ undergo deformation cycles (review in Misra and Mukherjee, 2015, Dasgupta and Mukherjee, 2017). Even present passive margins, which are tectonically inactive, also undergo some deformation.

To fulfil these goals, seismic interpreters construct and analyse the 3D geometric arrangement of geological structures, such as folds, faults, horizons, in some cases fractures, ductile- and brittle shear zones, and intrusives, for example salt diapirs/domes and so on. Faults and folds are the most basic secondary/deformation induced structures that are identified on seismic sections. The faults may be classified as normal, reverse, strike-slip etc. The folds may be compressional, ductile shear induced, salt-related or passive drapes over a structural high (Ramsay and Huber, 1987; van der Pluijm and Marshak, 2004; Twiss and Moores, 2007; Nemčok *et al.*, 2009; Mukherjee *et al.*, 2010, 2015; Davis *et al.*, 2012; Mukherjee and Mulchrone, 2012; Park, 2013; Rowland *et al.*, 2013; Fossen, 2016). Salt/volcanic intrusions can also be identified in seismic sections (van Gent *et al.*, 2011; Jackson *et al.*, 2014). Structures, thus interpreted, indicate the structural setting/style, for example either extensional, compressional, or strike-slip. As an integral part of structural analysis, the 3D geometry may be tested for admissibility and validity. This step is termed 'balancing' the seismic sections (Dahlstrom, 1969). Balancing sections validates the interpretation and helps the interpreter to deduce a sound geological history.

Structural analysis can be of the following types (Davis *et al.*, 2012; Rowland *et al.*, 2013):

- Descriptive analysis: here faults and horizons are mapped, and their attitude, geometry, shear senses and all other physical attributes of deformation described. Relative time relationships between sedimentation and deformation, and the link between one deformation structure and another, are studied in detail and described. The best methods involve mapping faults and horizons over the study area and studying their 3D architecture (Davis *et al.*, 2012, Part III).

Atlas of Structural Geological Interpretation from Seismic Images, First Edition. Edited by Achyuta Ayan Misra and Soumyajit Mukherjee.

- Kinematic analysis: here movements due to deformation, for example translation, rotation, distortion etc. are analysed. Description of the structures, for example attitude of horizons and faults, from descriptive analysis is used. Kinematic analysis (as in Mukherjee and Biswas 2014, 2015) gives an idea of the overall strain in the study area or a part of the study area. More importantly, it can be used to test 'kinematic compatibility' of a set of deformation structures. When structures are mutually kinematically incompatible, they most logically belong to different deformation events (Rowland *et al.*, 2013, Chapter 10).
- Dynamic analysis: the forces/stresses associated to deformations are analysed. Stress directions have been analysed based on data collected in the field, e.g., striated fault planes, stylollites, extensional joints, sheared dykes, brittle and ductile shear planes (Angelier 1979, 1984, 1989; Lisle, 1989; Srivastava *et al.* 1995; Žalohar and Vrabec, 2007; Rowland *et al.*, 2013, Chapter 10; Misra *et al.*, 2014; Misra and Mukherjee, 2017; Babar *et al.*, 2017; Kaplay *et al.*, 2017; Dasgupta and Mukherjee, 2017). Only recently it has also been performed on seismic data (van Gent *et al.*, 2009, 2010).

2.2 Advantages and Limitations of Seismic Data Analyses

Structural analysis has been traditionally done in the field by observing actual rocks and their deformation. The other approach has been to study analogue models (Mukherjee *et al.*, 2012). Reflection seismic technology, on the other hand, is now advancing to a stage where seismic sections will act as the virtual geological field. Other data sets, such as well and potential field data, are also used to unravel the geologic structure and history, most commonly to support reflection seismic data. Seismic data have a few advantages over other data sets:

- *Ease of use*: A significant logistic effort is required during seismic data acquisition and processing (Chapter 1). After the data are ready and loaded on workstations, however, it is comfortable for the interpreter to sit and view data from across the world. Geological fieldwork would be time-consuming. 3D visualization is easily possible with 3D imagery in software or a virtual reality set-up, in contrast to a field analysis.
- *Depth of observation*: also, during outcrop study, it is difficult to get data from depths beyond few hundred metres and practically impossible beyond one kilometre. Modern reflection seismic data can image sections deeper than 25 km in certain areas (Nemčok *et al.*, 2013, 2016; Misra *et al.*, 2015). Fieldwork on the surface or in mines, drilled wells and so on do not reach such depths. Compared to reflection seismic data, other geophysical techniques do not provide such resolution.
- *Three-dimensional nature*: seismic data are normally three-dimensional. Even 2D seismic data are a grid of a few lines, which are correlated and contoured to provide 3D time maps. Sometimes the lines are very close-spaced and can be processed and migrated as a 3D volume. Such data are often termed as a "poor man's 3D" (Wiederhold, 2005)!
- *Resolution*: the vertical and spatial resolution of seismic data adequately identify and map hydrocarbon reservoirs, traps and seals. The resolution is also enough to identify shallow hazards, carbon storage reservoirs etc.
- *Geological proxy*: reflection seismic sections have proven to be a close proxy to geology. For example, it is easy to test whether sedimentary sequences, certain sedimentary features, faults, unconformities etc. are actually present when visualized on a seismic section. This facilitates a good support and 3D perspective for field and well data. Time slices derived from 3D data mimic outcrops on a geological section and also show major structural features, while interpreted horizon slices show paleo-depositional environment as well as small-scale structural features. Wheeler diagrams are often made from vertical sections to identify erosion zones and periods of nondeposition, enabling conversion of seismic time to geological time.
- *Derivatives*: there can be numerous derivatives from reflections seismic data, for example seismic attributes (Chapter 1.4), which can be used to identify structures, sedimentary facies and so on.

However, there are some limitations of seismic data besides common pitfalls in interpretation. The most important are:

- *Resolution*: resolution is normally insufficient when small-scale structures are to be identified, for example foliations, fractures, and lineations such as slickensides. Even small scale sedimentary structures like crossbeds, ripples, groove marks, small channels (< 15–20 m thickness) are not possible to identify in seismic data.
- *Frequency*: during propagation of seismic waves into the Earth there is preferential absorption of high frequencies and, thus, the deeper events are predominantly imaged with lower frequencies and higher velocities. As a result, deeper events have far less resolution than the shallower events. Thus, mapping thin beds at a deeper level is a big challenge.
- *Imaging*: seismic imaging is excellent in most areas and numerous features can be interpreted with confidence. 3D Prestack depth migration can image steeply dipping faults and beds, including subsalt stratigraphies, that is the lithological layer beneath a salt canopy, using the latest seismic migration algorithms such as reverse

time migration provided velocity depth modelling is reasonable and calibrated with well velocity data in near vicinity. But the processes are expensive. Imaging at greater depths (>8 km) or below salt, basalt, coal, igneous/metamorphic basement is a challenge, as they absorb and reflect most of the energy from their top surface. Certain noise in the seismic band and strong multiples are often difficult to separate and remove. These affect imaging and resolving/detecting thin beds.

- *Difficult acquisition*: acquisition may be a challenge in many areas, such as mountainous regions, swamps, volcanic rocks dominated areas and so on, and may result in significant noise in the seismic data. Such data are difficult to interpret.
- *Time-migrated data*: this is one of the most important things to consider while interpreting seismic data. The basic seismic data are time-migrated and the vertical axis is time, in milliseconds, while the horizontal axis is in meters. This distorts the geometry due to vertical exaggeration along the depth of the section (Figure 2.1). Thus, the dips of fault planes appear gentle at deeper levels in time-migrated data.
- *Lateral velocity variation:* overlying lateral velocity variation can cause pseudo-structures in the time section in the underlying beds because of pull-ups and pull-downs. The depth picture could be in contrast to the time maps. Low velocity zones present in features like shale diapir, mud volcano, mass transport complex etc. result in poor imaging of beds/sequences lying beneath such features.
- *Vertical exaggeration*: display scales are not commonly 1:1, even though 1:1 sections are desirable during structural analyses. However, in some instances, for example while studying low-dipping normal faults or gently dipping, very low relief structures, vertically exaggerated sections may be advantageous. The interpreter must know when to use vertically exaggerated sections and remain cautious while interpreting (Figure 2.2).
- *Absence of direct evidence*: in a reflection seismic section geological features can be interpreted. However, predicting lithofacies, especially in a clastic environment, is intriguing using the P-wave seismology. Converted wave can help. A plot of *Vp*/*Vs* is normally a good indicator. In the absence of converted wave recording, which is normally the case, *Vp*/*Vs* is derived from AVO (amplitude variation with offset) attributes. Reliability of results depends on the quality of seismic data, and calibration with well and rock physics templates. Integration of other geoscientific data like sedimentological facies and petrophysics is carried out rigorously. Direct hydrocarbon indicators (DHIs), such as the AVO attribute (Ross, 2000), are merely analytical models indicating the possible presence of hydrocarbons. Such DHI attributes are used especially in case of high gas saturation having excellent data quality and there are well data to calibrate.

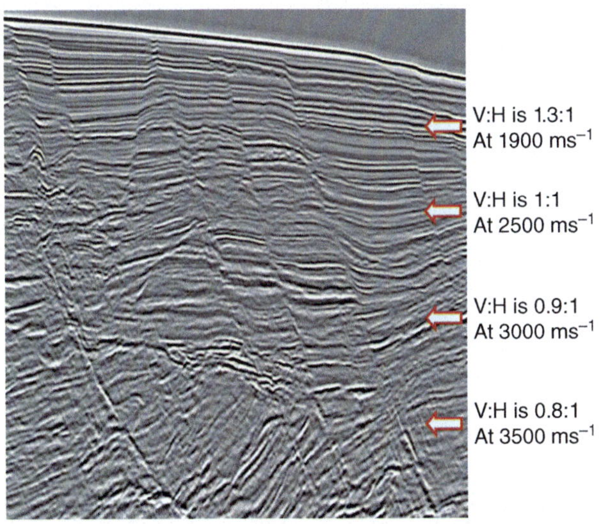

Figure 2.1 Variation of velocity with depth for a seismic time section leading to a varying vertical exaggeration. The depths in this seismic data range from 0 to 2500 ms. Seismic data courtesy Reliance Industries Ltd. Reproduced with permission from the Directorate General of Hydrocarbons (DGH), India.

While interpreting seismic data, faults and horizons must be analysed together as they compliment each other. For example, without mapping the marker horizons the throw along a series of faults is indeterminate. Again, if the faults are not marked on the seismic section, correlating horizons on either sides of the fault can be problematic.

2.3 Interpreting Faults

The essence of seismic structural analysis is interpreting faults on reflection seismic images. Detecting continuity of reflections and their termination lies at the core of interpreting faults on reflection seismic data. The spatial geometry of faults determines the structural setting of the study area and constitutes the base of structural analyses. Throws of faults deciphered on a reflection seismic image may scale m up to km. The faults with the largest throws are the regional faults, which indicate large-scale deformation that control basin architecture. The faults with smaller throws are mostly confined to certain layers and indicate small- to medium-scale deformation. These faults play an important role in reservoir parameters, such as fluid flow. All the faults in the seismic data may not be picked during a single analysis. The nature of analysis will determine the scale of faults to be interpreted. The quality of the seismic data determines how easily faults may be picked up on the reflection seismic profiles. Faults may be identified on profiles and maps,

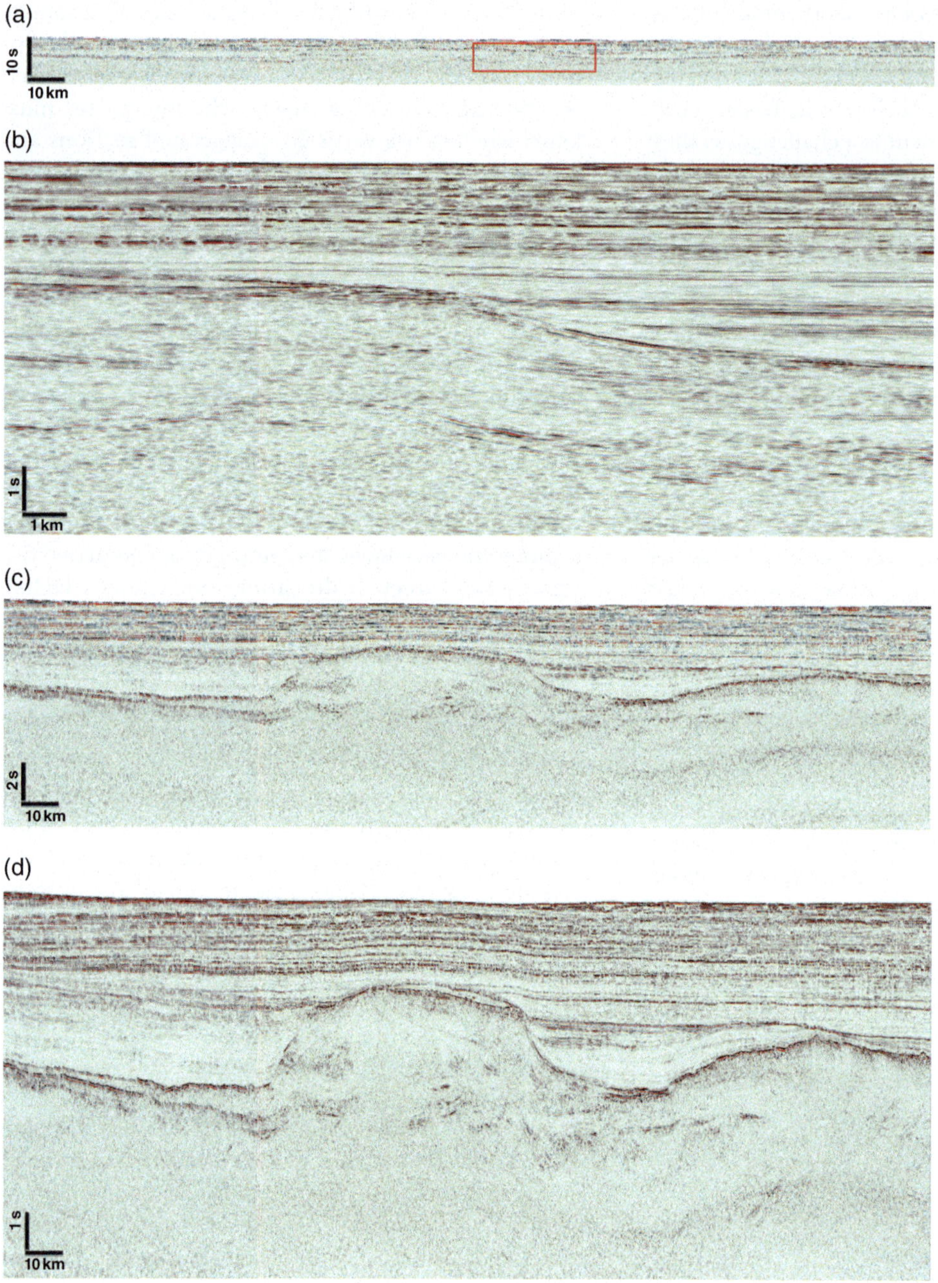

Figure 2.2 Part of a seismic section in two-way-time from the east Indian passive margin showing the 85° E Ridge. (a) 1:1 section, i.e. this section is not vertically exaggerated; (b) enlargement of the red rectangle in (a); (c) 5:1 and (d) 10:1 vertically exaggerated sections of the same line. The 85°E Ridge can be understood quite easily in (c) and the seismically transparent sedimentary cap atop the ridge becomes clear in (d). The 1:1 section is useful in observing the geometry of the clinoforms. Choosing the right vertical exaggeration (VE) is of prime importance in seismic interpretation. At regional scales a higher VE and at smaller scales a lower VE is commonly used. However, it is always better to change and use different VEs while interpreting a seismic profile. Seismic data courtesy Reliance Industries Ltd. Reproduced with permission from the Directorate General of Hydrocarbons (DGH), India.

along horizon or time-slices, of reflection seismic data or any of its derived attributes (Chapter 1.4) using the following criteria (Figures 2.3 and 2.4):

- *Offset in stratigraphic markers*: this is the easiest and the best way to identify faults. Displaced/slipped/offset stratigraphic markers deciphered in seismic image connote faults in most cases. In this connection, interpreters have to be careful for structures such as the oceanic fracture zones (Chapter 6.5).
- *Termination of reflections*: reflections may terminate due to a change in facies, for example an older sequence

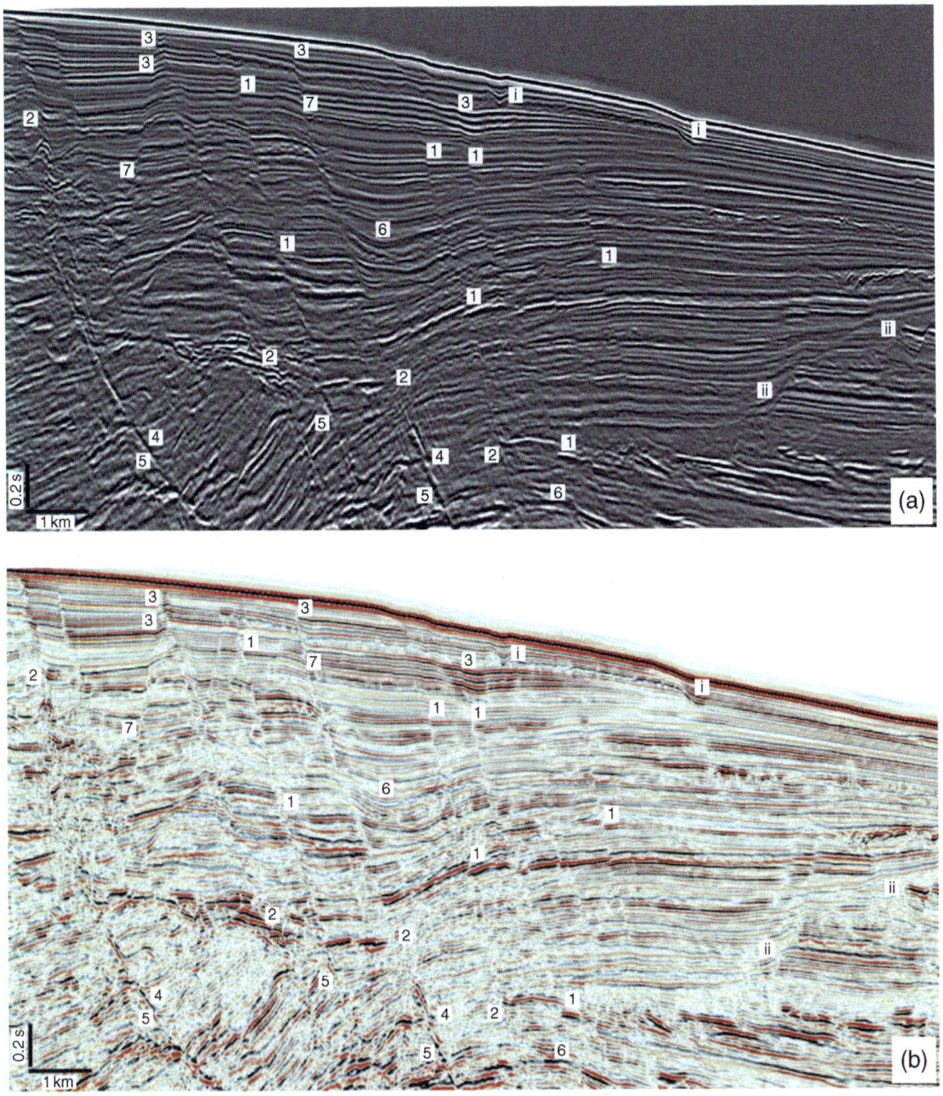

Figure 2.3 Criteria to identify faults on (a) grey scale and (b) coloured red-white-black sections. These different fault identification criteria are detailed in Section 2.3. Numbers correspond to those in Section 2.3. 1: offset in stratigraphic markers; 2: termination of reflections; 3: abrupt changes in dip; 4: abrupt changes in seismic patterns; 5: fault-plane reflections; 6: associated folding or sag; 7: fault shadow. Possible sedimentary features similar to fault identification criteria are: (i) abrupt change in dip due to a channel; (ii) termination of reflection due to canyon cuts. Seismic data courtesy Reliance Industries Ltd. Reproduced with permission from the the Directorate General of Hydrocarbons (DGH), India.

eroded and later filled by an 'incised valley fill'. Reflections may also terminate due to geophysical reasons, such as stacking problems. When such reasons have been ruled out, terminations of reflections would indicate faults. This is true when reflections on the hanging- or foot wall counterpart are unconformable. The interpreter, therefore, needs to use this criterion carefully (ii in Figure 2.3 are not faults).

- *Abrupt changes in dip*: parts of faults may not be interpretable on the seismic data because the total throw may remain undetected at a particular seismic resolution (Chapter 1.3.4). At such locations the reflections depict abruptly steepened dips. Almost invariably, there is an interpretable fault just below, that is in the deeper section of the reflector. If such a fault cannot be interpreted, the change in dip may not indicate a fault. Such a change in dip may also occur due to some sedimentological reason, such as slopes present in mass transport complexes (Moernaut *et al.*, 2009) (i in Figure 2.3).
- *Abrupt changes in seismic pattern/facies*: this is due to sedimentary or geophysical reasons. For example, seismic facies within an incised valley fill (IVF) differs remarkably from that outside it and a sharp boundary

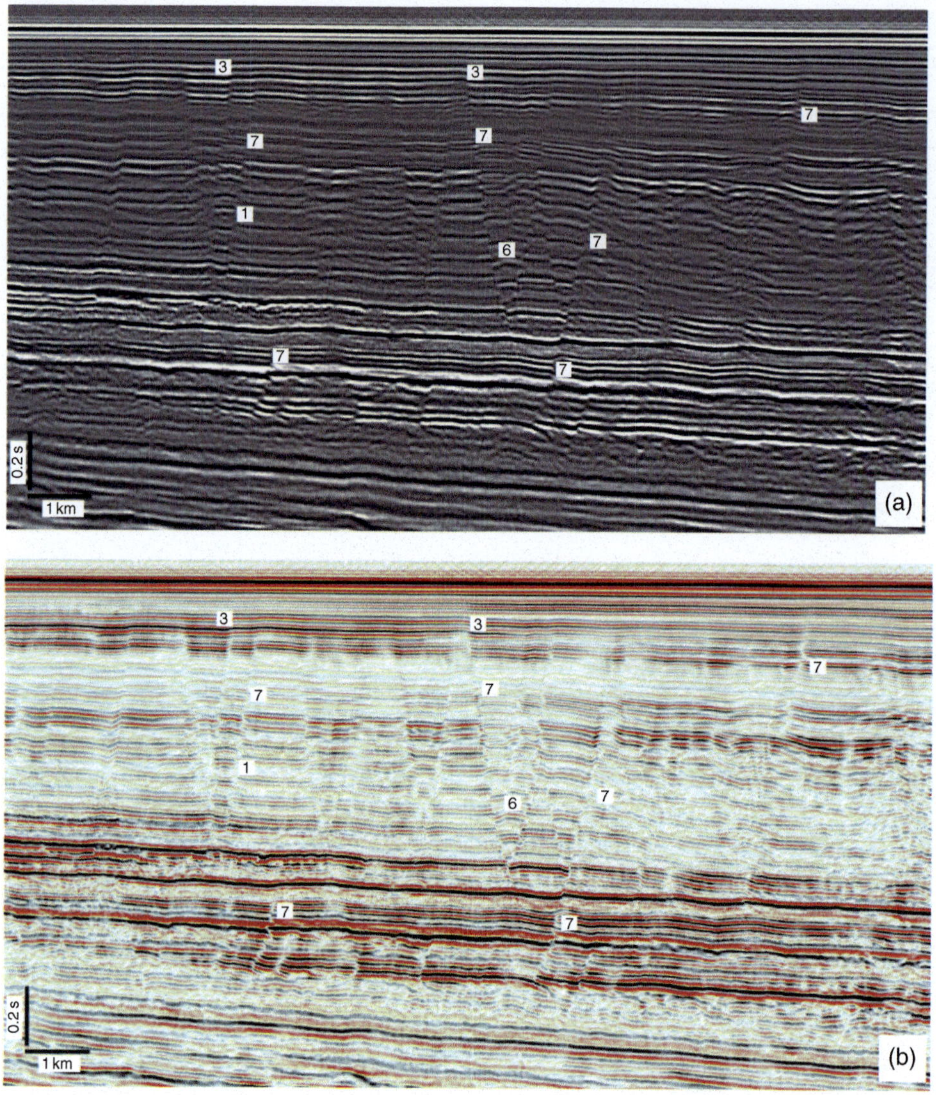

Figure 2.4 Criteria to identify faults on (a) grey scale and (b) coloured red-white-black sections. These different fault identification criteria are detailed in Section 2.3. The numbers in these figures correspond to those in Section 2.3. 1: offset in stratigraphic markers; 3: abrupt changes in dip; 6: associated folding or sag; 7: fault shadow. This seismic section shows good examples of fault shadows for fault identification. Seismic data courtesy Reliance Industries Ltd. Reproduced with permission from the the Directorate General of Hydrocarbons (DGH), India.

separates the IVF from the host facies. Nevertheless, incised valley fills can easily be identified from their own typical sedimentary seismic facies (Catuneanu, 2006; Nordfjord *et al.*, 2006; Chaumillon *et al.*, 2008; Green, 2009). When such possibilities are negated, abrupt changes in seismic facies would indicate faults.

- *Fault-plane reflections*: faults can be identified when they appear as reflections themselves. Fault-plane reflections appear due to (i) fluid movement/accumulation along fault planes (Haney *et al.*, 2005), (ii) pore fluid pressure differences between the hanging wall and the footwall or the fault zone itself from the surrounding rocks (Haney *et al.*, 2004, 2007), (iii) sedimentary injections along fault planes (Bureau *et al.*, 2013), and (iv) mineralizations along the fault plane (Drummond *et al.*, 2000). The first three points are common in sedimentary sections while the fourth one usually works within the basement.
- *Associated folding or sag*: folding and/or sagging of sediments is commonly associated with faulting (Passchier 2001; Mukherjee 2010a, 2010b, 2014b, 2017; Mukherjee and Koyi, 2009, 2010a, 2010b). Rollover antiforms commonly occur in hanging walls of listric growth faults (McClay, 1990; Magbagbeola and Willis, 2007; Ferrer *et al.*, 2008). Normal fault-related drag sags the hanging wall blocks juxtaposing the fault planes. Such sags are quite common along growth faults (e.g. Dasgupta and Mukherjee, 2017) (Chapter 3.2). Folds, both antiforms and synforms, are mostly associated with thrusts and can be visualized in reflection seismic

sections (Dong *et al.*, 2013; Schlische *et al.*, 2014; Zhang *et al.*, 2014; Li *et al.*, 2015). Folds on reflection seismic sections may, therefore, indicate faults. However, compactional folds, as opposed to drape folds, are not formed due to faults and, thus, many times are not related to faulting in the vicinity.

- *Fault shadow*: the amplitude may dim along a fault plane. This happens for two main reasons: (i) sharp lateral velocity variation across the two sides of the fault causing ray bending and (ii) diffractions created by the fault plane that may not have collapsed during migration.
- *Discontinuities*: Seismic discontinuities in lines (in 2D seismic) or in surfaces (in 3D seismic) indicate joints/fractures/fissures/ cracks or fault planes. Discontinuities are identified on vertical sections and orientation of the fault/fracture is seen in the horizon-based seismic attribute maps. This indication is often combined with other evidence to conclude the presence of a fault.
- *Convergence of evidence*: this is not a criterion in itself! When a number of criteria together indicate presence of faulting, while a single one fails, a fault may be present. Since interpretation, *per se*, is subjective, such a practice is common in seismic interpretation (Bally, 1983; Goudswaard and Jenyon, 1991; Tearpock and Bischke, 2002; Shaw *et al.*, 2006).

However, even after observing all the evidences above, minor faults, where the displacement is below seismic resolution, fractures and their orientation are often difficult to interpret in P-wave seismic data. Alfred rotation observed in converted shear waves can predict fracture orientation. Large fracture and fracture zones often show lowering of impedance and inverted data, especially Prestack, may be able to delineate fracture zones. All the faults marked in this atlas (refer to Parts I to IV) invariably use one or more of these types of evidences to identify faults.

The interpreted faults and folds must ultimately be analysed for a meaningful interpretation. Seismic data by themselves cannot unravel the subsurface geology and need to be complemented by other geoscientific data in almost invariably. Section 2.4 describes the most common studies on structural geology and tectonics usually performed on reflection seismic data. There are numerous other studies where a structural geological interpretation may apply.

2.4 Some Common Analyses

2.4.1 Geometry of Deformation Structures

This is a study aimed at the overall understanding of the deformation of the terrain. It is always carried out in a 3D seismic volume or on very close-spaced 2D lines processed as a 3D volume. The mapped fault networks and geometries are usually studied on maps to analyse the deformation style. Geometry can be studied on a scale from m up to km for different purposes. The objectives may vary widely: tsunami generation potential of faults (Moore *et al.*, 2007), basin formation studies (Okay *et al.*, 2000), deformation in hydrocarbon reservoirs (Jev *et al.*, 1993; Bahorich and Farmer, 1995; Chen and Sidney, 1997; Lonergan and Cartwright, 1999; Cacas *et al.*, 2001; Meldahl *et al.*, 2001) etc.

2.4.2 Crustal Architecture Analyses

The uniqueness of this study is that it can be performed on even a single 2D reflection seismic line subject to limitations. However, the required record length of the data often exceeds 12 seconds, cable lengths are large (>10 km) and the length of the seismic line surpasses 500 km. Examples of such seismic lines are the BABEL lines (Chapter 6.1–6.3), the BIRPS seismic lines around England (Klemperer and Hobbs, 1991), the SCREECH seismic lines in Canada (Funck *et al.*, 2003; Shillington *et al.*, 2006; Ranero and M Pérez-Gussinyé, 2010) and the DECORP-ECORS seismic data in Germany (Wenzel *et al.*, 1991; Brun and Gutscher, 1992). Additionally, such lines are also acquired for crustal architecture analysis for hydrocarbon exploration, for example the GXT India Span (Nemčok *et al.*, 2013; Misra, 2015; Misra *et al.*, 2015). Most of the present knowledge on rifts and passive margins has come from studies from such seismic sections (Manatschal, 2004; Péron-Pinvidic *et al.*, 2007; Jammes *et al.*, 2010; Bronner *et al.*, 2011; Sutra and Manatchal, 2012; Franke, 2013). The crustal architecture of compressional regimes is also studied on such lines (Jia *et al.*, 2006; Massoli *et al.*, 2006; Franke *et al.*, 2008; Marcaillou and Collot, 2008; Hesse *et al.*, 2010; Dong *et al.*, 2013; Rotevatn and Jackson, 2014). Submarine mounds, related oceanic crust and related tectonic structures are often studied using such data (Singh *et al.*, 2006a, 2006b; Autin *et al.*, 2010; Choudhuri *et al.*, 2014; Misra *et al.*, 2015). A large volume of other geophysical and geological data, such as seismic refraction, potential field, bore hole, outcrop data and so on, is required to support the analyses. These studies are often combined with gravity modelling to further understand the densities of the crustal domains and also magnetic modelling to find the ages of sea floor spreading, wherever present.

2.4.3 Progressive Deformation Analysis

Deformation progresses through time and fault planes reactivate through continuing stress. Such studies are carried out on very high resolution (~5–10 m), usually with 3D seismic data. Throw-depth (T-Z or Th-Z) plots are prepared to analyse the change of throw of the fault with time (Baudon and Cartwright, 2008 a, 2008b, 2008c; Pochat *et al.*, 2009). Throw-length (T-L) plots are prepared to study progress and linkage of faults (Tvedt *et al.*, 2013). To correlate the stratigraphy well data are required. Such studies are extremely important in hydrocarbon exploration, since trapping and sealing of faults reduce as faults reactivate. The timing of formation of traps vis-à-vis the timing of hydrocarbon generation is also matched to understand charging/ filling of the reservoir. Formation of structures and structural or combination traps can be studied here as well.

2.4.4 Strain Analysis

Strain and stress analyses are usually done using outcrop data. However, recently 3D seismic data have been used to analyse stress and strain (van Gent *et al.*, 2009, 2010). Very high resolution seismic data (~5–10 m) are required and the depth of imaging may be limited only down to the area of interest, that is mostly in the sedimentary section. The lithology interval for the analysis must be a brittle zone in the earlier mentioned study (van Gent *et al.*, 2009, 2010). This method of stress and strain analyses can only deduce the orientation of the principal stress axes and their relative magnitude for the last deformation episode.

2.5 Workflow

A question has often been asked whether a workflow exists for structural geological interpretation on reflection seismic data. Unfortunately, a workflow with check boxes to tick off while performing structural geological interpretation does not exist. The examples of some of the types of analyses that can be attempted on seismic data, which were presented in the previous sections, also reiterate that a single workflow cannot be devised for seismic structural analysis. It depends on the ultimate goal and the scale of observation. However, note that structural geological interpretation is inherently dynamic and must be done with the age-controlled horizon interpretation. Both of these compliment each other and cannot be, rather, *should not* be, performed in isolation. The usual pathway of structural geological interpretation is in harmony with geophysical data processing, sedimentology, stratigraphy, basin analysis (if available or possible) and structural geology. The structural geological interpreter needs to provide feedback to the all those research groups and receive their inputs to improve the interpretation.

The very first step must be to prepare a seismic 'fitness' map. A seismic fitness map may be a representation of the signal-to-noise ratio (Chapter 1.3.1) of the seismic data. It shows the areas of the data set that have good (or 'fit') and poor (or 'unfit') data quality. Such a map is required when a data set contains input from different vintages. This map will lead to an understanding about what sort of analysis can be performed with the seismic data available and how much improvement may be needed to solve the problem. In the next step, the aim or the problem is to be decided. This follows by structural and stratigraphic interpretation. At advanced stages drilled well data, petrophysics, bore hole geophysics etc. also involve to constrain the seismic interpretation. The seismic depth volume must be calibrated with velocity data from available drilled wells. This calibrated depth volume is to be used for structural and sequence stratigraphic interpretation. The structural geological interpretation is updated continuously with input of such data. It is important to note that an interpretation must always be validated for 'admissibility' and 'viability' by balancing the interpreted seismic section, palinspastic restorations and modelling. The time spent on validation to correctly understand the geology is always better than going through unwanted surprises during or after drilling. Structural interpretation of seismic images from offshore should be in conformity with the tectonics of the terrain as a whole as understood from onshore studies (Misra and Mukherjee 2017; Kaplay *et al.*, 2017; Babar *et al.*, 2017).

References

Angelier, J. 1979. Determination of the mean principal directions of stresses for a given fault population. *Tectonophysics*, **56**, T17–T26.

Angelier, J. 1984. Tectonic analysis of fault slip data sets. *Journal of Geophysical Research: Solid Earth*, **89**, 5835–5848.

Angelier, J. 1989. From orientation to magnitudes in paleostress determinations using fault slip data. *Journal of Structural Geology*, **11**, 37–50.

Autin, J., Leroy, S., Beslier, M.O. *et al.* 2010. Continental break-up history of a deep magma-poor margin based on seismic reflection data (northeastern Gulf of Aden margin, offshore Oman). *Geophysical Journal International*, **180**, 501–519.

Babar, M.D., Kaplay, R.D., Mukherjee, S. and Kulkarni, P.S. 2017. Evidences of deformation of dykes from Central Deccan Volcanic Province, Aurangabad, Maharashtra, India. In: Mukherjee, S., Misra, A.A., Calvès, G. and

Nemčok, M. (eds) *Tectonics of the Deccan Large Igneous Province: An Introduction*. Special Publications vol. 445, The Geological Society, London, pp. 337–353. doi: http://doi.org/10.1144/SP445.13.

Bally, A.W. 1983. *Seismic Expression of Structural Styles*. AAPG Studies in Geology, #15, 3 volumes.

Bahorich, M. and Farmer, S. 1995. 3-D seismic discontinuity for faults and stratigraphic features: The coherence cube. *The Leading Edge*, **14**, 1053–1058.

Baudon, C. and Cartwright, J. 2008a. 3D seismic characterisation of an array of blind normal faults in the Levant Basin, Eastern Mediterranean. *Journal of Structural Geology*, **30**, 746–760.

Baudon, C. and Cartwright, J. 2008b. Early stage evolution of growth faults: 3D seismic insights from the Levant Basin, Eastern Mediterranean. *Journal of Structural Geology*, **30**, 888–898.

Baudon, C. and Cartwright, J. 2008c. The kinematics of reactivation of normal faults using high resolution throw mapping. *Journal of Structural Geology*, **30**, 1072–1084.

Bronner, A., Sauter, D., Manatschal, G. *et al.* 2011. Magmatic breakup as an explanation for magnetic anomalies at magma-poor rifted margins. *Nature Geoscience*, **4**, 549–553.

Brun, J.P. and Gutscher, M.A. 1992. Deep crustal structure of the Rhine Graben from DEKORP-ECORS seismic reflection data: a summary. *Tectonophysics*, **208**, 139–147.

Bureau, D., Mourgues, R., Cartwright, J. *et al.* 2013. Characterisation of interactions between a pre-existing polygonal fault system and sandstone intrusions and the determination of paleo-stresses in the Faroe-Shetland basin. *Journal of Structural Geology*, **46**, 186–199.

Cacas, M.C., Daniel, J.M. and Letouzey, J. 2001. Nested geological modelling of naturally fractured reservoirs. *Petroleum Geoscience*, **7**, S43–S52.

Catuneanu, O. 2006. *Principles of Sequence Stratigraphy*. Elsevier.

Chaumillon, E., Proust, J.N., Menier, D. and Weber, N. 2008. Incised-valley morphologies and sedimentary-fills within the inner shelf of the Bay of Biscay (France): a synthesis. *Journal of Marine Systems*, **72**(1), 383–396.

Chen, Q. and Sidney, S. 1997. Seismic attribute technology for reservoir forecasting and monitoring. *The Leading Edge*, **16**, 445–448.

Choudhuri, M., Nemčok, M., Stuart, C. *et al.* 2014. 85° E Ridge, India – constraints on its development and architecture. *Journal of the Geological Society of India*, **84**, 513–530.

Dahlstrom, C.D.A. 1969. Balanced cross sections. *Canadian Journal of Earth Sciences*, **6**, 743–757.

Dasgupta, S., Mukherjee, S. 2017. Brittle shear tectonics in a narrow continental rift: asymmetric non-volcanic Barmer basin (Rajasthan, India). *The Journal of Geology*, **125**, 561–591.

Davis, G.H., Reynolds, S.J. and Kluth, C. 2012. *Structural Geology of Rocks and Regions*, 3rd edn. John Wiley & Sons. Hoboken, NJ.

Dong, S., Gao, R., Yin, A. *et al.* 2013. What drove continued continent-continent convergence after ocean closure? Insights from high-resolution seismic-reflection profiling across the Daba Shan in central China. *Geology*, **41**, 671–674.

Drummond, B.J., Goleby, B.R., Owen, A.J. *et al.* 2000. Seismic reflection imaging of mineral systems: Three case histories. *Geophysics*, **65**, 1852–1861.

Ferrer, O., Roca, E., Benjumea, B. *et al.* 2008. The deep seismic reflection MARCONI-3 profile: Role of extensional Mesozoic structure during the Pyrenean contractional deformation at the eastern part of the Bay of Biscay. *Marine and Petroleum Geology*, **25**, 714–730.

Fossen, H. 2016. *Structural Geology*. Cambridge University Press.

Franke, D. 2013. Rifting, lithosphere breakup and volcanism: Comparison of magma-poor and volcanic rifted margins. *Marine and Petroleum Geology*, **43**, 63–87.

Franke, D., Barckhausen, U., Heyde, I. *et al.* 2008. Seismic images of a collision zone offshore NW Sabah/Borneo. *Marine and Petroleum Geology*, **25**, 606–624.

Funck, T., Hopper, J.R., Larsen, H.C. *et al.* 2003. Crustal structure of the ocean-continent transition at Flemish Cap: Seismic refraction results. *Journal of Geophysical Research: Solid Earth*, **108**(B11), 2531.

Goudswaard, W.B. and Jenyon, M. (eds). 1991. *Seismic Atlas of Structural and Stratigraphic Features*. European Association of Exploration Geophysicists.

Green, A.N. 2009. Palaeo-drainage, incised valley fills and transgressive systems tract sedimentation of the northern KwaZulu-Natal continental shelf, South Africa, SW Indian Ocean. *Marine Geology*, **263**(1), 46–63.

Haney, M., Sheiman, J., Snieder, R. *et al.* 2004. Fault-plane reflections as a diagnostic of pressure differences in reservoirs: a case study from South Eugene Island Block 330. *Proceedings of the EAGE special session on Fault and Top Seals*. www.cwp.mines.edu/Documents/cwpreports/cwp456.pdf (last accessed 5 September2017).

Haney, M.M., Snieder, R., Sheiman, J. and Losh, S. 2005. Geophysics: A moving fluid pulse in a fault zone. *Nature*, **437**, 46.

Haney, M., Snieder, R., Ampuero, J.P. and Hofmann, R. 2007. Spectral element modelling of fault-plane reflections arising from fluid pressure distributions. *Geophysical Journal International*, **170**, 933–951.

Hesse, S., Back, S. and Franke, D. 2010. The structural evolution of folds in a deepwater fold and thrust belt – a case study from the Sabah continental margin offshore NW Borneo, SE Asia. *Marine and Petroleum Geology*, **27**, 442–454.

Jackson, C.A., Jackson, M.P., Hudec, M.R. and Rodriguez, C. 2014. Internal structure, kinematics, and growth of a salt wall: Insights from 3-D seismic data. *Geology*, **42**, 307–310.

Jammes, S., Tiberi, C. and Manatschal, G. 2010. 3D architecture of a complex transcurrent rift system: the example of the Bay of Biscay–Western Pyrenees. *Tectonophysics*, **489**, 210–226.

Jev, B.I., Kaars-Sijpesteijn, C.H., Peters, M.P.A.M. *et al.* 1993. Akaso Field, Nigeria: use of integrated 3-D seismic, fault slicing, clay smearing, and RFT pressure data on fault trapping and dynamic leakage. *AAPG Bulletin*, **77**, 1389–1404.

Jia, D., Wei, G., Chen, Z. *et al.* 2006. Longmen Shan fold-thrust belt and its relation to the western Sichuan Basin in central China: New insights from hydrocarbon exploration. *AAPG Bulletin*, **90**, 1425–1447.

Kaplay R.D., Babar M.D., Mukherjee S., Kumar T.V. 2017. Morphotectonic expression of geological structures in eastern part of south east Deccan volcanic province (around Nanded, Maharashtra, India) In: Mukherjee, S., Misra, A.A., Calvès, G. and Nemčok, M. (eds) *Tectonics of the Deccan Large Igneous Province: An Introduction.* Special Publications vol. 445, The Geological Society, London, pp. 317–335. doi: http://doi.org/10.1144/SP445.12

Klemperer, S.L. and Hobbs, R.W. 1991. *The BIRPS Atlas: Deep Seismic Reflections Profiles Around the British Isles.* Cambridge University Press, Cambridge.

Li, J., Dong, S., Yin, A. *et al.* 2015. Mesozoic tectonic evolution of the Daba Shan Thrust Belt in the southern Qinling orogen, central China: Constraints from surface geology and reflection seismology. *Tectonics*, **34**, 1545–1575. doi: 10.1002/2014TC003813.

Liner, C.L. 2004. *Elements of 3D Seismology.* 2nd edn. PennWell Books, Tulsa, OK.

Lisle, R.J. 1989. Paleostress analysis from sheared dike sets. *Geological Society of America Bulletin*, **101**, 968–972.

Lonergan, L. and Cartwright, J.A.,1999. Polygonal faults and their influence on deep-water sandstone reservoir geometries, Alba Field, United Kingdom central North Sea. *AAPG Bulletin*, **83**, 410–432.

Magbagbeola, O.A. and Willis, B.J. 2007. Sequence stratigraphy and syndepositional deformation of the Agbada Formation, Robertkiri field, Niger Delta, Nigeria. *AAPG Bulletin*, **91**, 945–958.

Manatschal, G. 2004. New models for evolution of magma-poor rifted margins based on a review of data and concepts from West Iberia and the Alps. *International Journal of Earth Sciences*, **93**, 432–466.

Marcaillou, B. and Collot, J.Y. 2008. Chronostratigraphy and tectonic deformation of the North Ecuadorian–South Colombian offshore Manglares forearc basin. *Marine Geology*, **255**, 30–44.

Massoli, D., Koyi, H.A. and Barchi, M.R. 2006. Structural evolution of a fold and thrust belt generated by multiple décollements: analogue models and natural examples from the Northern Apennines (Italy). *Journal of Structural Geology*, **28**, 185–199.

McClay, K.R. 1990. Extensional fault systems in sedimentary basins: a review of analogue model studies. *Marine and Petroleum Geology*, **7**, 206–233.

Meldahl, P., Heggland, R., Bril, B. and de Groot, P. 2001. Identifying faults and gas chimneys using multiattributes and neural networks. *The Leading Edge*, **20**, 474–482.

Misra, A.A. 2015. *Sheared Nature of the Indian Western Continental Margin around Mumbai, India: Onshore and Offshore Geoscientific Studies.* PhD Thesis, Indian Institute of Technology Bombay, India.

Misra, A.A. and Mukherjee, S. 2015. *Tectonic Inheritance in Continental Rifts and Passive Margins.* Springer.

Misra, A.A. and Mukherjee, S. 2017. Dyke-brittle shear relationships in the Western Deccan Strike Slip Zone around Mumbai (Maharashtra, India). In: Mukherjee, S., Misra, A.A., Calvès, G. and Nemčok, M. (eds) *Tectonics of the Deccan Large Igneous Province: An Introduction.* Special Publications vol. 445, The Geological Society, London, pp. 269–295. doi: 10.1144/SP445.4.

Misra, A.A., Bhattacharya, G., Mukherjee, S. and Bose, N. 2014. Near N-S paleo-extension in the western Deccan region in India: Does it link strike-slip tectonics with India-Seychelles rifting? *International Journal of Earth Sciences*, **103**, 1645–1680.

Misra, A.A., Sinha, N. and Mukherjee, S. 2015. Repeat ridge jumps and microcontinent separation: insights from NE Arabian Sea. *Marine and Petroleum Geology*, **59**, 406–428.

Moernaut, J., De Batist, M., Heirman, K. *et al.* 2009. Fluidization of buried mass-wasting deposits in lake sediments and its relevance for paleoseismology: results from a reflection seismic study of lakes Villarrica and Calafquén (South-Central Chile). *Sedimentary Geology*, **213**, 121–135.

Moore, G.F., Bangs, N.L., Taira, A. *et al.* 2007. Three-dimensional splay fault geometry and implications for tsunami generation. *Science*, **318**, 1128–1131.

Mukherjee, S. 2010a. Microstructures of the Zanskar Shear Zone. *Earth Science India*, **3**, 9–27.

Mukherjee, S. 2010b. Structures in Meso- and Micro-scales in the Sutlej section of the Higher Himalayan Shear Zone, Indian Himalaya. *e-Terra*, **7**, 1–27.

Mukherjee, S. 2011. Flanking microstructures from the Zanskar Shear Zone, NW Indian Himalaya. *YES Bulletin*, **1**, 21–29.

Mukherjee, S. 2013. *Deformation Microstructures in Rocks.* Springer, Berlin.

Mukherjee, S. 2014a. *Atlas of Shear Zone Structures in Meso-scale.* Springer.

Mukherjee, S. 2014b. Review of flanking structures in meso- and micro-scales. *Geological Magazine*, **151**, 957–974.

Mukherjee, S. 2015. *Atlas of Structural Geology*. Elsevier, Amsterdam.

Mukherjee, S. 2017. Review on symmetric structures in ductile shear zones. *International Journal of Earth Sciences*, **106**, 1453–1468.

Mukherjee, S. 2017. Airy's isostatic model: a proposal for a realistic case. Arabian *Journal of Geosciences*, **10**, 268.

Mukherjee, S. and Biswas, R. 2014. Kinematics of horizontal simple shear zones of concentric arcs (Taylor Couette flow) with incompressible Newtonian rheology. *International Journal of Earth Sciences*, **103**, 597–602.

Mukherjee, S. and Biswas, R. 2015. Biviscous horizontal simple shear zones of concentric arcs (Taylor Couette flow) with incompressible Newtonian rheology. In: Mukherjee S, Mulchrone KF (Eds) Ductile Shear Zones: From Micro – to Macro-scales. Wiley Blackwell. pp. 59–62.

Mukherjee, S. and Koyi, H. A. 2009. Flanking Microstructures. *Geological Magazine*, **146**, 517–526.

Mukherjee, S. and Koyi, H. A., 2010a. Higher Himalayan Shear Zone, Sutlej Section- Structural Geology & Extrusion Mechanism by Various Combinations of Simple Shear, Pure Shear & Channel Flow in Shifting Modes. *International Journal of Earth Sciences*, **99**, 1267–1303.

Mukherjee, S. and Koyi, H. A., 2010b. Higher Himalayan Shear Zone, Zanskar Section- Microstructural Studies & Extrusion Mechanism by a Combination of Simple Shear & Channel Flow. *International Journal of Earth Sciences*, **99**, 1083–1110.

Mukherjee, S., Koyi, H.A. and Talbot, C.J. 2012. Implications of channel flow analogue models in extrusion of the Higher Himalayan Shear Zone with special reference to the out-of-sequence thrusting. *International Journal of Earth Sciences*, **101**, 253–272.

Mukherjee, S. and Mulchrone, K. 2012. Estimating the viscosity and Prandtl number of the Tso Morari Gneiss Dome, western Indian Himalaya. *International Journal of Earth Sciences*, **101**, 1929–1947.

Mukherjee, S., Punekar, J., Mahadani, T. and Mukherjee, R., 2015. A review on intrafolial folds and their morphologies from the detachments of the western Indian Higher Himalaya. In: Mukherjee S, Mulchrone KF (Eds) Ductile Shear Zones: From Micro- to Macro-scales. Wiley Blackwell pp. 182–205.

Mukherjee, S., Talbot, C.J. and Koyi, H.A. 2010. Viscosity estimates of salt in the Hormuz and Namakdan salt diapirs, Persian Gulf. *Geological Magazine*, **147**, 497–507.

Nemčok, M., Schamel, S. and Gayer, R., 2009. *Thrustbelts: Structural architecture, thermal regimes and petroleum systems*. Cambridge University Press.

Nemčok, M., Sinha, S.T., Stuart, C.J. *et al.* 2013. East Indian margin evolution and crustal architecture: integration of deep reflection seismic interpretation and gravity modelling. In: Mohriak, W.U., Danforth, A. L.; Post, P. J. *et al.* (Eds.), *Conjugate Divergent Margins*. Special Publications vol. 369, The Geological Society, London, pp. 477–496.

Nemčok, M., Sinha, S.T., Doré, A.G. *et al.* 2016. Mechanisms of microcontinent release associated with wrenching-involved continental break-up; a review. In: Nemčok, M., Rybár, S., Sinha, S.T. *et al.* (eds.) *Transform Margins: Development, Controls and Petroleum Systems*. Special Publications vol. 431, The Geological Society, London, pp. 323–359. http://doi.org/10.1144/SP431.14

Nordfjord, S., Goff, J.A., Austin, J.A. and Gulick, S.P.S., 2006. Seismic facies of incised-valley fills, New Jersey continental shelf: implications for erosion and preservation processes acting during latest Pleistocene–Holocene transgression. *Journal of Sedimentary Research*, **76**(12), 1284–1303.

Okay, A.I., Kaşlılar-Özcan, A., Imren, C. *et al.* 2000. Active faults and evolving strike-slip basins in the Marmara Sea, northwest Turkey: a multichannel seismic reflection study. *Tectonophysics*, **321**, 189–218.

Park, R.G., 2013. *Foundation of Structural Geology*. Routledge.

Passchier, C.W. 2001. Flanking Structures. *Journal of Structural Geology*, **23**, 951–962.

Passchier, C.W. and Trouw, R.A. 2005. *Microtectonics*. 2nd Ed. Springer Verlag pp. 366.

Péron-Pinvidic, G., Manatschal, G., Minshull, T.A. and Sawyer, D.S. 2007. Tectonosedimentary evolution of the deep Iberia-Newfoundland margins: Evidence for a complex breakup history. *Tectonics*, **26**, TC2011.

Pochat, S., Castelltort, S., Choblet, G. and Van Den Driessche, J. 2009. High-resolution record of tectonic and sedimentary processes in growth strata. *Marine and Petroleum Geology*, **26**, 1350–1364.

Ramsay, J.G. and Huber, M.I. 1987. *The techniques of modern structural geology: Folds and Fractures* (Vol. **2**). Academic Press.

Ranero, C.R. and Pérez-Gussinyé, M. 2010. Sequential faulting explains the asymmetry and extension discrepancy of conjugate margins. *Nature*, **468**, 294–299.

Ross, C. 2000. Effective AVO cross-plot Modeling, *Geophysics*, **65**, 3.

Rotevatn, A. and Jackson, C.A.L. 2014. 3D structure and evolution of folds during normal fault dip linkage. *Journal of the Geological Society*, **171**, 821–829.

Rowland, S.M., Duebendorfer, E.M. and Schiefelbein, I.M. 2013. *Structural Analysis and Synthesis: A Laboratory Course in Structural Geology*, 3rd edn. John Wiley & Sons.

Schlische, R.W., Groshong, R.H., Withjack, M.O. and Hidayah, T.N. 2014. Quantifying the geometry, displacements, and subresolution deformation in thrust-ramp anticlines with growth and erosion: From models to seismic-reflection profile. *Journal of Structural Geology*, **69**, 304–319.

Shaw, J.H., Christopher, C. and Suppe, J. 2006. *Seismic Interpretation of Contractional Fault-Related Folds: An AAPG Seismic Atlas.* AAPG Studies in Geology #53, American Association of Petroleum Geologists, Tulsa, OK.

Shillington, D.J., Holbrook, W.S., Van Avendonk, H.J. *et al.* 2006. Evidence for asymmetric nonvolcanic rifting and slow incipient oceanic accretion from seismic reflection data on the Newfoundland margin. *Journal of Geophysical Research: Solid Earth*, **111**(B9).

Singh, S.C., Crawford, W.C., Carton, H. *et al.* 2006a. Discovery of a magma chamber and faults beneath a Mid-Atlantic Ridge hydrothermal field. *Nature*, **442**, 1029–1032.

Singh, S.C., Harding, A.J., Kent, G.M. *et al.* 2006b. Seismic reflection images of the Moho underlying melt sills at the East Pacific Rise. *Nature*, **442**, 287–290.

Srivastava, D.C., Lisle, R.J. and Vandycke, S. 1995. Shear zones as a new type of palaeostress indicator. *Journal of Structural Geology*, **17**, 663–676.

Sutra, E. and Manatschal, G. 2012. How does the continental crust thin in a hyperextended rifted margin? Insights from the Iberia margin. *Geology*, **40**(2), 139–142.

Tearpock, D.J. and Bischke, R.E. 2002. *Applied Subsurface Geological Mapping with Structural Methods.* Prentice Hall.

Twiss, R.J. and Moores, E.M. 2007. *Structural Geology.* W.H. Freeman.

Tvedt, A.B., Rotevatn, A., Jackson, C.A.L. *et al.* 2013. Growth of normal faults in multilayer sequences: a 3D seismic case study from the Egersund Basin, Norwegian North Sea. *Journal of Structural Geology*, **55**, 1–20.

van der Pluijm, B. and Marshak, S. 2004. *Earth structures: An Introduction to Structural Geology and Tectonics*, 2nd edn. W.W. Norton & Co.

van Gent, H., Back, S., Urai, J.L. and Kukla, P. 2010. Small-scale faulting in the Upper Cretaceous of the Groningen block (The Netherlands): 3D seismic interpretation, fault plane analysis and regional paleostress. *Journal of Structural Geology*, **32**, 537–553.

van Gent, H., Urai, J.L. and De Keijzer, M. 2011. The internal geometry of salt structures–A first look using 3D seismic data from the Zechstein of the Netherlands. *Journal of Structural Geology*, **33**, 292–311.

van Gent, H.W., Back, S., Urai, J.L. *et al.* 2009. Paleostresses of the Groningen area, the Netherlands – Results of a seismic based structural reconstruction. *Tectonophysics*, **470**, 147–161.

Wiederhold, H., 2005. Poor Man's 3D – A Simple Approach to 3D Seismic Surveying: A Case History. In: Butler, D.K. (ed) *Near-Surface Geophysics.* SEG Investigations in Geophysics Series No. 13, Society of Exploration Geophysicists, Tulsa, OK, pp. 695–706.

Wenzel, F., Brun, J.P. and ECORS-DEKORP Working Group. 1991. A deep reflection seismic line across the Northern Rhine Graben. *Earth and Planetary Science Letters*, **104**(2), 140–150.

Žalohar, J. and Vrabec, M. 2007. Paleostress analysis of heterogeneous fault-slip data: The Gauss method. *Journal of Structural Geology*, **29**, 1798–1810.

Zhang, S., Gao, R., Li, H. *et al.* 2014. Crustal structures revealed from a deep seismic reflection profile across the Solonker suture zone of the Central Asian Orogenic Belt, northern China: An integrated interpretation. *Tectonophysics*, **612**, 26–39.

Part I

Faults

Faults are blocks of rocks moving past each other. In this section, examples are given of:

- Normal faults (Chapters 3, 4, 6, 8, 15, 16, 18, 19, 20, 24)
- Reverse faults/Thrusts (Chapter 5)
- Strike-slip faults (Chapters 17, 18, 23)
- Fault linkages/transfers/relay ramps (Chapter 21)
- Reactivation and inversion (Chapters 7, 9, 10, 12, 13, 22)
- Sediment interactions with normal fault (Chapters 3, 4, 14, 21, 22, 24)

3

Normal Faults in the Crawford-Skipper Basin, East Shetland Platform, United Kingdom

*William Reid and Stefano Patruno**

Petroleum Geo-Services (PGS), Weybridge, UK
** Stefano.Patruno@pgs.com*

The area between the East Shetland Platform and the western edge of the Viking Graben, called the Crawford-Skipper Basin (Patruno & Reid, 2017) (Figure 3.1), has undergone significant pre-Tertiary tectonics, with different phases of active extension and at least one event of compressional inversion, resulting in complex patterns of fault reactivations and formation of new faults. Extension started in the Devonian Period (see Patruno & Reid, 2016, 2017; Patruno, 2017; Patruno *et al.*, in press). A major syn-rift wedge can be observed in the Devonian unit (e.g., in the half-graben formed by Fault 1), which is known to have resulted from an extensional collapse of the Caledonian orogen (Coward *et al.*, 1989; Seranne, 1992; Ziegler, 1992; Marshall and Hewett, 2003; Wilson *et al.*, 2010). The second extensional event is Carboniferous to Rotliegend in age (Stemmerik *et al.*, 2000; Kombrink *et al.*, 2010) and is highlighted by Fault 5 and by a minor reactivation of Fault 1, forming a minor Carboniferous/Rotliegend syn-rift wedge. Subsequently, the regional Late Permian to Triassic rifting event (Glennie, 1995; Balson *et al.*, 2002; Goldsmith *et al.*, 2003; Wilson *et al.*, 2010; Whipp *et al.*, 2014; Patruno & Reid, 2017; Patruno, 2017; Patruno *et al.*, in press) resulted in the structuration of most of the fault blocks observed in this section. Newly-formed or reactivated Permo-Triassic extensional faults include faults 1–7, together with the relatively minor faults situated to the west of Fault 1. In this particular location, the seismic stratigraphic relationships reveal that most of the Permo-Triassic syn-depositional fault movement occurred in the Triassic.

This is reflected by Triassic syn-rift wedges (e.g., to the hangingwall of Fault 1, faults 3–4 and 6–7), with only minor possible Zechstein thickening (e.g., Fault 4). In particular, the Triassic fill to the hangingwall of faults 1 and 2 represents the vast majority of the total Mesozoic sediments preserved along this section of the East Shetland Platform. Low-magnitude Jurassic faulting can be inferred in the grabens and half-grabens delimited by faults 1 and 6–7, where a thin Jurassic fill has been preserved (Figure 3.2a). Major upper Jurassic syn-rift wedges are observed on the Viking Graben (Ziegler, 1992; Fraser *et al.*, 2003; Turner *et al.*, in press). Footwall uplift of the Viking Graben border faults might be responsible for the lack of upper Jurassic strata in the wells penetrating this part of the East Shetland Platform and Crawford Spur area. Cretaceous reflectors gradually onlap the Viking Graben tectonic scarp and a thin to very thin upper Cretaceous veneer is observed over the platform.

A relatively significant compressional inversion of the main Triassic syn-rift wedge can be noticed in the centres of the diagrams in Figure 3.2a, with 'Base Jurassic', 'Base Cretaceous' and 'Base Tertiary' seismic horizons all subject to antiformal folding. This compression is probably of 'Early Alpine' age, as constrained by: (i) sub-horizontal lower Paleocene reflectors onlapping the antiformal structure and (ii) Cretaceous strata becoming anomalously thin towards the antiformal culmination, which corresponds to the Triassic hangingwall depocentres of faults 1 and 2. Alpine age compressions might be as early as Aptian in age (c.f., Patruno *et al.*, 2015; Unida & Patruno, 2016).

Furthermore, on the structural culmination of the Crawford Spur (horst bounded by Faults 5 and 6), two angular unconformities are visible. In particular, the uppermost Devonian is truncated by the 'Base Permian Unconformity' and the Rotliegend itself is partly truncated by the Base Zechstein. This could be the result of combined compressional uplift in Variscan times (Middle Carboniferous to Middle Permian) (Coward *et al.*, 1989; Seranne, 1992; Corfield *et al.*, 1996; Balson *et al.*, 2002; Zanella and Coward, 2003; Patruno, 2017; Patruno & Reid, 2016, 2017; Patruno *et al.*, in press) and

Atlas of Structural Geological Interpretation from Seismic Images, First Edition. Edited by Achyuta Ayan Misra and Soumyajit Mukherjee.

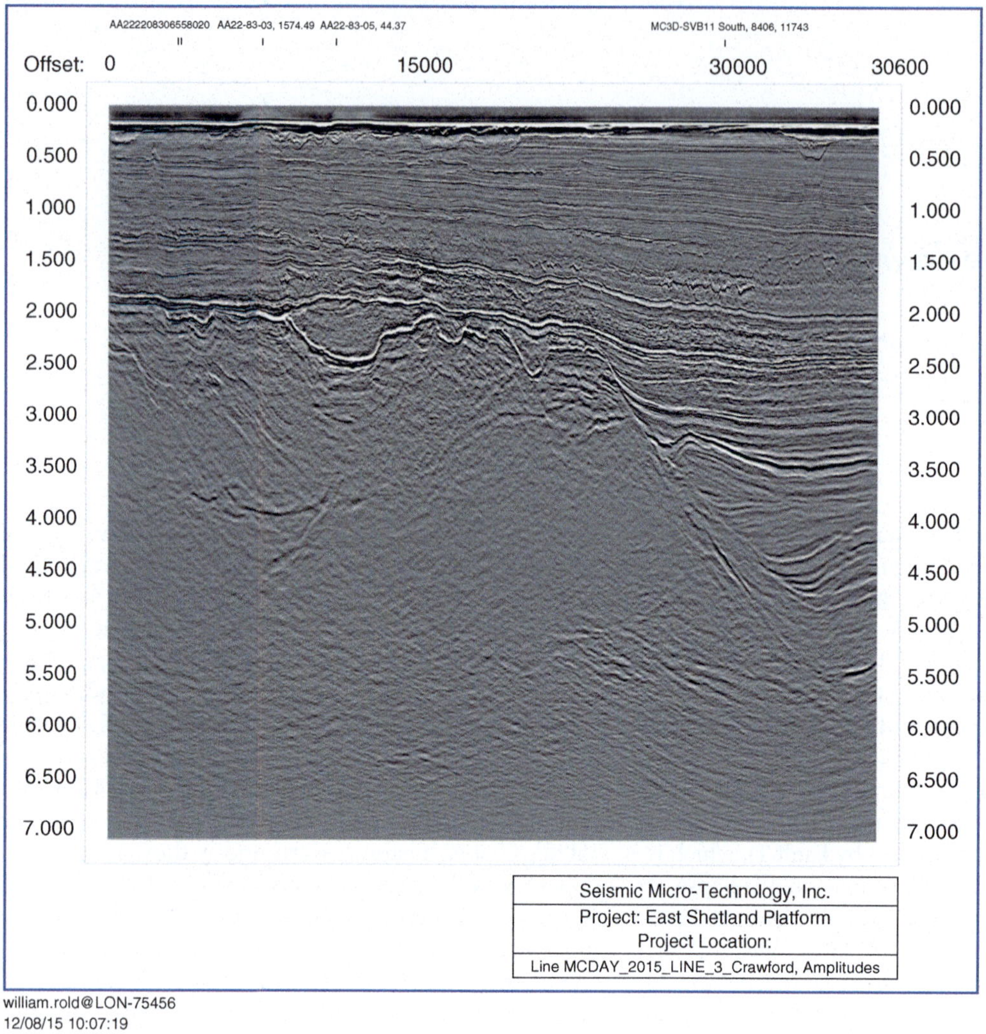

Figure 3.1 West–east oriented uninterpreted seismic cross-section along the easternmost East Shetland Platform, Crawford Spur and the western edge of the Viking Graben (UKCS Quadrant 9).

footwall uplift during the various Carboniferous to Jurassic extensional events.

In the Triassic main reservoir interval (Heron Group), the Crawford wells recorded a dip of over 40° (Yaliz, 1991). This dip is difficult to distinguish when looking at the Triassic reflectors in the full-stack data (Figure 3.2b(C)). The well dipmeter data are only properly understood by examining the relative acoustic impedance section (Figure 3.2b(D)). The relative acoustic impedance section extracted from broadband GeoStreamer® data shows a clear distinction between the Jurassic interval and the Triassic package, and it reveals westerly-dipping Triassic reflectors associated with at least three minor listric faults, detached from the top of the Zechstein (Figure 3.2b(D)).

The Crawford Field is situated within this heavily compartmentalized interval on top of a pre-existing Paleozoic horst. This oil and gas field contains a combination of fault- and unconformity-related traps and dip closures, with reservoir components within Triassic sandstones (Skagerrak and Cormorant Formations, Heron Group), Bajocian-Bathonian sandstones (Hugin Formation, Pentland Formation), Campanian limestones (Shetland Group) and upper Paleocene Heimdal sandstones (Yaliz, 1991). The Crawford Field contains 20.6 million barrels of oil and 38.1 billion cubic feet of gas as ultimately recoverable reserves and proves the importance of understanding and imaging the subtle fault-related structures of the East Shetland Platform.

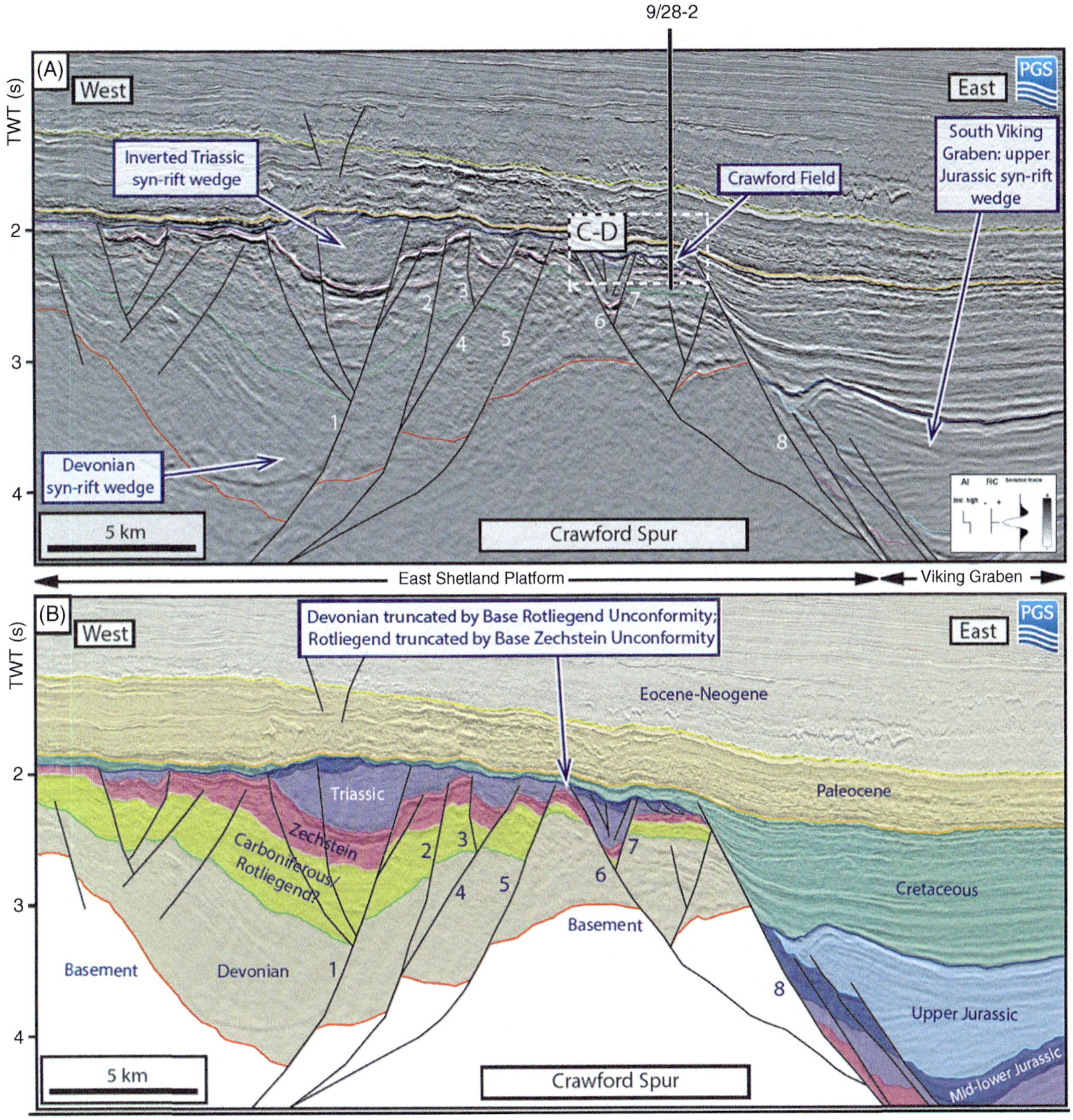

Figure 3.2a West–east oriented interpreted seismic cross-section (A) and geoseismic section (B) across the eastern-most East Shetland Platform, Crawford Spur and the western edge of the Viking Graben (UKCS Quadrant 9). Adapted from Patruno & Reid 2017. Courtesy of PGS GeoStreamer® MC3D-Q16-2013 3D survey.

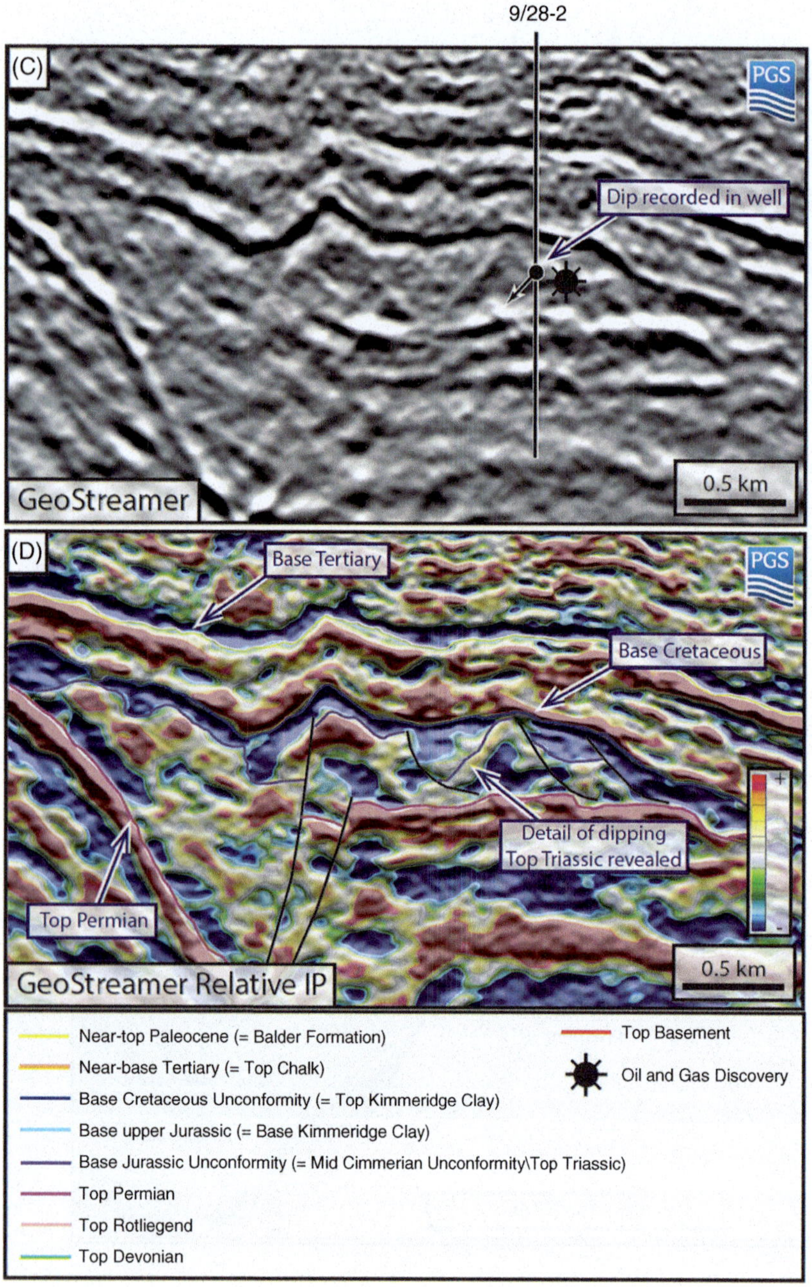

Figure 3.2b (C) An expanded view of the Jurassic and Triassic reservoir interval of the Crawford discovery is shown (post-stack) and (D) (relative acoustic impedance).

References

Balson, P., Butcher, A., Holmes, R. *et al.* 2002. North Sea Geology. Strategic Environmental Assessment, SEA2&3. Technical Report 008 Rev1, Geology. British Geological Survey.

Corfield, S.M., Gawthorpe, R.L., Gage, M. *et al.* 1996. Inversion tectonics of the Variscan foreland of the British Isles. *Journal of the Geological Society of London*, **153**, 17–32.

Coward, M.P., Enfield, M.A. and Fischer, M.W. 1989. Devonian basins of Northern Scotland: extension and inversion related to Late Caledonian – Variscan tectonics. In: Cooper, M.A. and Williams, G.D. (eds) *Inversion Tectonics*, Special Publications vol. 44, The Geological Society, London, pp. 275–308.

Fraser, S., Robinson, A., Johnson, H. *et al.* 2003. Upper Jurassic. In: Evans, D., Graham, C., Armour, A. and

Bathurst, P. (eds) *The Millennium Atlas: Petroleum Geology of the Central and Northern North Sea*, The Geological Society, London, pp. 157–189.

Glennie, K.W. 1995. Permian and Triassic rifting in northwest Europe. In: Boldy, S.A.R. (ed) *Permian and Triassic Rifting in Northwest Europe*, Special Publications vol. 91, The Geological Society, London, pp. 1–5.

Goldsmith, P.J., Hudson, G. and Van Veen, P. 2003. Triassic. In: Evans, D., Graham, C., Armour, A. and Bathurst, P. (eds) *The Millennium Atlas: Petroleum Geology of the Central and Northern North Sea*, The Geological Society, London, pp. 105–127.

Kombrink, H., Besly, B.M., Collinson, J.D. *et al.* 2010. Carboniferous. In: Doornenbal, J.C. and Stevenson, A.G. (eds) *Petroleum Geological Atlas of the Southern Permian Basin Area*, EAGE Publications bv, Houten, The Netherlands, pp. 81–99.

Marshall, J.E.A. and Hewett, A.J. 2003. Devonian. In:Evans, D., Graham, C., Armour, A. and Bathurst, P. (eds) *The Millennium Atlas: Petroleum Geology of the Central and Northern North Sea*, The Geological Society, London, pp. 65–81.

Patruno, S. and Reid., W. 2017. New plays on the Greater East Shetland Platform (UKCS Quadrants 3, 8-9, 14-16) – part 2: Newly reported Permo-Triassic intra-platform basins and their influence on the Devonian-Paleogene prospectivity of the area. *First Break*, **35**, 59–69.

Patruno, S. 2017. A new look at the geology and prospectivity of a North Sea frontier area with modern seismic: the East Shetland Platform. 79th EAGE Conference and Exhibition 2017, 15 June 2017, At Paris, France (Expo Porte de Versailles), Volume: Oral Presentation Session: Exploration Plays, Prospects and Prospects Evaluation II, Abstract Th C2 09. Extended abstract and oral presentation. Extended abstract available online at: http://www.earthdoc.org/publication/publicationdetails/?publication=88346

Patruno, S. and Reid, W. 2016. New plays on the Greater East Shetland Platform (UKCS Quadrants 3, 8-9, 14-16) – part 1: Regional setting and a working petroleum system. *First Break* (peer-reviewed 'Technical Article' sections), **34**, 33–45.

Patruno, S., Reid, W. and Berndt, C. (in press). Regional tectono-stratigraphic architecture, evolution and hydrocarbon prospectivity of the Greater East Shetland Platform and Mid North Sea High: a comparative synopsis for the under-explored platforms of the eastern UK Continental Shelf. In: Palaeozoic Plays of NW Europe. Geological Society Special Publications, The Geological Society, London, United Kingdom, v. x, p. xxxx. in review.

Patruno, S., Turner, C., Cronin, B. *et al.* (in press). The South Viking Graben; Overview of upper Jurassic rift geometry, biostratigraphy, extent of submarine fan systems and rift analogues. Memoir of the Brae Play conference.

Patruno, S., Triantaphyllou, M.V., Erba, E. *et al.* 2015. The Barremian and Aptian stepwise development of the 'Oceanic Anoxic Event 1a' (OAE 1a) crisis: integrated benthic and planktic high-resolution palaeoecology along the Gorgo a Cerbara stratotype section (Umbria-Marche Basin, Italy). *Palaeogeography, Palaeocology, Palaeoecology*, **424**, 147–182. DOI:10.1016/j.palaeo.2015.01.031.

Patruno, S. and Unida, S. 2016. The palynostratigraphy of the upper Maiolica, Selli Level and the lower Marne a Fucoidi units in the proposed Barremian / Aptian (Lower Cretaceous) GSSP stratotype at Gorgo a Cerbara, Umbria-Marche Basin, Italy. *Palynology*, **40**(2), 230–246.

Seranne, M. 1992. Devonian extensioinal tectonics versus Carboniferous inversion in the northern Orcadian basin. *Journal of the Geological Society*, **149**, 27–37.

Stemmerik, L., Ineson, J.R. and Mitchell, J.G. 2000. Stratigraphy of the Rotliegend Group in the Danish part of the Northern Permian Basin, North Sea. *Journal of the Geological Society*, **157**, 1127–1136.

Whipp, P.S., Jackson, C. A.-L., Gawthorpe, R.L. *et al.* 2014. Normal fault array evolution above a reactivated rift fabric; a subsurface example from the northern Horda Platform, Norwegian North Sea. *Basin Research*, **26**, 523–549.

Wilson, R.W., Holdsworth, R.E., Wild, L.E. *et al.* 2010. Basement-influenced rifting and basin development: a reappraisal of post-Caledonian faulting patterns from the North Coast Transfer Zone, Scotland. In: Law, R.D., Butler, R.W.H., Holdsworth, R.E. *et al.* (eds) *Continental Tectonics and Mountain Building: The Legacy of Peach and Horne*, Special Publications vol. 335, The Geological Society, London, pp. 795–826.

Yaliz, A. 1991. The Crawford Field, Block 9/28a, UK North Sea. In: Abbotts, I.L. (ed) *United Kingdom Oil and Gas Fields, 25 Years Commemorative Volume*, Memoirs No. 14, The Geological Society, London, pp. 287–294.

Zanella, E. and Coward, M.P. 2003. Structural framework. In: Evans, D., Graham, C., Armour, A. and Bathurst, P. (eds) *The Millennium Atlas: Petroleum Geology of the Central and Northern North Sea*. The Geological Society of London, pp. 42–59.

Ziegler, P.A., 1992. North Sea rift system. *Tectonophysics*, **208**, 55–75.

4

Normal Faults Bounding Rift Grabens in Krishna-Godavari Basin, India

Achyuta Ayan Misra

Reliance Industries Ltd, Mumbai, India
Achyuta.Misra@ril.com; achyutaayan@gmail.com

The Krishna-Godavari Basin occurs in the central part of the Eastern Continental Margin (ECMI) of India (Figure 4.1) in the state of Andhra Pradesh. It formed as Antarctica rifted from India in Late Jurassic – Early Cretaceous (Rao, 2001; Nemčok *et al*., 2013). This part of the ECMI shows classical normal fault-bound rift grabens with prominent sediment growth packages (Figures 4.2 and 4.3). The seismic sections presented here show one large rifted graben in the proximal margin. (The definition of 'proximal margin' can be found in Manatschal (2004) and Manatschal *et al*. (2007)). In both these seismic sections early synrift, peak synrift, late synrift and postrift sediments can be easily identified, while prerift is doubtful (Figures 4.2 and 4.3).

Prerift sediments are identified based on their seismic facies. Reflections in the prerift sediments characteristically parallel the basement top and do not show growth of sediments. '*Growth*' of sediments is their expansion and for normal faults it occurs towards the fault. The synrift sediment package, which shows growth in the entire interval, can be separated into primarily three subpackages, which are early synrift, peak synrift and late synrift. Angular unconformities probably separate these subpackages. In the sections here (Figure 4.2b and 4.3b), the pack of sediments overlying the basement with prerift-like seismic signature is very early synrift, as confirmed from biostratigraphic data from one well drilled by the Reliance Industries Ltd. on the rift shoulder.

Stretching started during *early synrift* and faults grew slowly. This sedimentary package separates from the prerift by an angular unconformity. The early synrift sedimentary sequence shows minimal growth of sediments and may onlap on the top of prerift sediment sequence. They may even subparallel the prerift top. Thus, it may sometimes be very difficult to separate prerift from early synrift sediment package. Biostratigraphy from well cutting or cores might be helpful in such situations.

Peak synrift or *peak rift* stage is the phase when there is maximum stretching and fault movement compared to the prior two rift stages. This stage is identified easily by noting the most prolific growth of sediments.

Late synrift is the stage where the deformation slows and the prolific growth is no longer observed. However, in some cases the difference between the growths in the peak- and late synrift may be too little to be separated. An unconformity within the synrift may help to separate out the two phases. This marks the end of tectonic faulting. However, a few faults may continue growing due to differences in compaction on either side of the fault. The top of late synrift sequence is also called a breakup unconformity.

Postrift sedimentary sequences are identified from their downlapping on top of the synrift. The postrift shows a sagging of the basin due to the thermal cooling of the lithosphere after rifting. Marine incursion in the rifted grabens may occur much earlier than the postrift phase. Thus, it is difficult using biostratigraphy alone to ascertain a postrift sequence. Seismic facies analysis is more helpful in such a situation.

In the synrift sequence, normal dragging (Mukherjee, 2014) of the sedimentary sequence can be observed along the fault plane (red rectangles in Figure 4.2b). These drags are commonly observed along many rifted grabens, which may occur due to (i) faulting dragged existing sediments and/or (ii) sediments that have fallen off the eroded footwall forming an edge of sediments. During stretching many faults initiate, out of which only the weakest continue to the final stages of the stretching phase to bound the rifted grabens. Such faults may abort at any stage during rifting, for example the fault marked by dotted lines in Figures 4.2b and 4.3b.

Atlas of Structural Geological Interpretation from Seismic Images, First Edition. Edited by Achyuta Ayan Misra and Soumyajit Mukherjee.

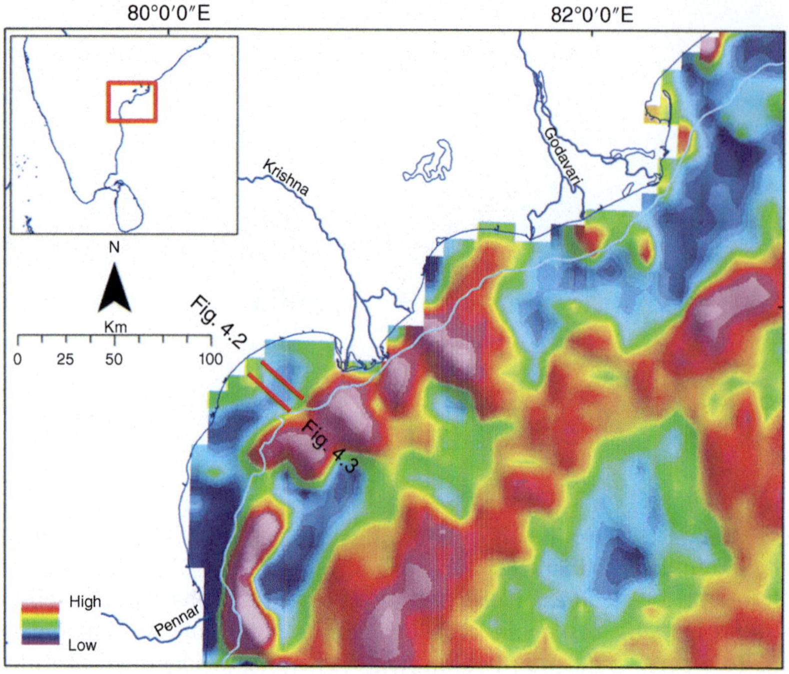

Figure 4.1 Krishna-Godavari Basin along the ECMI overlaid on overlaid on hill-shaded, upward-continued (10 km) Bouguer gravity anomaly data (see the colour scale). Blue line: Shelf-slope break. Red lines: indicate respective seismic sections.

Figure 4.2a Uninterpreted section in the stretching zone of Krishna-Godavari Basin. Seismic data courtesy Reliance Industries Ltd. Reproduced with permission from the Directorate General of Hydrocarbons (DGH), India.

Footwalls very often erode and are redeposited as sediments in the graben because they are subaerially exposed or very near the water surface (Ravnås and Steel, 1998). Large parts of the footwall may erode (Figure 4.3d). The seismic reflection of the basement top becomes rugged and a fault scarp forms due to the erosion of the fault plane (fault F2 in Figure 4.3d). The seismic facies of the basement also shows chaotic reflections indicating fracturing and erosion (for fault F2 in Figure 4.3d).

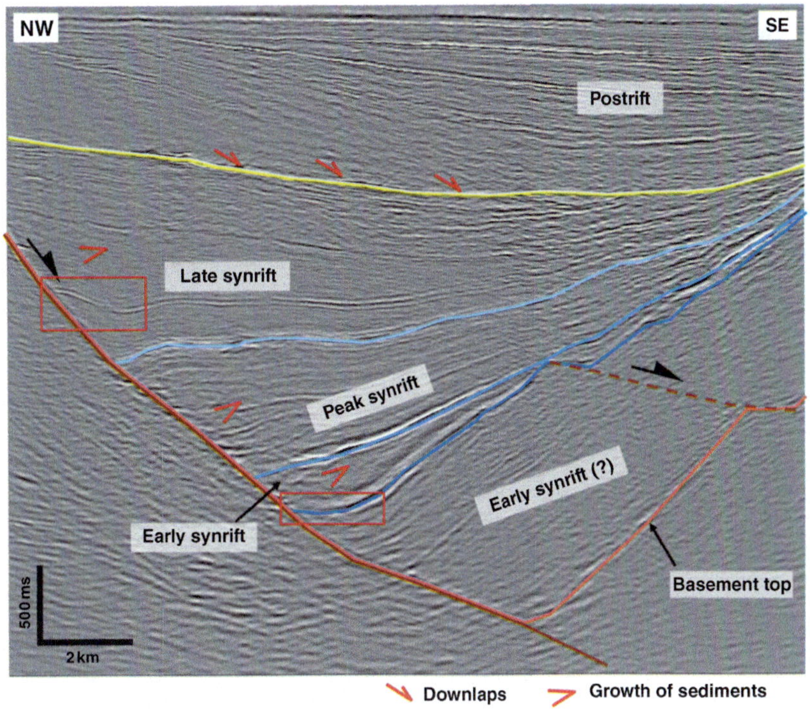

Figure 4.2b Interpretation of the seismic section in Figure 4.2a showing the various elements of a rifted graben. Note the difference in growth of the sediments for the different stages of the rift. Yellow line: break-up or top-synrift unconformity. Red rectangles: normal dragging of sediments along the fault plane. Seismic data courtesy Reliance Industries Ltd. Reproduced with permission from the Directorate General of Hydrocarbons (DGH), India.

Figure 4.3a Uninterpreted section in the stretching zone of Krishna-Godavari Basin. Note the noise content is more in this section compared to the one in Figure 4.2. Seismic data courtesy Reliance Industries Ltd. Reproduced with permission from the Directorate General of Hydrocarbons (DGH), India.

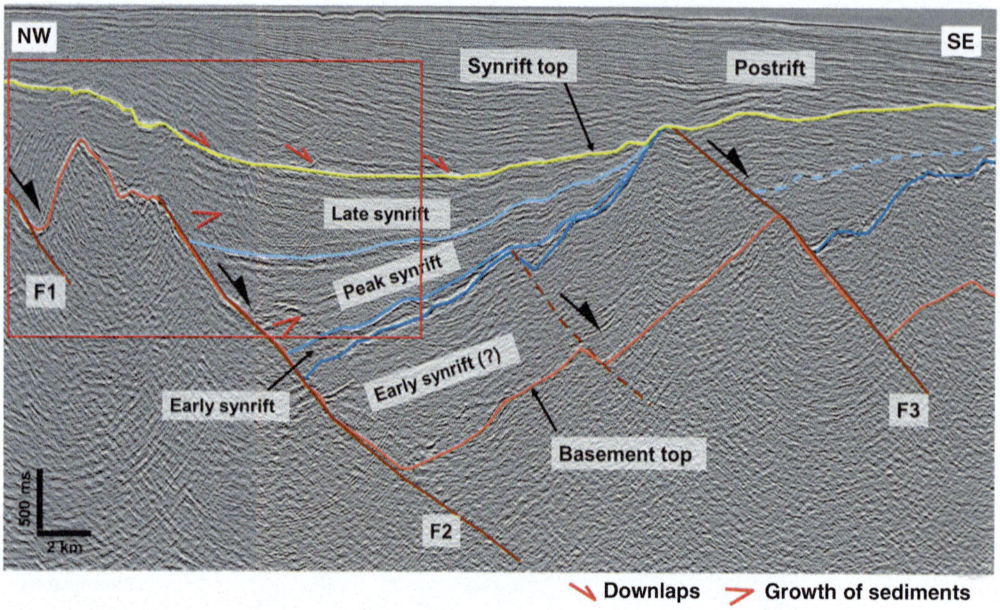

Figure 4.3b Interpretation of the seismic section in Figure 4.3a. Note the difference in growth of the sediments for the different stages of the rift. The higher reaches of the fault F2 eroded and a fault scarp remains. Yellow line: break-up or top-synrift unconformity. Red rectangle: enlarged section in Figures 4.3c and 4.3d (extent marked as red rectangle). Seismic data courtesy Reliance Industries Ltd. Reproduced with permission from the Directorate General of Hydrocarbons (DGH), India.

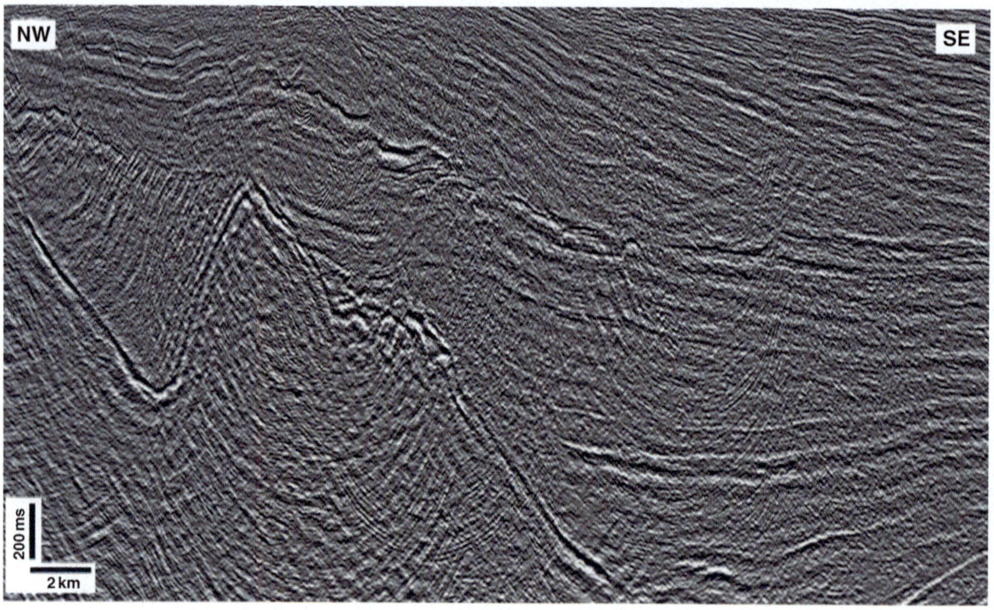

Figure 4.3c Uninterpreted section: enlarged part shown within red rectangle in Figure 4.3b. Seismic data courtesy Reliance Industries Ltd. Reproducedwith permission from the Directorate General of Hydrocarbons (DGH), India.

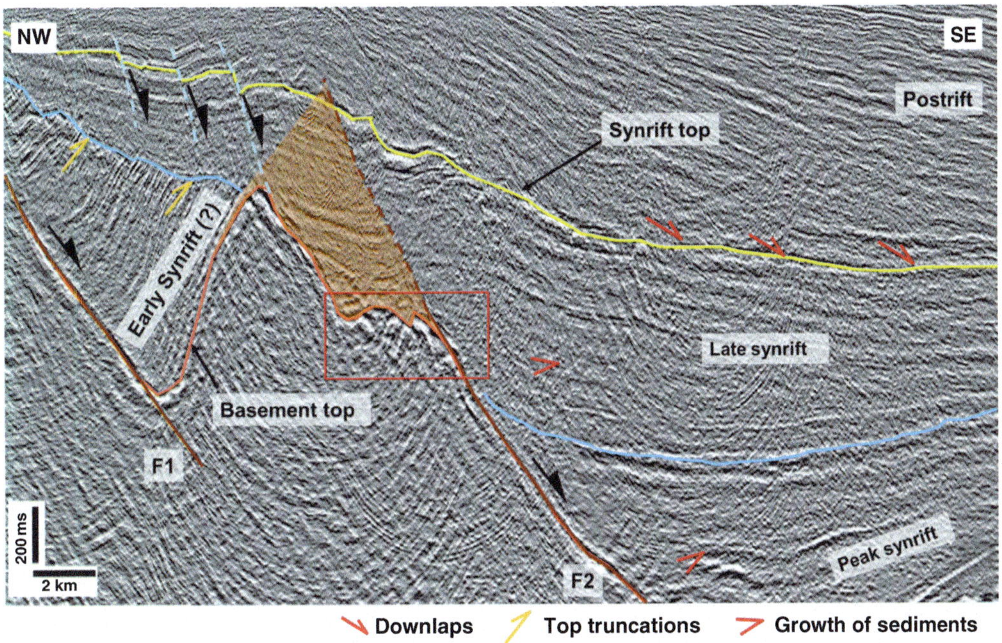

Figure 4.3d Interpretation of the seismic section in Figure 4.3c. Brown dotted line: possible original fault F2, presently a compaction related fault is present at the same location. Brown transparent polygon: possible original basement configuration. The synrift sediments on the NW graben belong to early to peak synrift. Note the other compaction related faults (in blue dotted line). Yellow line: breakup or top-synrift unconformity. Seismic data courtesy Reliance Industries Ltd. Reproduced with permission from the Directorate General of Hydrocarbons (DGH), India.

References

Manatschal, G., 2004. New models for evolution of magma-poor rifted margins based on a review of data and concepts from West Iberia and the Alps. *International Journal of Earth Sciences*, **93**, 432–466.

Manatschal, G., Müntener, O., Lavier, L.L. *et al.* 2007. Observations from the Alpine Tethys and Iberia–Newfoundland margins pertinent to the interpretation of continental breakup. In: Karner, G.D., Manatschal, G. and Pinheiro, L.M. (eds) *Imaging, Mapping and Modelling Continental Lithosphere Extension and Breakup*, Special Publications vol. 282, The Geological Society, London, pp. 291–324.

Mukherjee, S. 2014. Review of flanking structures in meso- and micro-scales. *Geological Magazine*, **151**, 957–974.

Nemčok, M., Sinha, S.T., Stuart, C.J. *et al.* 2013. East Indian margin evolution and crustal architecture: integration of deep reflection seismic interpretation and gravity modelling. In: Mohriak, W.U., Danforth, A., Post, P.J. *et al. Conjugate Divergent Margins*, Special Publications vol. 369, The Geological Society, London, pp. 477–496.

Rao, G.N. 2001. Sedimentation, stratigraphy, and petroleum potential of Krishna-Godavari basin, East Coast of India. *AAPG Bulletin*, **85**, 1623–1643.

Ravnås, R. and Steel, R.J. 1998. Architecture of marine rift-basin successions. *AAPG Bulletin*, **82**, 110–146.

5

Bitsminda Anticline: South Kura Foreland Fold and Thrust Belt, Georgia

Victor Alania[1], A. Chabukiani[1,2], O. Enukidze[1], K. Gogrichiani[2], Z. Mikeladze[2], and G. Tatishvili[2]*

[1] *I. Javakhishvili State University, Institute of Geophysics, Tbilisi, Georgia*
[2] *State Agency of Oil and Gas, Ministry of Energetics, Tbilisi, Georgia*
* *v_alania2002@yahoo.com*

Structural interpretation of S–N trending seismic profile used fault-related folding theories (Suppe and Medwedeff 1990; Shaw *et al.*, 2006). Identification of stratigraphic units at a depth for seismic profiles was based on outcrop correlations and Bitsminda-1 borehole data. Interpreted seismic reflection profiles show that the Bitsminda anticline is a shallow fault-propagation fold with the front limbs broken by thrust faults. The fold shows a strongly normal dragged (Mukherjee 2014) hanging wall block, and a less dragged footwall block. Syntectonic (or synorogenic) Middle and Upper Miocene strata are represented by shallow marine, predominantly, terrigenic fine clastics, limestones and marls with rare conglomeratic interbeds and sandstones and thick continental sediments (alternation of conglomerates, clays and sandstones; 1000–2000 m thick) (Alania *et al.*, 2015) (Figure 5.1).

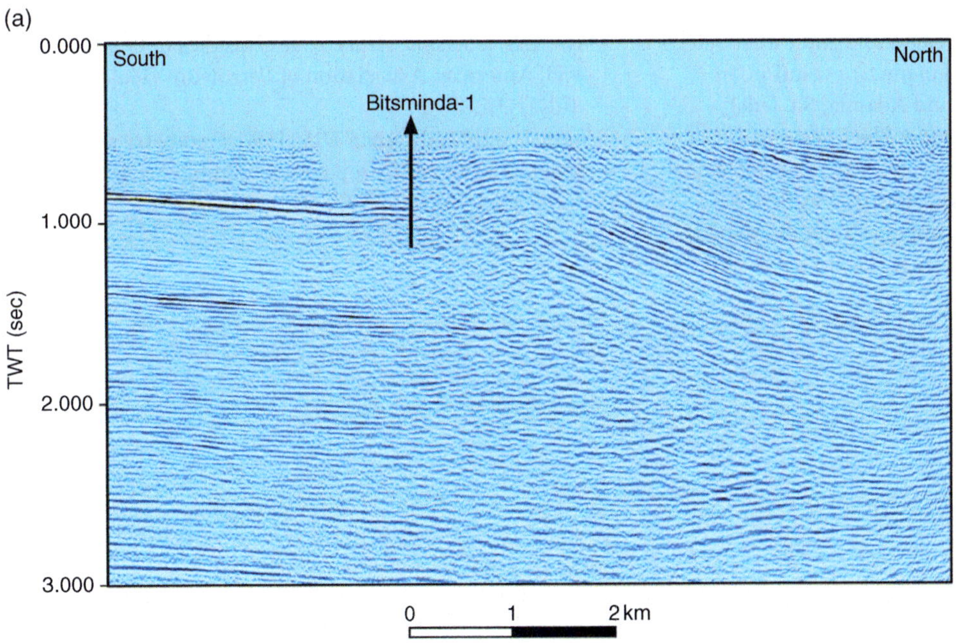

Figure 5.1 Uninterpreted (a) and interpreted (b) 2D time-migrated seismic profile through the Bitsminda anticline: southern part of the Kura foreland fold and thrust belt (Georgia). Abbreviations: F_1 and F_2 is south-vergent thrust; F_3 is backthrust; P_3 –Oligocene; N_1^1 – Lower Miocene; N_2^2 – Middle Miocene; N_2^3S – Upper Miocene (Sarmatian); N_2^3 m-p – Upper Miocene (Meotis – Pontian).

Atlas of Structural Geological Interpretation from Seismic Images, First Edition. Edited by Achyuta Ayan Misra and Soumyajit Mukherjee.

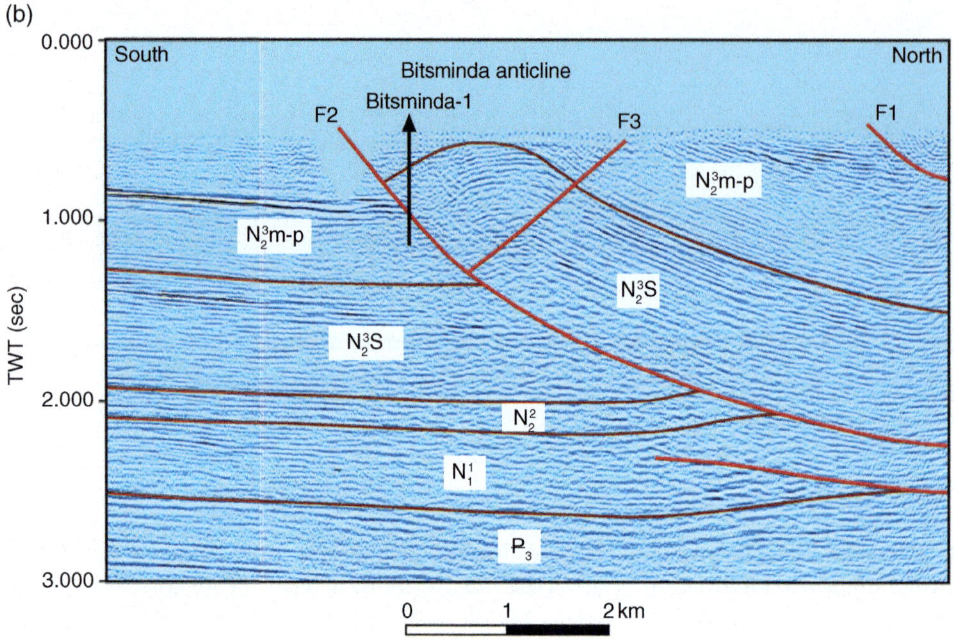

Figure 5.1 (Cont'd)

References

Alania, V., Chabukiani, A., Chagelishvili, R. *et al.* 2015. Growth structures, piggyback basins and growth strata of Georgian part of Kura foreland fold and thrust belt: implication for Late Alpine kinematic evolution. In Sosson, M., Stephenson, R. and Adamia, Sh. (eds) *Tectonic Evolution of the Eastern Black Sea and Caucasus*. Special Publications vol. 428, The Geological Society, London, pp. 171–185. doi:10.1144/SP428.5.

Mukherjee, S. 2014. Review of flanking structures in meso- and micro-scales. *Geological Magazine*, **151**, 957–974.

Shaw, J.H., Christopher, C. and Suppe, J. 2006. *Seismic Interpretation of Contractional Fault-Related Folds: An AAPG Seismic Atlas*. AAPG Studies in Geology #53, American Association of Petroleum Geologists, Tulsa, OK.

Suppe, J. and Medwedeff, D.A. 1990. Geometry and kinematics of fault-propagation folding. *Eclogae Geologicae Helvetiae*, **83**, 409–454.

6

Diffraction and Refraction Occurring in a Fault Zone in Corinth Gulf, Greece as Seen by Vertical Seismic Profiling

Joachim Place[1], Charles Naville[2], and Isabelle Moretti[3]*

[1] *Formerly at Department of Earth Sciences, Uppsala University, Sweden*
[2] *IFP, Division Géologie - Géochimie, Rueil-Malmaison, France*
[3] *GDF SUEZ, Courbevoie, France*
* *joachim.geo@gmail.com*

Vertical seismic profiling (VSP) is a field measurement procedure in which the seismic source is activated at a fix surface position and the seismic signal is recorded by sensors located in a well at successive depth levels (Coppens and Mari, 1991). The first objective of VSP is the vertical velocity survey. Additionally, it is possible to discern downgoing and upgoing events and to obtain calibrated seismic reflections tied with depth. If the source is located near the well, the survey is called a Zero Offset VSP; if the source is significantly far from the well, it is called an Offset VSP (Coppens and Mari, 1991).

When VSP measurements are carried out in the vicinity of fault zones, singular arrivals have to be interpreted in detail as the expression of the fault zone presence. The example of Offset-VSP data detailed here was recorded in the Aig-10 borehole intersecting the Aigion Fault Corinth Gulf, Greece (Figure 6.1). The Corinth Rift is one of the most seismically active areas in Europe. This activity is related to N–S extension strain, which commenced during the Early Miocene. A series of WNW–ESE striking faults has developed to the southern shoulder of the rift (Figure 6.1). One of them, the Aigion Fault, has shown some recent seismic activity and is studied here (Figure 6.2).

A four-component tool (three orthogonal geophones and a hydrophone) was used to record the data. For simplicity, only the vertical component is shown in Figure 6.3. The depth interval is represented in Figure 6.4, along with the source (P-vibrator) position by the scarp formed by the active fault (see also Figure 6.2).

Apart from the direct arrival (1), two arrivals stand out, because of either an amplitude higher than that of the direct arrival (2) or an early arrival time (3). Calculation of reoriented components allowed a detailed analysis of the polarization of the seismic arrivals along with their travel time. Arrival (2) is interpreted as P-waves travelling in the hangingwall, propagating subparallel to the fault zone, then diffracting off a corner formed by the fault zone offsetting the radiolarite and limestone unit. Arrival (3) is interpreted as a refraction occurring along the fault zone. Arrival times of events (2) and (3) have been used to refine the geometry of the fault, especially to estimate its fault throw (about 200 m).

The low amplitude of the direct arrival (1) indicates that the medium between the source and receivers strongly attenuates the seismic waves, which is in agreement with the presence of poorly consolidated conglomerates (Figure 6.4). The low amplitude of the arrival (3) is likely related to the damage of the medium within and in the vicinity of the fault zone that is encountered by the refracted waves. The detailed approach based on four-component data is available elsewhere (Place *et al.*, 2007b) and also includes results of full waveform 2D finite-difference elastic modelling that support the interpretations.

Atlas of Structural Geological Interpretation from Seismic Images, First Edition. Edited by Achyuta Ayan Misra and Soumyajit Mukherjee.

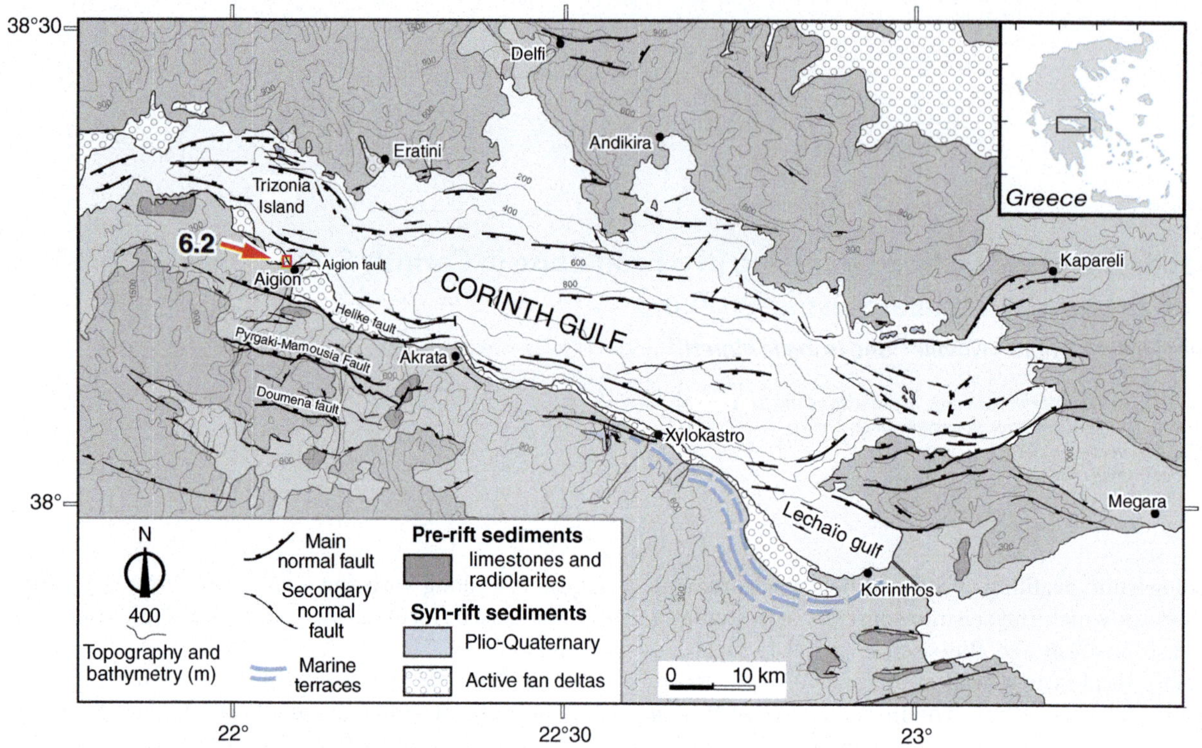

Figure 6.1 Simplified geological structural map of the Gulf of Corinth. The red framed area indicates the location shown in Figure 6.2. Place 2007a. Reproduced with permission of Elsevier.

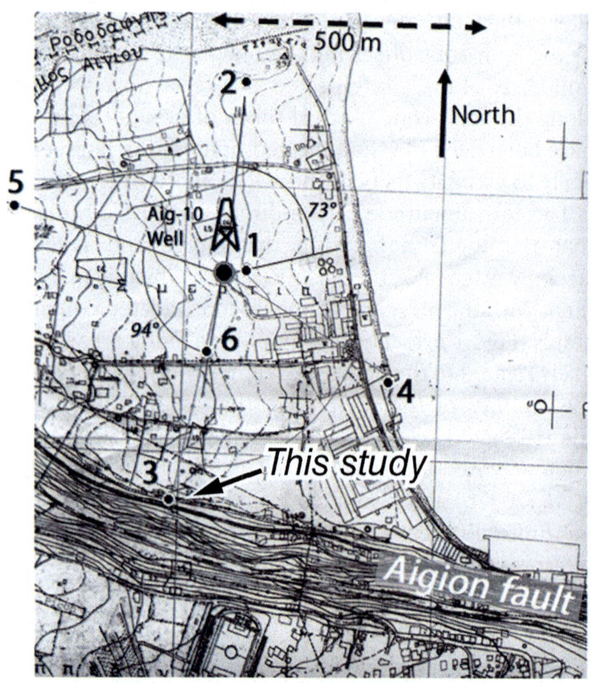

Figure 6.2 Position of the Aig-10 well and VSP sources (only the profile recorded using source 3 is studied here). Note the scarp of the Aigion Fault to the south. Place 2007b. Reproduced with permission of Elsevier.

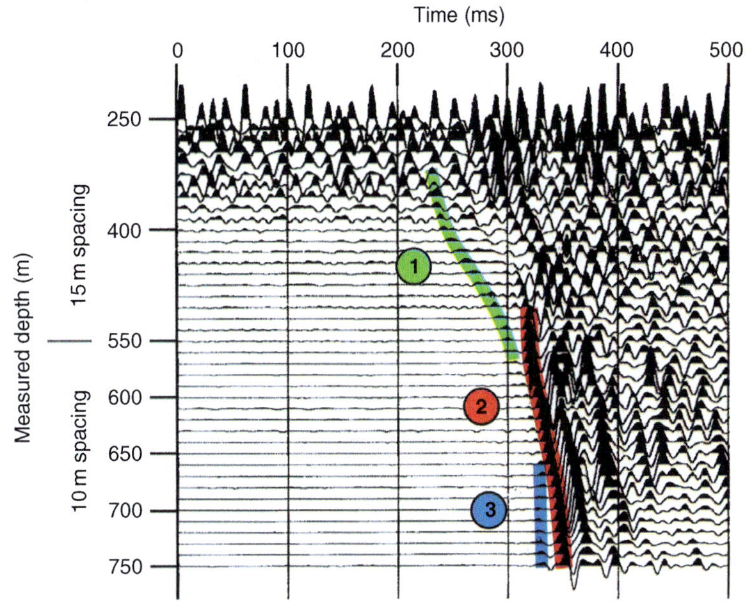

Figure 6.3 Vertical seismic profile recorded with source 3 (Figure 6.2). Place 2007b. Reproduced with permission of Elsevier.

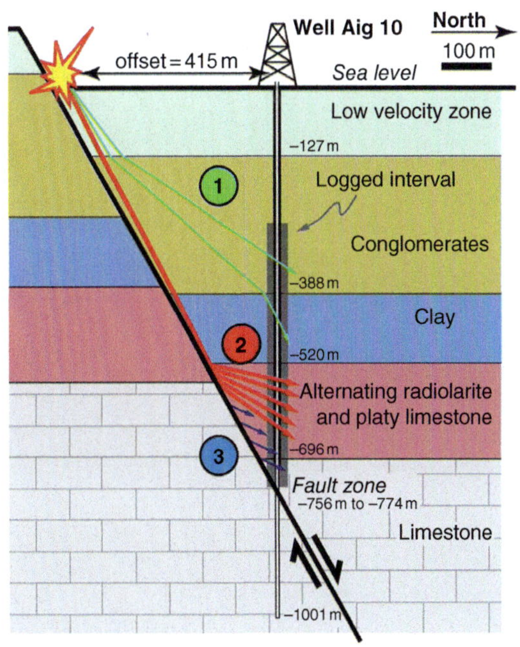

Figure 6.4 Sketch showing the interpreted ray paths of arrivals identified in Figure 6.1. Place 2007b. Reproduced with permission of Elsevier.

References

Coppens, F. and Mari, J.L. 1991. Seismic well surveying. Technip editions, pp. 128.

Place, J., Géraud, Y., Diraison, M. and Warr, L., 2007a. North–south transfer zones and paleo-morphological reconstruction of the Xylokastro area (Corinth Gulf, Greece). *Tectonophysics*, **440**, 121–139.

Place, J., Naville, C. and Moretti, I. 2007b. Fault throw determination using 4 component VSP: Aigion fault (Greece) case study. *Tectonophysics*, **440**, 141–158.

7

Horst Inversion Within a Décollement Zone During Extension Upper Rhine Graben, France

Joachim Place[*1], M. Diraison[2], Y. Géraud[3], and Hemin A. Koyi[4]*

[1] Formerly at Department of Earth Sciences, Uppsala University, Sweden
[2] Institut de Physique du Globe de Strasbourg (IPGS), Université de Strasbourg/EOST, Strasbourg, France
[3] Université de Lorraine, Vandoeuvre-lès-Nancy, France
[4] Department of Earth Sciences, Uppsala University, Sweden
*joachim.geo@gmail.com

The Merkwiller–Pechelbronn oil field of the Upper Rhine Graben has been a target for hydrocarbon exploration for over a century. The occurrence of the hydrocarbons is thought to be related to the noticeably high geothermal gradient of the area. Its magnitude peaks at about 100 °C.km^{-1} a few kilometres east of the oil field, in the vicinity of the Soultz-sous-Forêts horst, which has been the target of geothermal exploration in recent decades (Figure 7.1). Seismic investigations in the 40 km wide Upper Rhine Graben reveal significant details. We have chosen a 5 km time-migrated section of the PHN84J profile that focuses on the eastern flank of the Soultz-sous-Forêts horst to illustrate the complexity of basement-sedimentary cover deformation patterns involving décollements (Figure 7.2). This is a slightly revised version of the original interpretation of Place *et al.* (2010), where the deformation in the décollement level was more emphasized than in the present interpretation.

The section shows a couple of major normal faults bounding a graben within the upper part of the sedimentary cover, rooted in décollements of Triassic anhydrite or clay series (top of Buntsandstein to Middle Muschelkalk, and Keuper). These faults are interpreted to have been active during the Oligocene when the Upper Rhine Graben opened (Roussé, 2006). The seismic profile also reveals the presence of a deeply-located fault-bounded block labelled (1) within the décollement zone. This block, which we call 'inverted horst', has formed in two main stages:

Firstly, extension in the basement (along at least one basement fault), decoupled from the cover by décollement level(s), resulted in the formation of a large-scale monocline

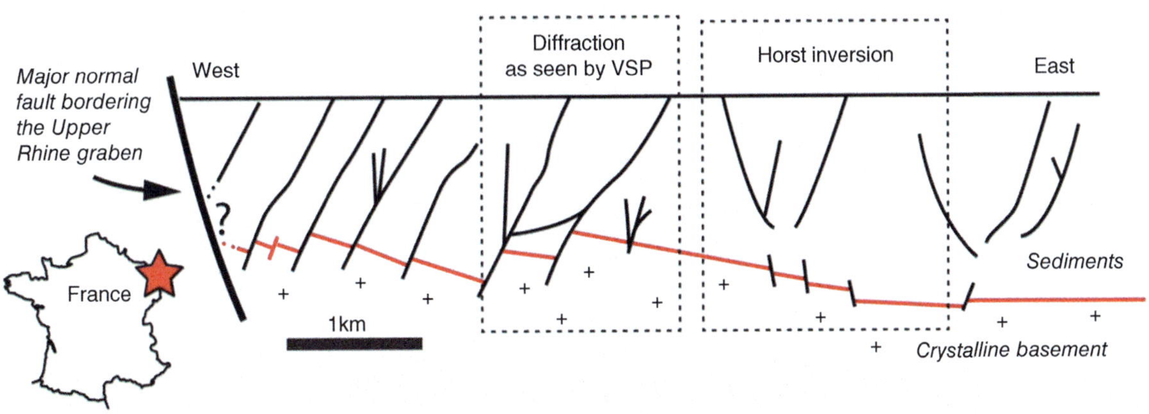

Figure 7.1 Regional geologic section showing the study areas for Chapters 7 and 8.

Atlas of Structural Geological Interpretation from Seismic Images, First Edition. Edited by Achyuta Ayan Misra and Soumyajit Mukherjee.

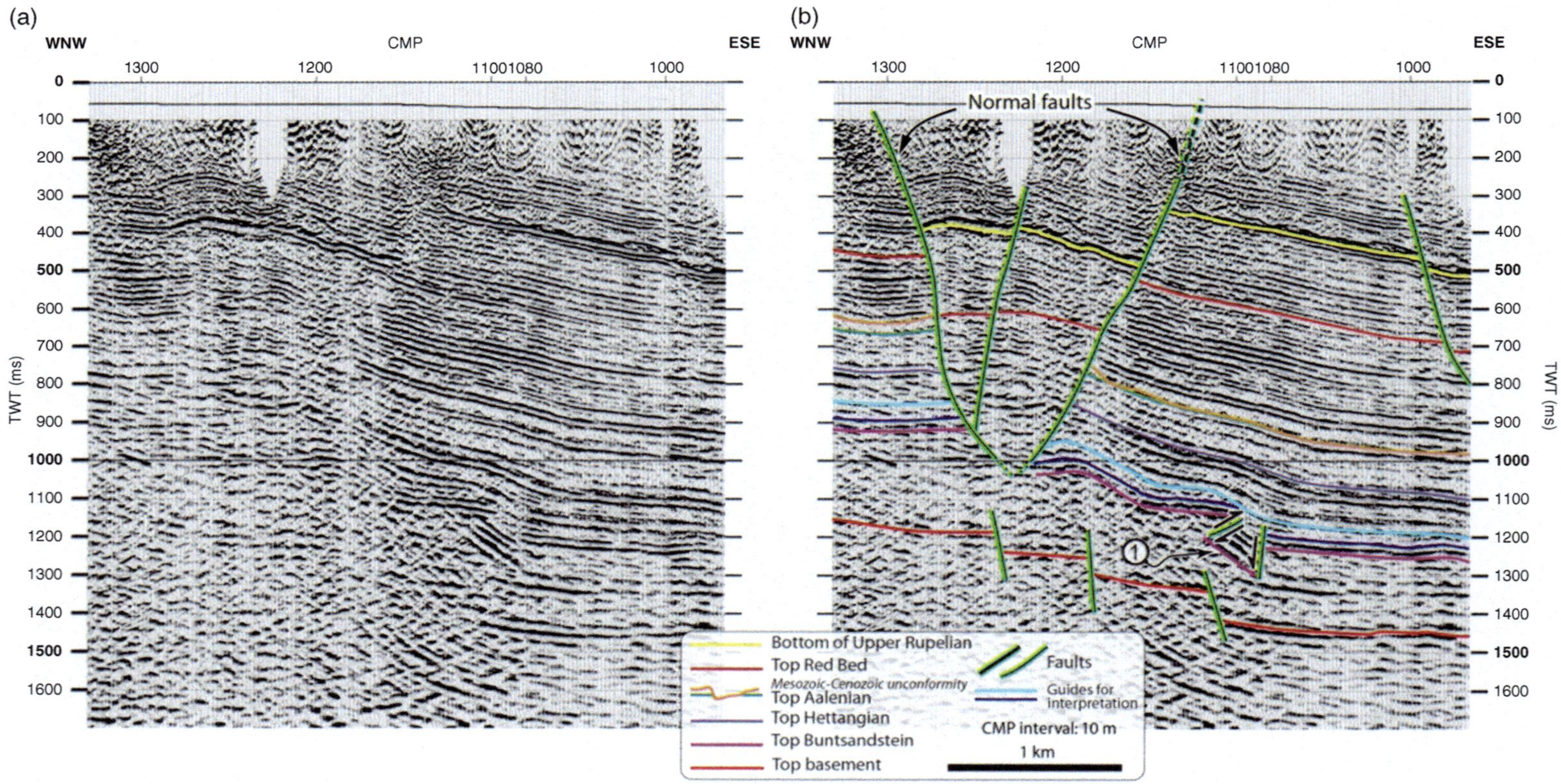

Figure 7.2 (a) Uninterpreted and (b) interpreted reflection seismic section showing the normal faults and the inverted horst block. Place *et al.* 2010. Reproduced with permission of Elsevier. Copyright © 2010, published by Elsevier Masson SAS. All rights reserved.

and a graben in the cover units (Figure 7.3a). The upper reflectors within the graben exhibit some thickening and bending towards the western fault (synthetic to basement faults), denoting a syn-tectonic deposition, and not any drag that sometimes associate with faults (Mukherjee, 2014 and Koyi *et al.*, 1993). The extensional regime also formed a detached deeply-located horst near the hangingwall hinge of the monocline (as obtained by Withjack and Callaway (2000) using analogue modelling) (Figures 7.2 and 7.3a).

Secondly, continued extension resulted in the eastern shoulder of the graben (about CMP 1120 and lower) to decouple from the basement and glide eastward on the ductile material. This ductile material, most likely made of Triassic clay and salt, also facilitated the popping down of the horst (1) alongside its inversion and clockwise rotation (Figure 7.3b). Concomitant bending of the reflectors above this inverted horst (1) attenuates upwards away from it (Figure 7.2). A slight clockwise rotation of the graben is also caused by the gliding of its eastern shoulder.

Interpretation of other seismic profiles indicates that the graben strikes around N170°E, which makes it prone to undergo Chattian compression (the maximum horizontal stress being NE–SW), and possible transpression when exposed to the Miocene to present Alpine push (oriented NW–SE) (Schumacher, 2002; Bourgeois *et al.*, 2005). These two tectonic events might have bent the upper layers in the graben (about 200–400 ms and CMP 1260) as well as causing out-of-plane flow of ductile material within the décollement zones.

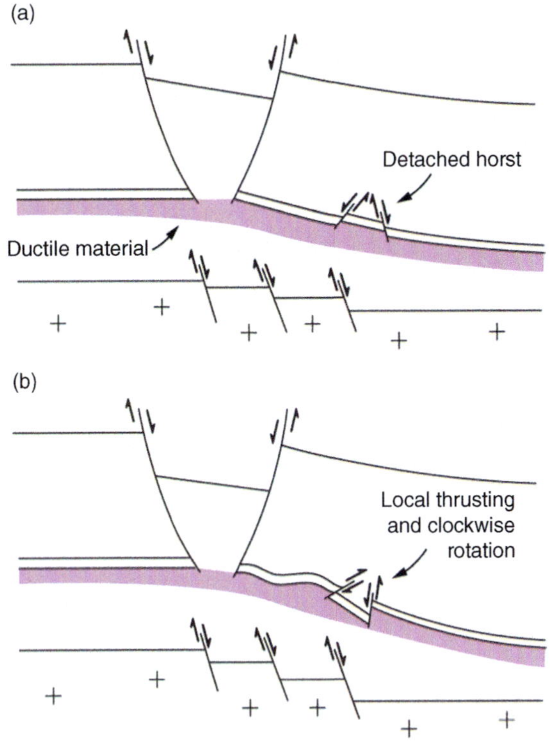

Figure 7.3 Schematic cross-section showing the two stage process for formation of the inverted horst block. (a) Stage 1 and (b) Stage 2.

References

Bourgeois O., Ford M., Diraison M. *et al.* 2007. Separation of rifting and lithospheric folding signatures in the NW-Alpine foreland. *International Journal of Earth Sciences*, **96**, 1003–1031.

Koyi, H. A., Jenyon, M.K. and Petersen K. 1993. The effect of basement faulting on diapirism. *Journal of Petroleum Geology*, **16**(3), 285–312.

Mukherjee, S. 2014. Review of flanking structures in meso- and micro-scales. *Geological Magazine*, **151**, 957–974.

Place J., Diraison M., Naville C. *et al.* 2010. Decoupling of deformation in the Upper Rhine Graben sediments. Seismic reflection and diffraction on 3-component Vertical Seismic Profiling (Soultz-sous-Forêts area). *Comptes Rendus Geoscience*, **342**, 575–586.

Roussé, S. 2006. Architecture et dynamique des séries marines et continentales de l'Oligocène Moyen et Supérieur du Sud du Fossé Rhénan: évolution des milieux de dépôt en contexte de rift en marge de l'avant pays alpin. PhD Thesis, University of Strasbourg, France.

Schumacher, M.E. 2002. Upper Rhine Graben: Role of preexisting structures during rift evolution. *Tectonics*, **21**(1), 6-1–6-17. doi: 10.1029/2001TC900022.

Withjack, M.O. and Callaway, S., 2000. Active normal faulting beneath a salt layer: an experimental study of deformation patterns in the cover sequence. *AAPG Bulletin*, **84**, 627–651.

8

Interpretation of a Vertical Seismic Profile Diffraction Occurring at a Basement–Sedimentary Cover Interface in the Rhine Graben, France

Joachim Place[1] and Charles Naville[2]*

[1] *Formerly at Department of Earth Sciences, Uppsala University, Sweden*
[2] *IFP, Division Géologie - Géochimie, Rueil-Malmaison, France*
* *joachim.geo@gmail.com*

Geothermal prospects in the Rhine Graben have focused on the crystalline basement as well as the lower part of the sediments filling the graben. This example presents a case study from the Soultz-sous-Forêts geothermal project in the Rhine Graben (France), with an aim to delineate of the sediment–basement interface using the complementarity between surface and downhole (VSP) seismic surveys (see location in Figure 7.1). Decollement occurring in Triassic units are evidenced by complex fault patterns, such as that of a fault connected to the Soultz-sous-Forêts fault (Place *et al*., 2010), and further emphasize the need for understanding the coupling between the sediments and the basement. A more comprehensive data set, references and description of the approach summarized here, can be found elsewhere (Place *et al*., 2010; Place, 2010).

On one hand, seismic reflection data were acquired within the frame of hydrocarbon exploration projects, with relatively shallow reflectors for targets. As a result, the delineation of the base of the sedimentary filling of the graben remains unclear in some places, as illustrated in Figure 8.1a. The low frequency response of the top basement is denoted near the centre of the section, as confirmed from the downhole data collected in GPK1, but is more difficult to identify further away from the borehole. This becomes problematic when trying to trace the Soultz-sous-Forêts fault down to the basement, as the top basement cannot be identified with confidence in the hangingwall.

On the other hand, several P- and S-wave VSP surveys have been conducted within the basement to characterize its geothermal potential, in particular in the GPK1 vertical borehole. In the data set acquired across the basement-sediment interface (highlighted in red in Figure 8.1), a very energetic diffraction occurs around 0.9 s with an apex at 1745 m. Even though no horizontal component data are available, the fact that the signal is highly energetic on the vertical component suggests a horizontal propagation of vertically polarized S-waves. Distance from the borehole to the diffractor was analysed using the curvature of the hyperbola (i.e. relative travel times) as was the absolute arrival time (see Place *et al.* (2007) for analytical developments). In the latter case, S-source VSP data recorded at greater depth indicate that the downgoing direct S-wave train is too slow to generate the diffraction (in yellow in Figure 8.2). Two possibilities therefore exist regarding the propagation mode of the downgoing waves generating the diffraction: (i) P-waves propagating at a lower average velocity than that recorded by zero-offset VSP in GPK1 due to the presence of the Soultz-sous-Forêts fault zone (Figure 8.1), or (ii) P–S conversion occurring in the sedimentary sequence, as such converted waves are potentially highly energetic (in green in Figure 8.1).

Despite limited constrains on the processes causing the diffraction, it is thought that such an energetic signal can be generated by a structure whose dimensions and contrast in physical properties should be detectable in seismic reflection data. As no azimuth information is available due to the lack of horizontal components, two possible locations for the diffractor have been plotted on Figure 8.1. The option to the east of the borehole cannot be retained as there is no evidence for the basement to host a structure of any kind capable of generating a strong diffraction. However, a better candidate is found to the west of the borehole, as the Soultz-sous-Forêts fault creates contact between the sedimentary materials and basement rocks. In this way, VSP data analysis brought crucial information to delineate with more confidence the top basement within the hanging wall of the Soultz-sous-Forêts fault in the seismic reflection section.

Atlas of Structural Geological Interpretation from Seismic Images, First Edition. Edited by Achyuta Ayan Misra and Soumyajit Mukherjee.

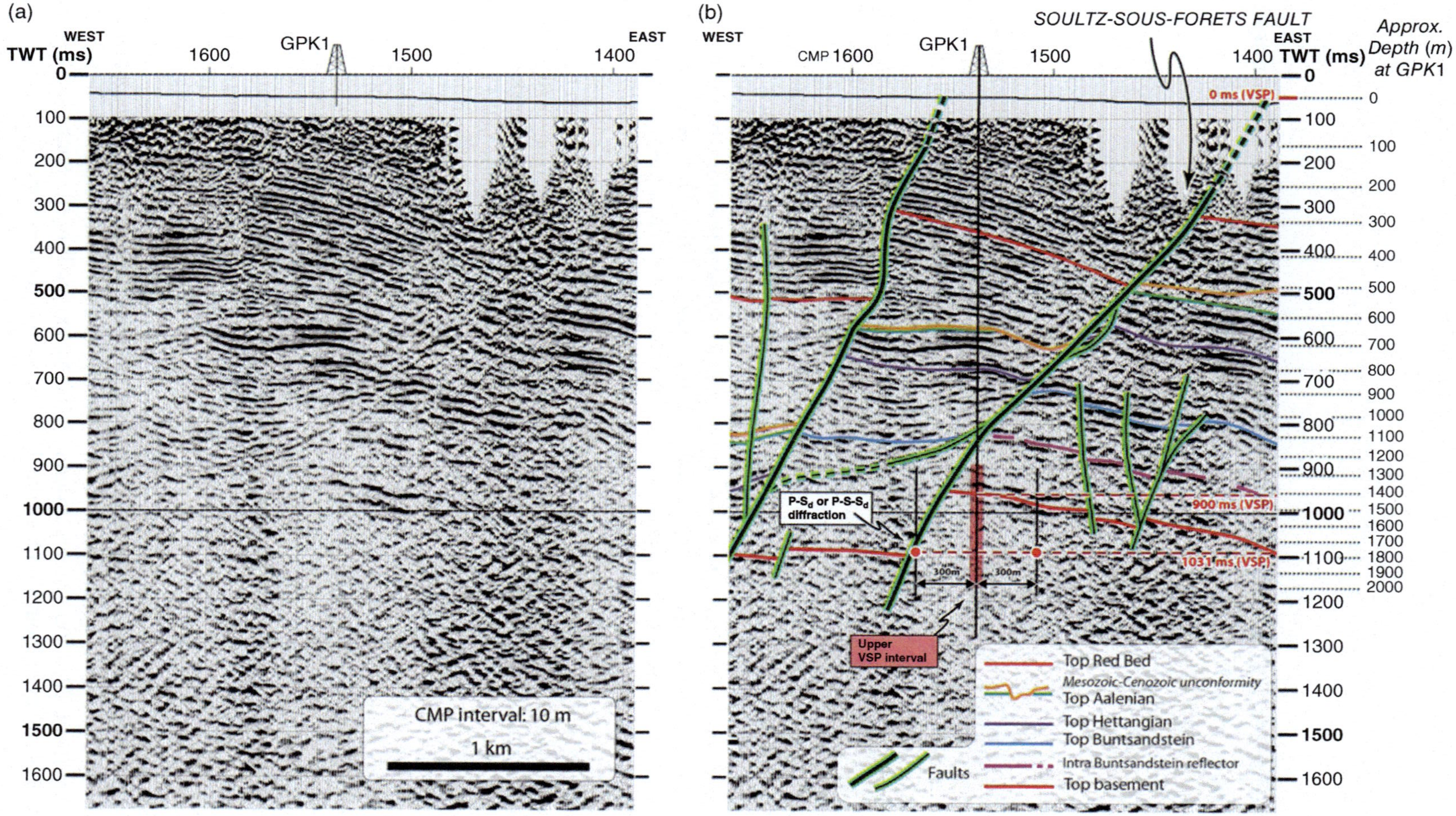

Figure 8.1 (a) Uninterpreted and (b) interpreted reflection seismic profile. Modified after Place *et al*. 2010. Reproduced with permission of Elsevier. Copyright © 2010, published by Elsevier Masson SAS. All rights reserved.

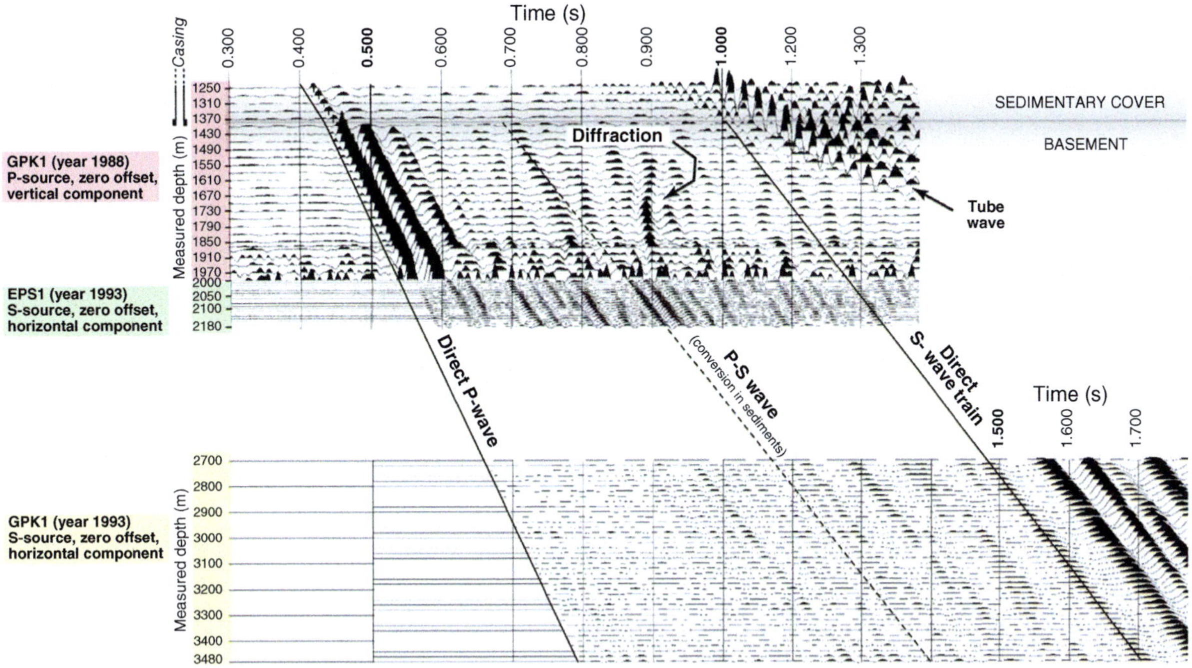

Figure 8.2 Consolidation of three vertical seismic profiles. Modified after Place *et al.* 2010. Reproduced with permission of Elsevier.

References

Place, J. 2010. Characterization of fluid flow paths within a fractured granitic batholith. Application to the Soultz-sous-Forêts Enhanced Geothermal System. University of Strasbourg, France (in French). http://scd-theses.u-strasbg.fr/2045/; last accessed 27 November 2017.

Place, J., Naville, C. and Moretti, I. 2007. Fault throw determination using 4-component VSP: Aigion fault (Greece) case study. *Tectonophysics*, **440**, 141–158.

Place J., Diraison M., Naville C. *et al.* 2010. Decoupling of deformation in the Upper Rhine Graben sediments. Seismic reflection and diffraction on 3-component Vertical Seismic Profiling (Soultz-sous-Forêts area). *Comptes Rendus Geoscience*, **342**, 575–586.

9

Clara Inversion Anticline, Adriatic Sea, Italy

Paolo Pace

GEPlan Consulting, Petroleum Geosciences, Ferrara (FE), Italy
Department of Engineering and Geology, 'Gabriele d'Annunzio' University, Chieti Scalo (CH), Italy
paolo.pace@geplan.it; p.pace@unich.it

The Clara structure occupies the NW sector of the Zone-B regional seismic survey covering the central Adriatic Sea of Italy (Figure 9.1). It is a compressional structure developed within the Adriatic foreland ahead of the Northern Apennine thrust belt (Figure 9.2).

The Adriatic is a foreland domain belonging to the Adria continental block that, at present, extends NW–SE including from north to south of the Po Plain, Istria, the Adriatic Basin, the Gargano Promontory and the Apulian Peninsula. It is quite unique since it represents a common foreland area shared by three different thrust belts that developed in distinct times: the Apennines to the west-southwest, the Southern Alps to the north-northwest and the Dinarides-Albanides to the east-northeast. The Adriatic region records a multiphase tectonic history with repeated events of normal faulting associated with basin formation followed by compressional deformation induced by the development of the surrounding thrust belts. Permian-Triassic and Jurassic rifting-drifting phases caused a complex paleogeography over the Adria crust, which was dominated by persistent carbonate platforms (Apulian and Adriatic platforms) and deep fault-bounded pelagic basins (Mesozoic Adriatic and Umbria-Marche Basins). Starting from the Cretaceous onwards, convergence between Africa and Europe switched deformation from extension to contraction inducing basin inversion and positive normal fault reactivation in the Adriatic region. During the Neogene, the present-day Adriatic area progressively became a foreland domain to the advancing Apennine thrust belt experiencing syn-orogenic flexure-induced subsidence partly accommodated by normal faulting. A thick, wedge- shaped sequence of Pliocene-Quaternary siliciclastic foredeep deposits progressively buried the carbonate successions (Scisciani and Calamita, 2009; Pace *et al.*, 2015). The Italian sector of the Adriatic is a long-established hydrocarbon province that has recorded in the past a successful exploration history. There are two main proven plays in the area; these consist of biogenic gas in the shallow Pliocene-Pleistocene foredeep sequence and medium-to-heavy oil and subordinate thermogenic gas in the underlying Jurassic-Miocene carbonates (Bertello *et al.*, 2010; Casero and Bigi, 2013).

The Clara anticline is the result of the reverse reactivation of a Mesozoic normal fault bounding a pre-existing half-graben basin (Figure 9.3). This inversion anticline involves the Paleozoic-Lower Jurassic sediments, the Triassic-Miocene carbonate sequence and the Pliocene foredeep deposits (Pace *et al.*, 2015). On seismic sections the structure appears as a markedly east-vergent, asymmetric, harpoon-shaped anticline facing its footwall and riding upon a high-angle reverse fault (Figure 9.4). The Messinian reflector (Me) is gently folded and elevated over its regional level. At depth, the high amplitude low frequency signals, corresponding to the top of the Lower Triassic (LT), show a normal offset across the reverse fault directly underneath the crest of the anticline. The top of the Lower Jurassic (LJ) that is within the syn-rift sequence shows no offset across the reverse reactivated normal fault, resembling the null point of inversion (Cooper and Williams, 1989). The growth of the thrust-related inversion anticline occurred during the Upper Pliocene-Lower Pleistocene interval, as testified by low-amplitude high frequency reflections progressively onlapping towards the fold limbs (yellow half-arrows). These strata postdate the earlier parts of the siliciclastic sequences that onlap the foreland ramp (blue half-arrows) of the Messinian reflector. Prograding clinoforms within

Atlas of Structural Geological Interpretation from Seismic Images, First Edition. Edited by Achyuta Ayan Misra and Soumyajit Mukherjee.

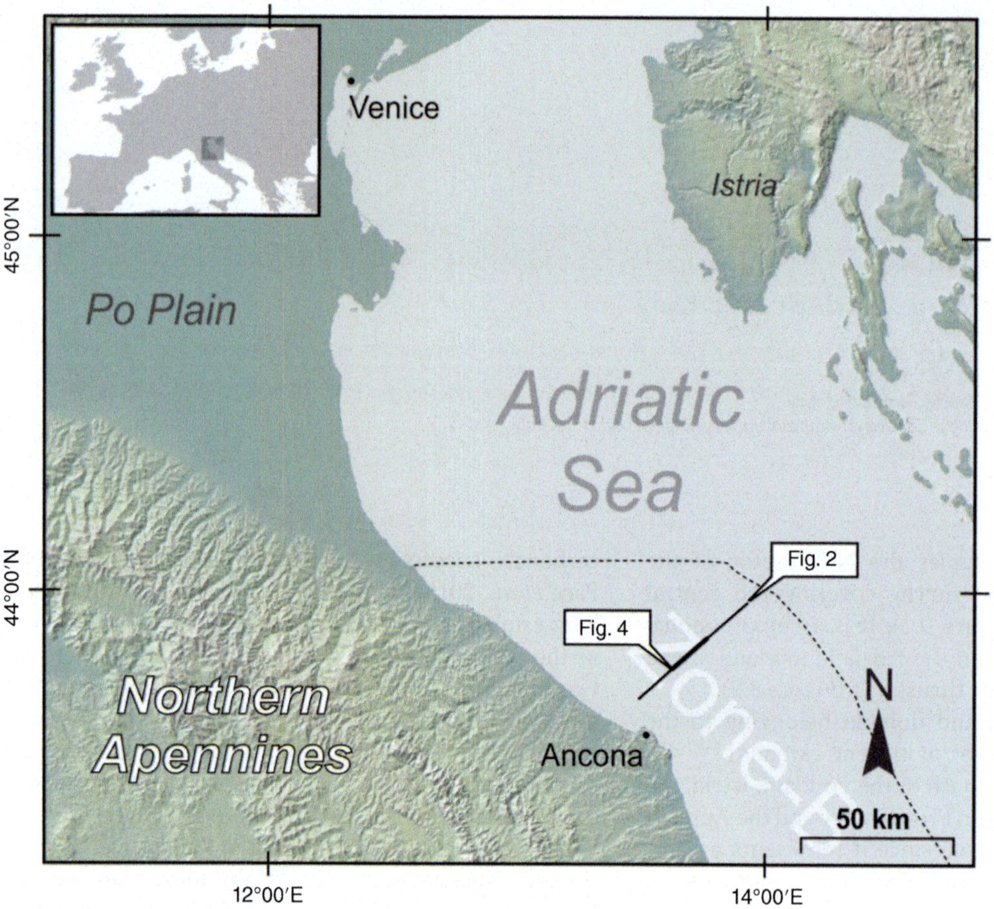

Figure 9.1 Location of the regional cross-section and of the Clara inversion anticline within the Adriatic Sea of Italy.

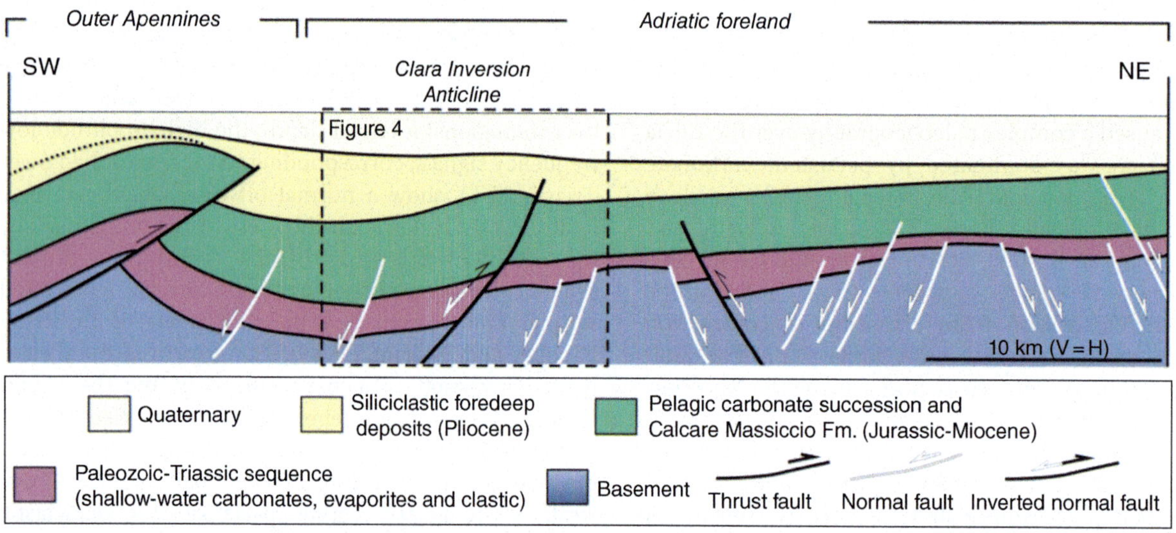

Figure 9.2 Regional cross-section showing the structural position of the Clara inversion anticline within the Adriatic foreland domain ahead of the Northern Apennine thrust belt of Italy. Adapted from Pace *et al*., 2015.

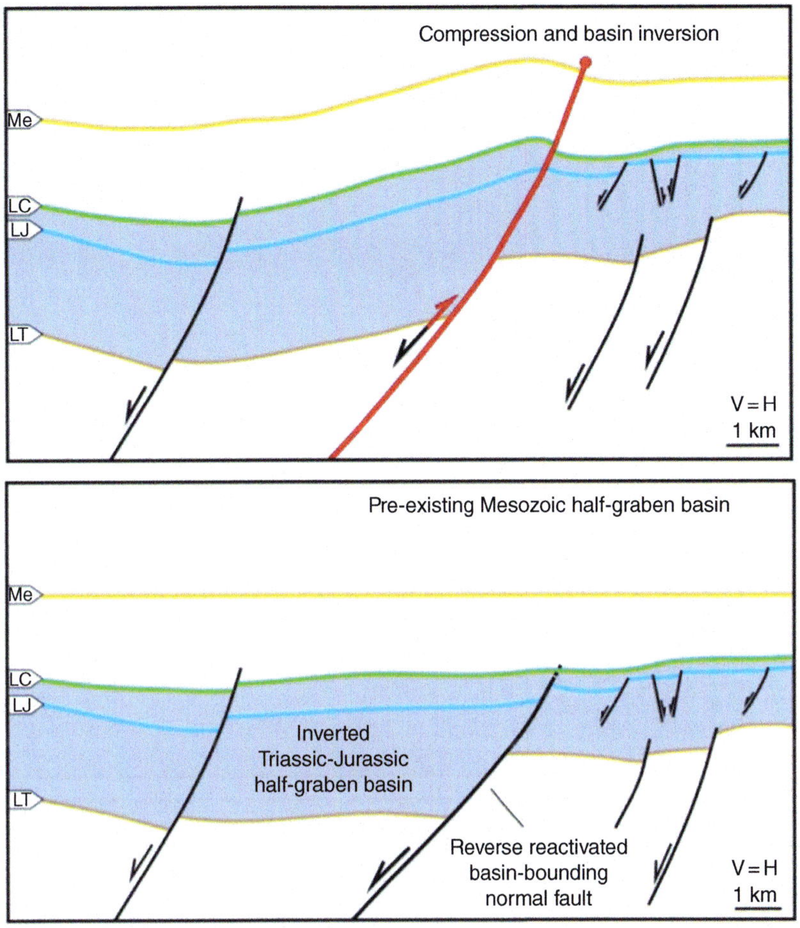

Figure 9.3 Schematic representation of the Clara anticline development resulting from the Pliocene-Quaternary reverse reactivation of a Mesozoic normal fault bounding a pre-existing half-graben basin. Adapted from Pace *et al.*, 2015.

the Quaternary interval are undeformed and postdate the growth of the inversion structure. Selective reactivation of the Mesozoic extensional faults is likely related to their orientation with respect to the Cenozoic compressional stress field during inversion. Normal faults that are orthogonal to the compressional stress are locked whereas ones that are oriented oblique are prone to be reworked as high-angle reverse faults (Cooper and Williams, 1989; Buchanan and Buchanan, 1995).

Seismic data are courtesy of Spectrum Ltd from the reprocessed 2D seismic survey in the Italian Adriatic Sea (www.spectrumgeo.com). The vintage seismic data have been reprocessed as part of a renewed interest of the hydrocarbon potential of the region (Peace *et al.*, 2012) and some recent works have generated a better understanding of the tectonostratigraphic setting and evolution of the Adriatic foreland (Casero and Bigi, 2013; Wrigley *et al.*, 2014, 2015; Pace *et al.*, 2015).

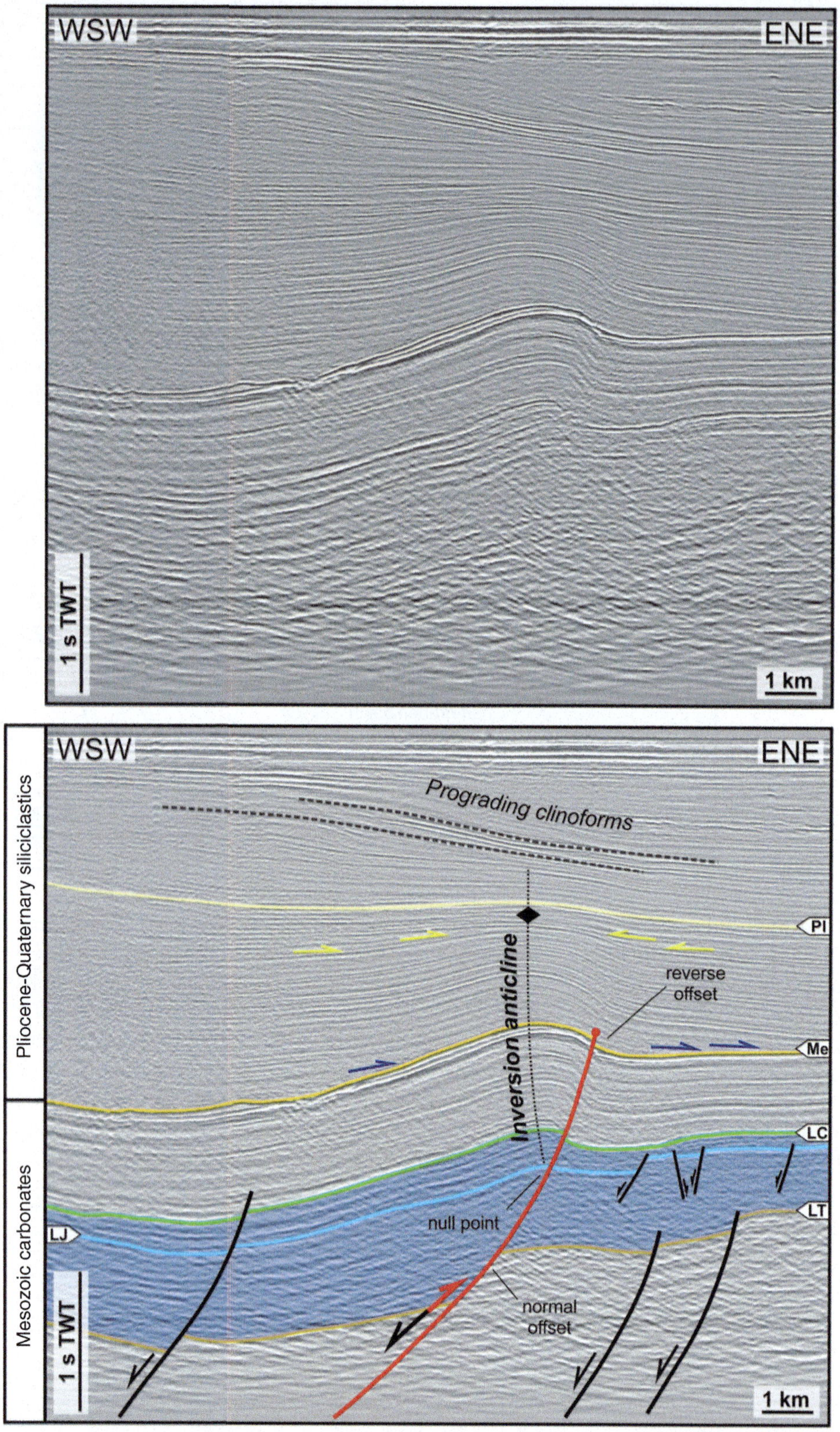

Figure 9.4 Uninterpreted and interpreted time seismic section across the Clara inversion anticline in the Adriatic Sea of Italy. Keys to reflectors, which were tied by the available wells in the area: top Pliocene (Pl); Messinian (Me) – Evaporites; top Lower Cretaceous (LC) Marne a Fucoidi FM; top Lower Jurassic (LJ) Calcare Massiccio Fm; top Lower Triassic (LT) Verrucano Fm. Adapted from Pace *et al.*, 2015. Courtesy of Spectrum.

Acknowledgements

The interpretation benefited from stimulating discussions with V. Scisciani, R.W.H. Butler and D. Iacopini. Spectrum Ltd. is gratefully thanked for providing access to the seismic data library of the Italian Adriatic Sea. Seismic interpretation was carried out with IHS Kingdom and Petrel. Structural modeling and section restoration were performed using Move and the Midland Valley Academic Software Initiative and is kindly acknowledged. Seismic presented in this paper is also available for inspection and interpretation on the Virtual Seismic Atlas (www.seismicatlas.org).

The Editors, Achyuta Ayan Misra and Soumyajit Mukherjee, are gratefully acknowledged for the editorial handling and for inviting to contribute in this atlas.

References

Buchanan, J.G. and Buchanan, P.G. 1995. *Basin Inversion*. Special Publications vol. 88, The Geological Society, London.

Bertello, F., Fantoni, R., Franciosi, R. *et al.* 2010. From thrust-and-fold belt to foreland: hydrocarbon occurrences in Italy. In: Vining, B.A. and Pickering, S.C. (eds) *Petroleum Geology: From Mature Basins to New Frontiers*, Petroleum Geology Conference series, vol. **7**, pp. 113–126.

Casero, P. and Bigi, S. 2013. Structural setting of the Adriatic basin and main related petroleum exploration plays. *Marine and Petroleum Geology*, **42**, 135–147. doi: 10.1016/j.marpetgeo.2012.07.006.

Cooper, M.A. and Williams, G.D. 1989. *Inversion Tectonics*. Special Publications vol. 44, The Geological Society, London.

Pace, P., Scisciani, V., Calamita, F. *et al.* 2015. Inversion structures in a foreland domain: Seismic examples from the Italian Adriatic Sea. *Interpretation*, **4**, SAA161–SAA176.

Peace, D., Brown, D., Zappaterra, E. *et al.* 2012. The Italian Adriatic Sea: Bringing new life to an established region. *The Leading Edge*, **31**, 802–809.

Scisciani, V. and Calamita, F. 2009. Active intraplate deformation within Adria: examples from the Adriatic region. *Tectonophysics*, **476**, 57–72.

Wrigley, R., Marszalek, A., Rodriguez, K. and Hodgson, N. 2014. Offshore Croatia – Hunting "Big Oil" in the centre of Europe, *First Break*, **32**, 61–68.

Wrigley, R., Hodgson, N., Esestime, P. 2015. Petroleum geology and hydrocarbon potential of the Adriatic basin, offshore Croatia. *Journal of Petroleum Geology*, **38**, 301–316.

10

Emma Inversion Structure, Adriatic Sea, Italy

Paolo Pace

GEPlan Consulting, Petroleum Geosciences, Ferrara (FE), Italy
Department of Engineering and Geology, 'Gabriele d'Annunzio' University, Chieti Scalo (CH), Italy
paolo.pace@geplan.it; p.pace@unich.it

The Emma structure is located within the central sector of the Zone-B regional seismic survey covering the central Adriatic Sea of Italy (Figure 10.1). It is a compressional structure developed within the Adriatic foreland ahead of the Central–Northern Apennine thrust belt (Figure 10.2).

The Adriatic is a foreland domain belonging to the Adria continental block that, at present, extends NW–SE including from north to south of the Po Plain, Istria, the Adriatic Basin, the Gargano Promontory and the Apulian Peninsula. It is quite unique, as it represents a common foreland area shared by three different thrust belts that developed in distinct times: the Apennines to the W–SW, the Southern Alps to the N–NW and the Dinarides-Albanides to the E–NE. The Adriatic region records a multiphase tectonic history with repeated normal faulting associated with basin formation followed by compression induced by the development of the surrounding thrust belts (Cooper and Williams, 1989; Buchanan and Buchanan, 1995). Permian-Triassic and Jurassic rifting-drifting phases caused a complex paleogeography over the Adria crust, which was dominated by persistent carbonate platforms (Apulian and Adriatic platforms) and deep fault-bounded pelagic basins (Mesozoic Adriatic and Umbria-Marche Basins). Starting from the Cretaceous onwards, convergence between Africa and Europe caused a switch in deformation from extension to contraction, inducing basin inversion and positive normal fault reactivation in the Adriatic region. During the Neogene, the present-day Adriatic area progressively became a foreland domain to the advancing Apennine thrust belt experiencing syn-orogenic flexure-induced subsidence partly accommodated by normal faulting. A thick, wedge-shaped sequence of Pliocene-Quaternary siliciclastic foredeep deposits progressively buried the carbonate successions (Scisciani and Calamita, 2009; Pace *et al.*, 2015). The Italian sector of the Adriatic is a long-established hydrocarbon province that has recorded in the past a successful exploration history. There are two main proven plays in the area that consist of biogenic gas in the shallow Pliocene-Pleistocene foredeep sequence and medium-to-heavy oil and subordinate thermogenic gas in the underlying Jurassic-Miocene carbonates (Bertello *et al.*, 2010; Casero and Bigi, 2013).

The Emma structure results from the positive inversion of a composite rift-faulted basin in the Adriatic foreland of Italy (Figure 10.2). The overall inversion structure is composed by an array of conjugate high-angle reverse faults and related anticlines (Figures 10.2 and 10.3). The Messinian reflector is mostly gently folded together with the overlying lower section of the Pliocene siliciclastic wedge and rests above its regional level, which is inclined towards the Apennine thrust belt (SW). The structure is cored by Upper Triassic/Lower Jurassic carbonates and shales (BLJ) deposited in an intraplatform euxinic basin (Emma Mesozoic Basin) (Zappaterra, 1994; Cazzini *et al.*, 2015). In both eastward and westward footwall blocks of the Emma inversion structure, the Upper Triassic/Lower Jurassic is represented by the platform carbonates (PLJ) of the Calcare Massiccio Fm. These thickness and facies variations across the structure testify that the Emma uplift resulted from the positive inversion of the pre-existing Emma Mesozoic Basin (Pace *et al.*, 2015). The Mesozoic black-shales of the Mesozoic Emma intraplatform anoxic basin represent the source rock for the oil discoveries in the adjacent fields (Casero and Bigi, 2013; Cazzini *et al.*, 2015). The interpretation of the base of the basinal and platform carbonate successions or top of the Triassic evaporites (TE) is problematic at depth (dotted line) due to migration hyperbolas mostly underneath the culmination of the Emma inversion structure (Figure 10.3).

Atlas of Structural Geological Interpretation from Seismic Images, First Edition. Edited by Achyuta Ayan Misra and Soumyajit Mukherjee.

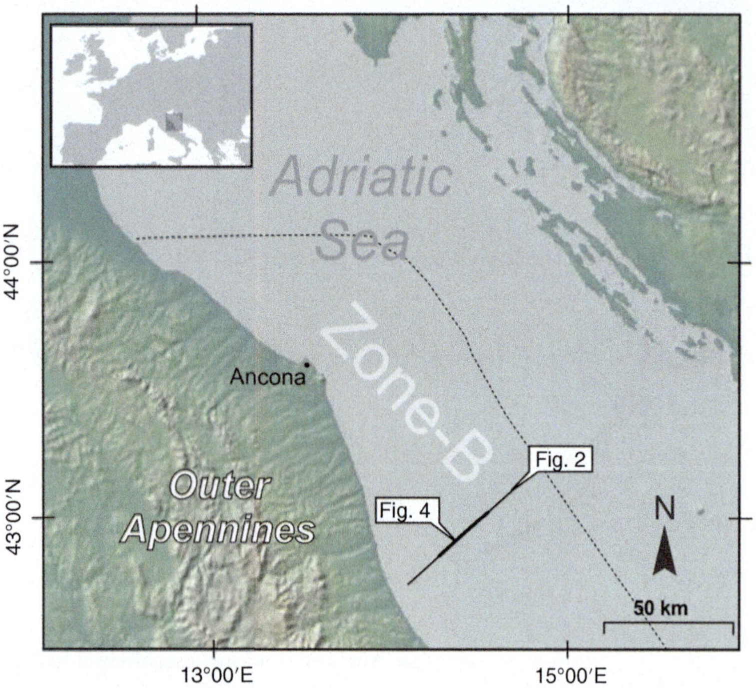

Figure 10.1 Location of the regional cross-section and of the Emma inversion structure within the Adriatic Sea of Italy.

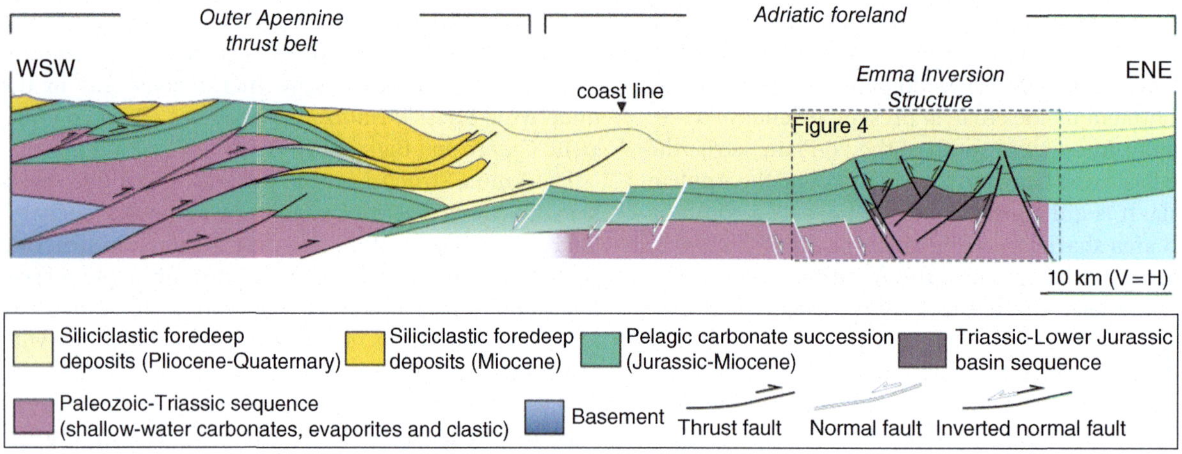

Figure 10.2 Regional cross-section showing the structural position of the Emma inversion structure within the Adriatic foreland domain ahead of the Central-Northern Apennine thrust belt of Italy. Adapted from Pace *et al.*, 2015.

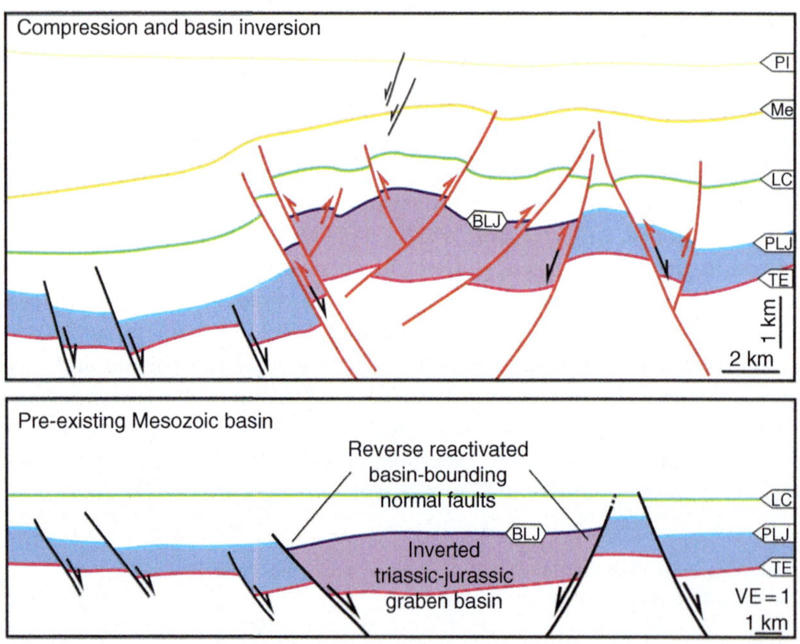

Figure 10.3 Schematic representation showing the development of the Emma structure resulting from the Cenozoic positive inversion of an inherited Mesozoic Basin. Adapted from Pace *et al.*, 2015.

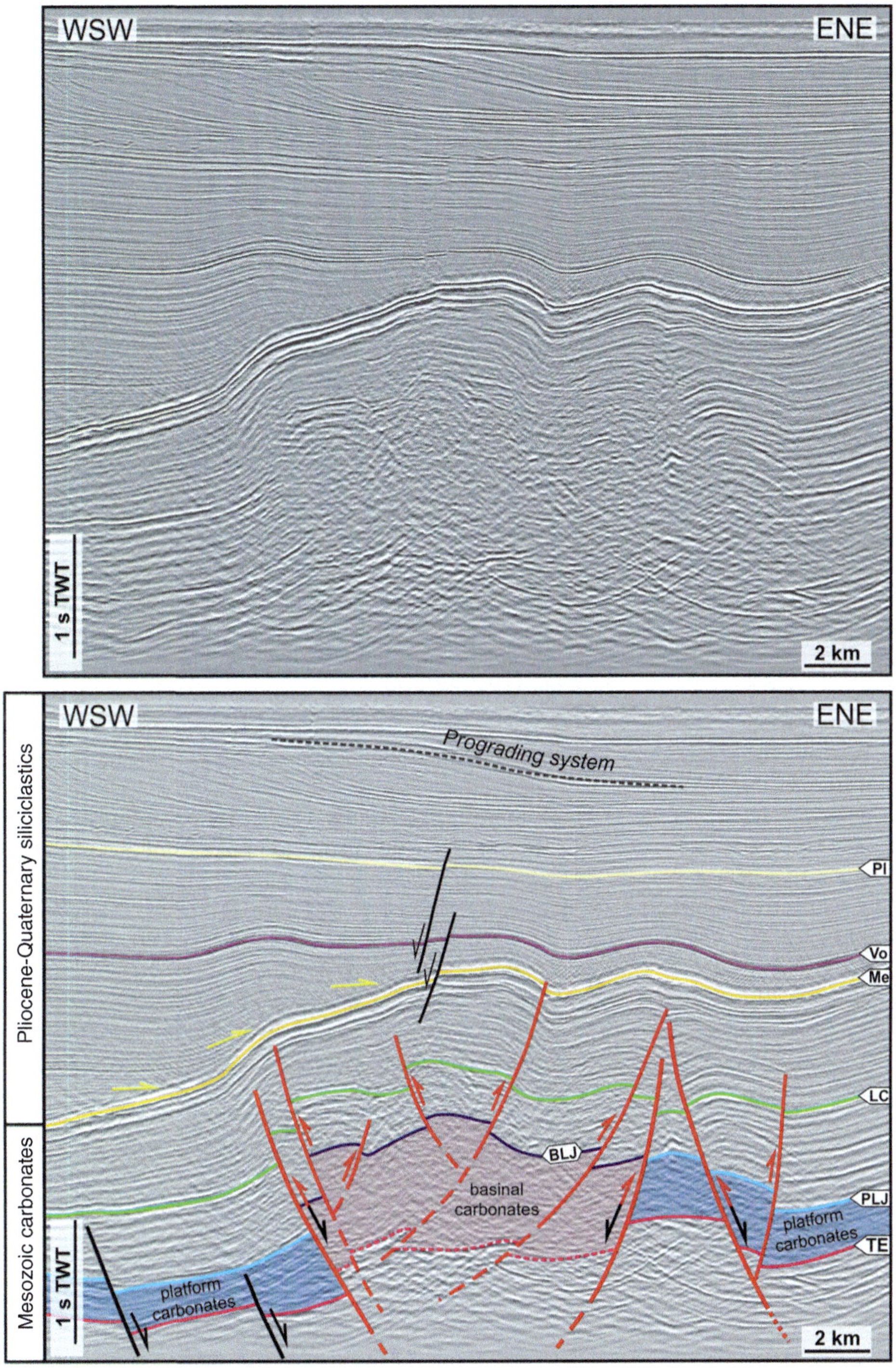

Figure 10.4 Uninterpreted and interpreted time seismic section across the Emma inversion structure in the Adriatic Sea of Italy. Keys to reflectors, which were tied by the available wells in the area: top Pliocene (Pl); Messinian (Me) – Evaporites; Pliocene volcanoclastic horizon (Vo); top Lower Cretaceous (LC); top Lower Jurassic platform carbonates (PLJ) – Calacare Massiccio Fm; top Lower Jurassic basinal carbonates (BLJ) – Emma Limestones; top Triassic evaporites (TE) – Burano Fm. Adapted from Pace *et al.*, 2015. Courtesy of Spectrum.

The most recent phase of growth of the Emma inversion structure can be charted during the Upper Pliocene/Lower Pleistocene as delineated by onlaps over the Messinian reflector (yellow half-arrows) and by thickness variations within the high frequency signals of the siliciclastic wedge. Well-rendered prograding clinoformal systems within the Quaternary package remain undeformed and postdate this event of positive inversion (Figure 10.4).

Seismic data are courtesy of Spectrum Ltd from the reprocessed 2D seismic survey in the Italian Adriatic

Sea (www.spectrumgeo.com). The vintage seismic data have been reprocessed as part of a renewed interest of the hydrocarbon potential of the region (Peace *et al.*, 2012) and some recent works have generated a better understanding of the tectonostratigraphic setting and evolution of the Adriatic foreland (Casero and Bigi, 2013; Wrigley *et al.*, 2014, 2015; Pace *et al.*, 2015).

Acknowledgements

The interpretation benefited from stimulating discussions with V. Scisciani, R.W.H. Butler and D. Iacopini. Spectrum Ltd. is gratefully thanked for providing access to the seismic data library of the Italian Adriatic Sea. Seismic interpretation was carried out with IHS Kingdom and Petrel. Structural modeling and section restoration were performed using Move and the Midland Valley Academic Software Initiative and is kindly acknowledged. Seismic presented in this paper is also available for inspection and interpretation on the Virtual Seismic Atlas (www.seismicatlas.org).

The Editors, Achyuta Ayan Misra and Soumyajit Mukherjee, are gratefully acknowledged for the editorial handling and for inviting to contribute in this atlas.

References

Buchanan, J.G. and Buchanan, P.G. 1995. *Basin Inversion.* Special Publications vol. 88, The Geological Society, London.

Bertello, F., Fantoni, R., Franciosi, R. *et al.* 2010. From thrust-and-fold belt to foreland: hydrocarbon occurrences in Italy. In: Vining, B.A. and Pickering, S.C. (eds) *Petroleum Geology: From Mature Basins to New Frontiers*, Petroleum Geology Conference series, vol. 7, pp. 113–126.

Casero, P. and Bigi, S. 2013. Structural setting of the Adriatic basin and main related petroleum exploration plays. *Marine and Petroleum Geology*, **42**, 135–147. doi: 10.1016/j.marpetgeo.2012.07.006.

Cazzini, F., Dal Zotto, O., Fantoni, R. *et al.* 2015. Oil and gas in the Adriatic foreland, Italy. *Journal of Petroleum Geology*, **38**, 255–279.

Cooper, M.A. and Williams, G.D. 1989. *Inversion Tectonics.* Special Publications vol. 44, The Geological Society, London.

Pace, P., Scisciani, V., Calamita, F. *et al.* 2015. Inversion structures in a foreland domain: Seismic examples from the Italian Adriatic Sea. *Interpretation*, **4**, SAA161–SAA176.

Peace, D., Brown, D., Zappaterra, E. *et al.* 2012. The Italian Adriatic Sea: Bringing new life to an established region. *The Leading Edge*, **31**, 802–809.

Scisciani, V. and Calamita, F. 2009. Active intraplate deformation within Adria: examples from the Adriatic region. *Tectonophysics*, **476**, 57–72

Wrigley, R., Marszalek, A., Rodriguez, K. and Hodgson, N. 2014. Offshore Croatia – Hunting "Big Oil" in the centre of Europe, *First Break*, **32**, 61–68.

Wrigley, R., Hodgson, N., Esestime, P. 2015. Petroleum geology and hydrocarbon potential of the Adriatic basin, offshore Croatia. *Journal of Petroleum Geology*, **38**, 301–316.

Zappaterra, E. 1994. Source-rock distribution model of the Periadriatic region. *AAPG Bulletin*, **78**, 333–354.

11

Inversion in the Umiat Basin (Part 1) National Petroleum Reserve Alaska (NPRA), North Slope, Alaska

Paolo Pace

GEPlan Consulting, Petroleum Geosciences, Ferrara (FE), Italy
Department of Engineering and Geology, 'Gabriele d'Annunzio' University, Chieti Scalo (CH), Italy
paolo.pace@geplan.it; p.pace@unich.it

The Umiat Basin extends in the eastern sector of the National Petroleum Reserve in Alaska (NPRA) in the North Slope of northwestern Artic Alaska (Bird, 1981). The NPRA is bound by Chukchi Sea westward, by the Beaufort Sea northward, by the Colville River eastward and by the Brooks Range southward (Figure 11.1).

The NPRA is located within the foreland basin ahead of the Brooks Range, which is known as the Colville Trough. The Umiat Basin represents a minor sub-basin of the Colville Trough (Figure 11.1). Across Arctic Alaska three megasequences were deposited over the metamorphosed basement (Franklinian): the Ellesmerian (Carboniferous-Jurassic), the Beaufortian (Jurassic-Cretaceous) and the Brookian (Cretaceous to Cenozoic) (Hubbard *et al.*, 1987).

The Franklinian sequence includes pre-Mississippian low-grade metamorphic rocks and some minor igneous rocks with granite intrusions and is commonly considered as the basement (Kulander *et al.*, 2005). The Ellesmerian sequence, including the Endicott, Lisburne and Sadlerochit groups, the Shublik Formation and the Sag River Sandstone, is composed of Mississippian to Triassic marine carbonate and nonmarine to shallow marine siliciclastic strata deposited on a passive continental margin. The Beaufortian sequence is a complex of Jurassic to Lower Cretaceous siliciclastic sediments recording erosional and depositional effects of rifting associated with the opening of the Arctic Ocean Basin (Hubbard *et al.*, 1987). The Brookian sequence consists of Cretaceous to Tertiary syn-orogenic siliciclastic sedimentary deposits derived from the Brooks Range orogenic belt and sedimented northward in deep-marine to nonmarine systems in the Colville foreland basin and on the continental margin north of the Barrow arch (Bird, 2001; Houseknecht and Bird, 2004). The sedimentary succession thickens into the deep, E–W trending Colville trough in the central part of the NPRA. Moving southward the sedimentary filling of the Colville foredeep becomes progressively more deformed being incorporated into the north-vergent fold-and-thrust belt of the Brooks Range.

The region has undergone a complex tectonic history, including several distinct orogenic events and significant anticlockwise rotation relative to the North American craton (Embry, 1990). The Brooks Range orogeny initiated in the Jurassic with sediments coming from the north filling the accommodation space in the foreland (Houseknecht and Bird, 2004). Subsequently, the Arctic microplate rotated counterclockwise into its current position and the Colville Basin was formed as a result of loading from the Brooks Range and rifting in the Beaufort (Schultz, 2013).

Basin subsidence in the region is retained to have initiated as the Early-Mid Devonian orogeny concluded (Hubbard *et al.*, 1987; Sherwood *et al.*, 2002). Other workers suggested a two-stage basin evolution characterized by early rifting in the Lower Paleozoic basement with subsequent Endicott Group deposition (Lower Ellesmerian Sequence), followed by a brief episode of northward compression (Mauch, 1987). In this scenario, stress field inverted locally induced by distal compressional tectonics.

Newly reprocessed 2D regional seismic profiles extending through the entire NPRA provide a clearer resolution imaging of the basin subsurface (Miller *et al.*, 2000). Their structural and stratigraphic interpretation confirm the presence of local compressional features, such as reverse faults, thrust-related folds and synkinematic thinning of strata, within the Umiat Basin (Fulk 2010; Schultz, 2013) (Figure 11.2).

This first seismic example shows the reverse reactivation of a pre-existing extensional fault system within the Umiat Basin (Figure 11.2) with both north- and

Atlas of Structural Geological Interpretation from Seismic Images, First Edition. Edited by Achyuta Ayan Misra and Soumyajit Mukherjee.

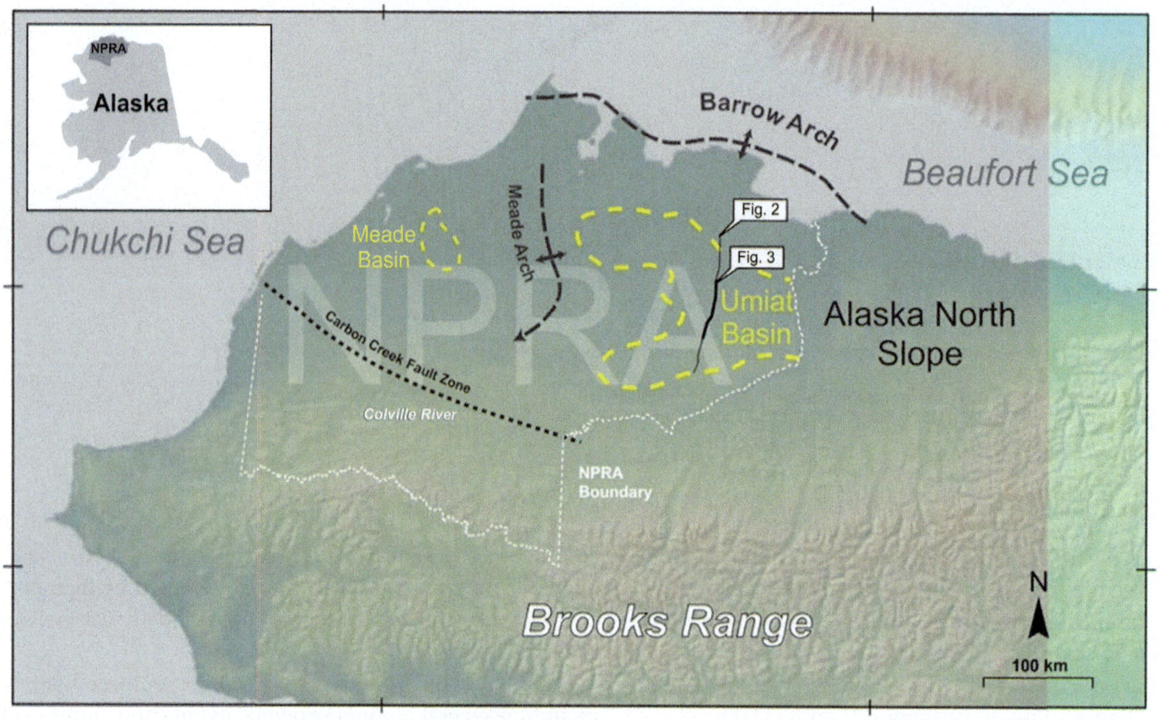

Figure 11.1 Location of the seismic sections across the Umiat Basin in the eastern sector of the National Petroleum Reserve in Alaska (NPRA), North Slope of NW Artic Alaska. Adapted from Kulander *et al.*, 2005.

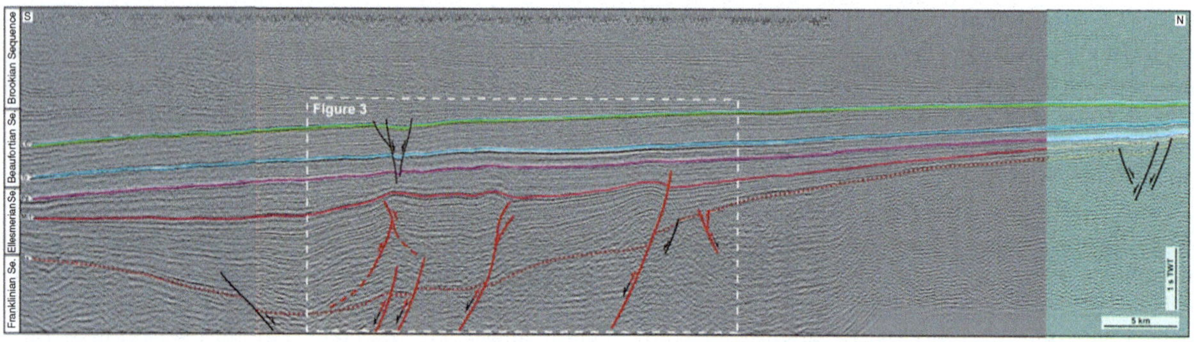

Figure 11.2 Regional seismic section across the Umiat Basin showing the occurrence of both compressional and extensional structural features. Seismic data from the USGS reprocessed 2D regional seismic data set in the NPRA (Miller *et al.*, 2000).

south-dipping normal faults rooted in the Franklinian basement. The positive inversion process caused reactivation in terms of reverse faulting of some of these basement-rooted normal faults (i.e. thick-skinned basement-involved style of deformation). Inversion of the pre-existing south-dipping normal faults along the foreland basement ramp have generated folding of the overlaying Mississippian sedimentary strata (lower section of the Ellesmerian Sequence) developing gentle, markedly asymmetric harpoon-shaped inversion anticlines (Figure 11.3). The reverse reactivated normal fault producing the inversion anticline in the central zone of the seismic section of Figure 11.3 exhibit a change in geometry from down-dip listric to up-dip anti-listric that is a typical geometry resulting from positive inversion (Cooper and Williams, 1989; Yamada and McClay, 2004). The multiple reactivations of the pre-existing basement normal faults is accompanied by complex folding and thrusting within the overlaying sedimentary sequence constituting the lower section of the Ellesmerian Sequence (left side of Figure 11.3). Normal faulting localized above that anticline (left side of Figure 11.3) is most likely related to collapse affecting the Pennsylvanian-Lower Cretaceous sequence. The weak shaley layers at

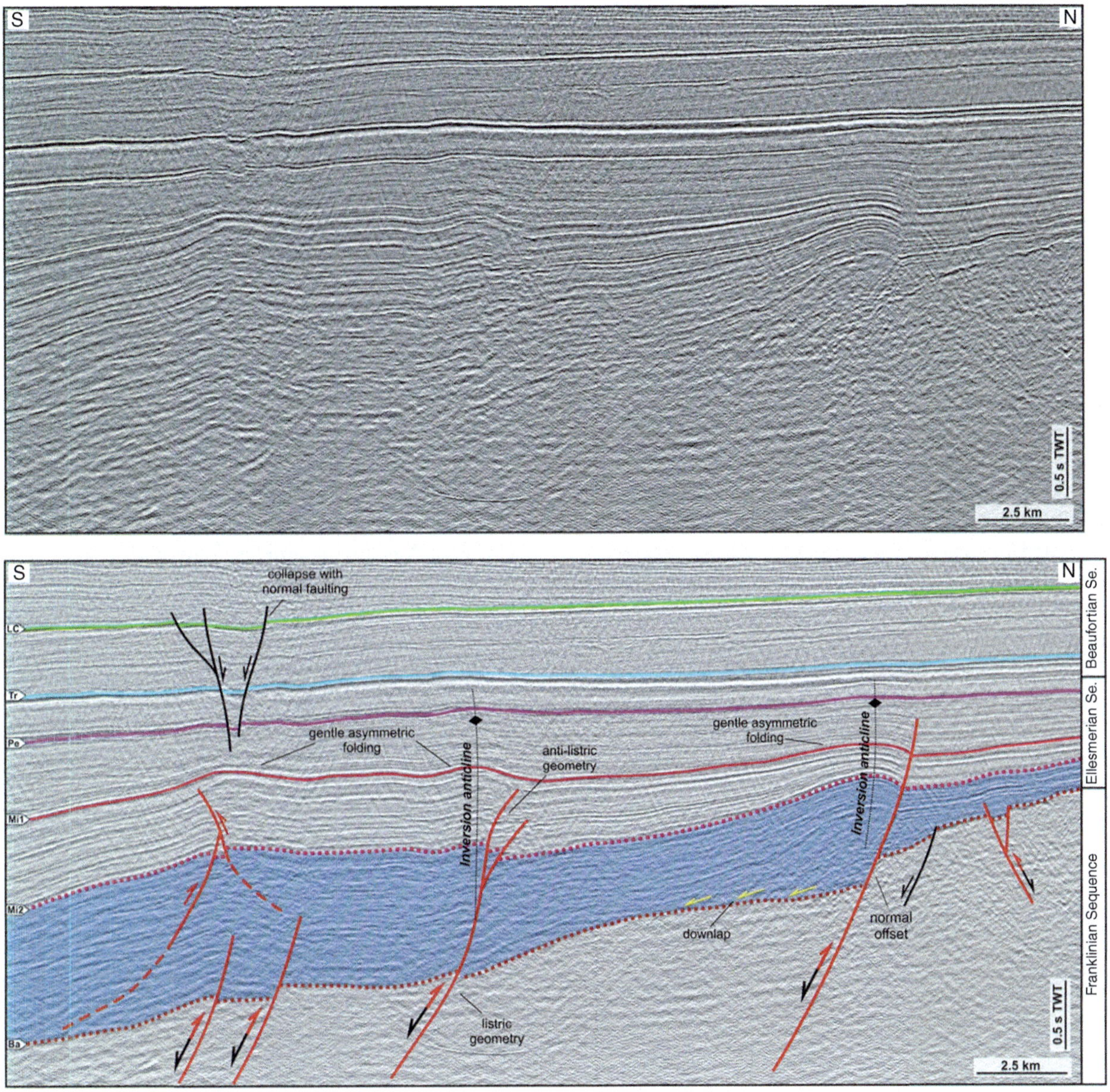

Figure 11.3 Uninterpreted and interpreted time seismic section showing the characteristics of positive inversion within the Umiat Basin. Seismic data from the USGS reprocessed 2D regional seismic data set in the NPRA (Miller *et al.*, 2000).

the top Ellesmerian/base Beaufortian interface (Shublik Formation and Kingak Shale) may have accommodated decoupling between the lower compression/inversion and the upper extension. The presence of all of these compressional structures testify that contractional reworking and inversion occurred within the Umiat Basin and it is likely related to distal compressional stresses transmitted through the basin.

Acknowledgements

The interpretation benefited from stimulating discussions with V. Scisciani. Seismic data are publicly available from the USGS reprocessed 2D regional seismic dataset in the NPRA (Miller *et al.*, 2000).

The Editors, Achyuta Ayan Misra and Soumyajit Mukherjee, are gratefully acknowledged for the editorial handling and for inviting to contribute in this atlas.

References

Bird, K.J. 1981. Petroleum exploration of the North Slope in Alaska, USA. Open-File Report 81-227, United States Geological Survey.

Bird, K.J. 2001. Framework geology, petroleum systems, and play concepts of the National Petroleum Reserve-Alaska. In: Houseknecht, D.W. (ed) *NPRA Core Workshop – Petroleum Plays and Systems in the National Petroleum Reserve*, SEPM Core Workshop 21, Alaska, pp. 5–17.

Cooper, M.A. and Williams, G.D. 1989. *Inversion Tectonics*. Special Publications vol. 44, The Geological Society, London.

Embry, A.F. 1990. Geological and geophysical evidence in support of the hypothesis of anticlockwise rotation of northern Alaska. *Marine Geology*, **93**, 317–329.

Fulk, B.R. 2010. Seismic analysis of the Ikpikpuk-Umiat basin, NPRA, North Slope, Alaska. MSc Thesis, San Diego State University.

Hubbard, R.J., Edrich, S.P. and Rattey, R.P. 1987. Geologic evolution and hydrocarbon habitat of the 'Arctic Alaska microplate.' *Marine and Petroleum Geology*, **4**, 2–34.

Houseknecht, D.W. and Bird, K.J. 2004. Sequence stratigraphy of the Kingak Shale(Jurassic–Lower Cretaceous), National Petroleum Reserve in Alaska. *AAPG Bulletin*, **88**, 279–302.

Kulander, C.S., Grow, J.A., Potter, C.J. *et al.* 2005. Interpreted Regional Seismic Reflection Lines, National Petroleum Reserve in Alaska. Open-File Report 2005-1402, United States Geological Survey.

Mauch, E.A. 1987. Foredeep: Stratigraphic and structural interpretation of selected seismic lines, Ikpikpuk-Umiat Basin, National Petroleum Reserve, Alaska. In: Bally A.W. (ed) *Atlas of Seismic Stratigraphy*, AAPG Studies in Geology vol. 3, American Association of Petroleum Geologists, Tulsa, OK.

Miller, J.J., Agena, W.F., Lee, W.L. *et al.* 2000. Regional seismic lines reprocessed using post-stack processing techniques: National Petroleum Reserve – Alaska (NPRA). Open File Report 00-286, United States Geological Survey.

Schultz, L.E. 2013. Sequence Stratigraphic and Structural Interpretation, Modeling and Restoration in the National Petroleum Reserve in Alaska (NPRA). BSc Thesis, Washington and Lee University.

Sherwood, K.W., Johnson, P.P., Craig, J.D. *et al.* 2002. Structure and stratigraphy of the Hanna Trough, U.S. Chukchi Shelf, Alaska. *Geological Society of America Special Papers*, **360**, 39–66.

Yamada, Y. and McClay, K.R. 2004. 3-D Analog modelling of inversion thrust structures. In: McClay, K.R. (ed) *Thrust Tectonics and Hydrocarbon Systems*. American Association of Petroleum Geologists Memoir vol. 82, pp. 276–302.

12

Inversion in the Umiat Basin (Part 2) National Petroleum Reserve Alaska (NPRA), North Slope, Alaska

Paolo Pace

GEPlan Consulting, Petroleum Geosciences, Ferrara (FE), Italy
Department of Engineering and Geology, 'Gabriele d'Annunzio' University, Chieti Scalo (CH), Italy
paolo.pace@geplan.it; p.pace@unich.it

The Umiat Basin extends in the eastern sector of the National Petroleum Reserve in Alaska (NPRA) in the North Slope of northwestern Artic Alaska (Bird, 1981). The NPRA is bound by Chukchi Sea westward, by the Beaufort Sea northward, by the Colville River eastward and by the Brooks Range southward (Figure 12.1).

The NPRA is located within the foreland basin ahead of the Brooks Range, which is known as the Colville Trough. The Umiat Basin represents a minor sub-basin of the Colville Trough (Figure 12.1). Across Arctic Alaska three megasequences were deposited over the metamorphosed basement (Franklinian): the Ellesmerian (Carboniferous-Jurassic), the Beaufortian (Jurassic-Cretaceous) and the Brookian (Cretaceous to Cenozoic) (Hubbard *et al.*, 1987).

The Franklinian sequence includes pre-Mississippian low-grade metamorphic rocks and some minor igneous rocks with granite intrusions and is commonly considered as the basement (Kulander *et al.*, 2005). The Ellesmerian sequence, including the Endicott, Lisburne and Sadlerochit groups, the Shublik Formation and the Sag River Sandstone, is composed of Mississippian to Triassic marine carbonate and nonmarine to shallow marine siliciclastic strata deposited on a passive continental margin. The Beaufortian sequence is a complex of Jurassic to Lower Cretaceous siliciclastic sediments recording erosional and depositional effects of rifting associated with the opening of the Arctic Ocean Basin (Hubbard *et al.*, 1987). The Brookian sequence consists of Cretaceous to Tertiary syn-orogenic siliciclastic sedimentary deposits derived from the Brooks Range orogenic belt and sedimented northward in deep-marine to nonmarine systems in the Colville foreland basin and on the continental margin north of the Barrow arch (Bird, 2001; Houseknecht and Bird, 2004). The sedimentary succession thickens into the deep, E–W trending Colville trough in the central part of the NPRA. Moving southward the sedimentary filling of the Colville foredeep becomes progressively more deformed being incorporated into the north-vergent fold-and-thrust belt of the Brooks Range.

The region has undergone a complex tectonic history, including several distinct orogenic events and significant anticlockwise rotation relative to the North American craton (Embry, 1990). The Brooks Range orogeny was initiated in the Jurassic with sediments coming from the north filling the accommodation space in the foreland (Houseknecht and Bird, 2004). Subsequently, the Arctic microplate rotated counterclockwise into its current position and the Colville Basin was formed as a result of loading from the Brooks Range and rifting in the Beaufort (Schultz, 2013).

Basin subsidence in the region is retained to have initiated as the Early-Mid Devonian orogeny concluded (Hubbard *et al.*, 1987; Sherwood *et al.*, 2002). Other workers suggested a two-stage basin evolution characterized by early rifting in the Lower Paleozoic basement with subsequent Endicott Group deposition (Lower Ellesmerian Sequence), followed by a brief episode of northward compression (Mauch, 1987). In this scenario, a local stress field inversion induced by distal compressional tectonics occurred.

Newly reprocessed 2D regional seismic profiles extending through the entire NPRA provide a clearer resolution imaging of the basin subsurface (Miller *et al.*, 2000). Their structural and stratigraphic interpretation confirm local compressional features, such as reverse faults, thrust-related folds and syn-kinematic thinning of strata, within the Umiat Basin (Fulk 2010; Schultz, 2013) (Figure 12.2).

This second seismic example across the Umiat Basin clearly shows a very mild positive inversion of

Atlas of Structural Geological Interpretation from Seismic Images, First Edition. Edited by Achyuta Ayan Misra and Soumyajit Mukherjee.

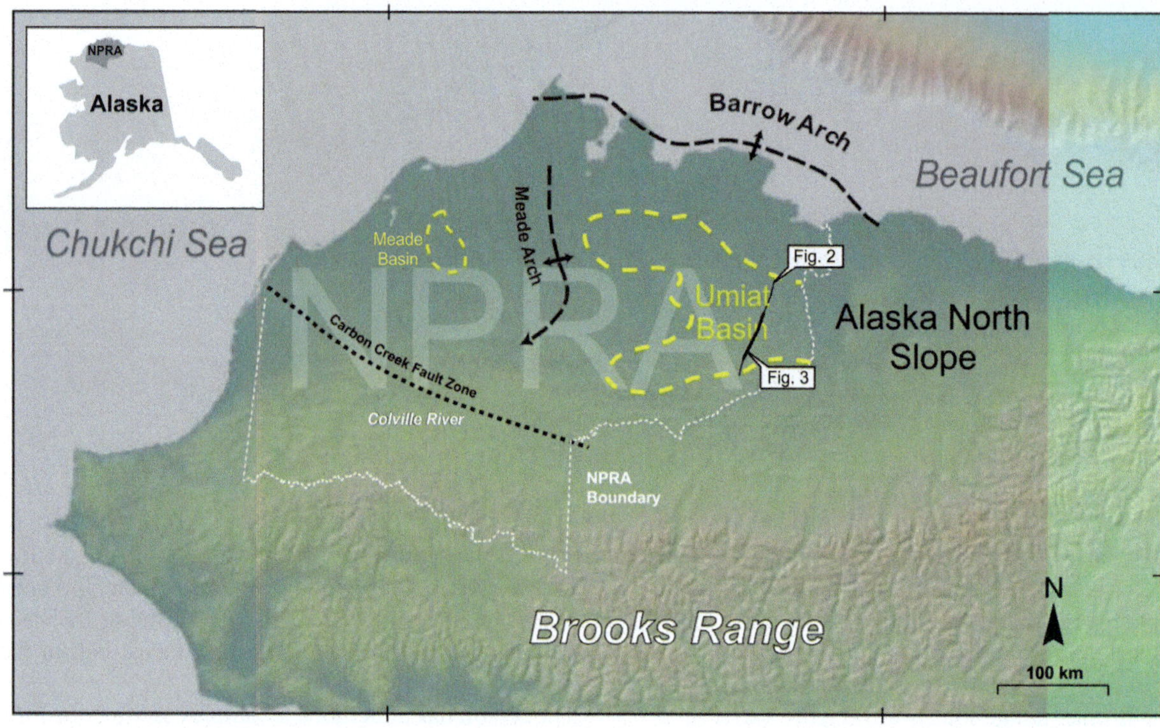

Figure 12.1 Location of the seismic sections across the Umiat Basin in the eastern sector of the National Petroleum Reserve in Alaska (NPRA), North Slope of northwestern Artic Alaska. Adapted from Kulander *et al.*, 2005.

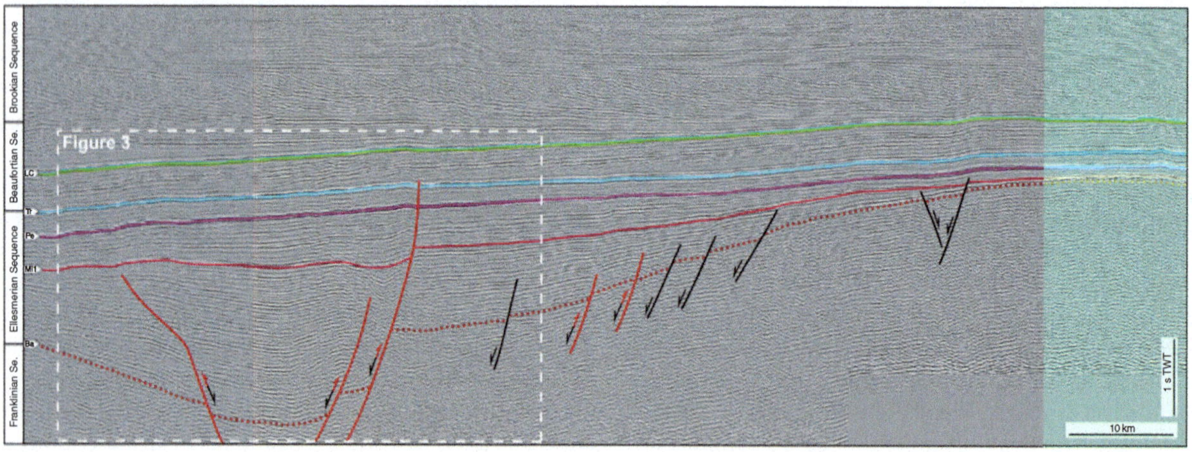

Figure 12.2 Regional seismic section across the Umiat Basin showing the occurrence of both compressional and extensional structural features. Seismic data from the USGS reprocessed 2D regional seismic dataset in the NPRA (Miller *et al.*, 2000).

a pre-existing basin (Cooper and Williams, 1989; Yamada and McClay, 2004). The basin, related to a conjugate normal fault system, accommodated a thicker sedimentary succession of the lower part of the Ellesmerian Sequence (Figure 12.3). Both the north- and south-dipping normal faults that are rooted into the Franklinian basement reactivated with reverse movement during basin inversion. Gentle asymmetric anticlines are developed in the hangingwall of the inverted faults. Some minor extensional faults are localized in the crest zone of one of the inversion anticlines and are probably related to crestal collapse during inversion (central area of Figure 12.3). The positive inversion of this minor basin as suggested by thickness variations and anticline geometries further testifies that contractional reworking and inversion occurred within the Umiat Basin and it is likely related to distal compressional stresses transmitted through the basin.

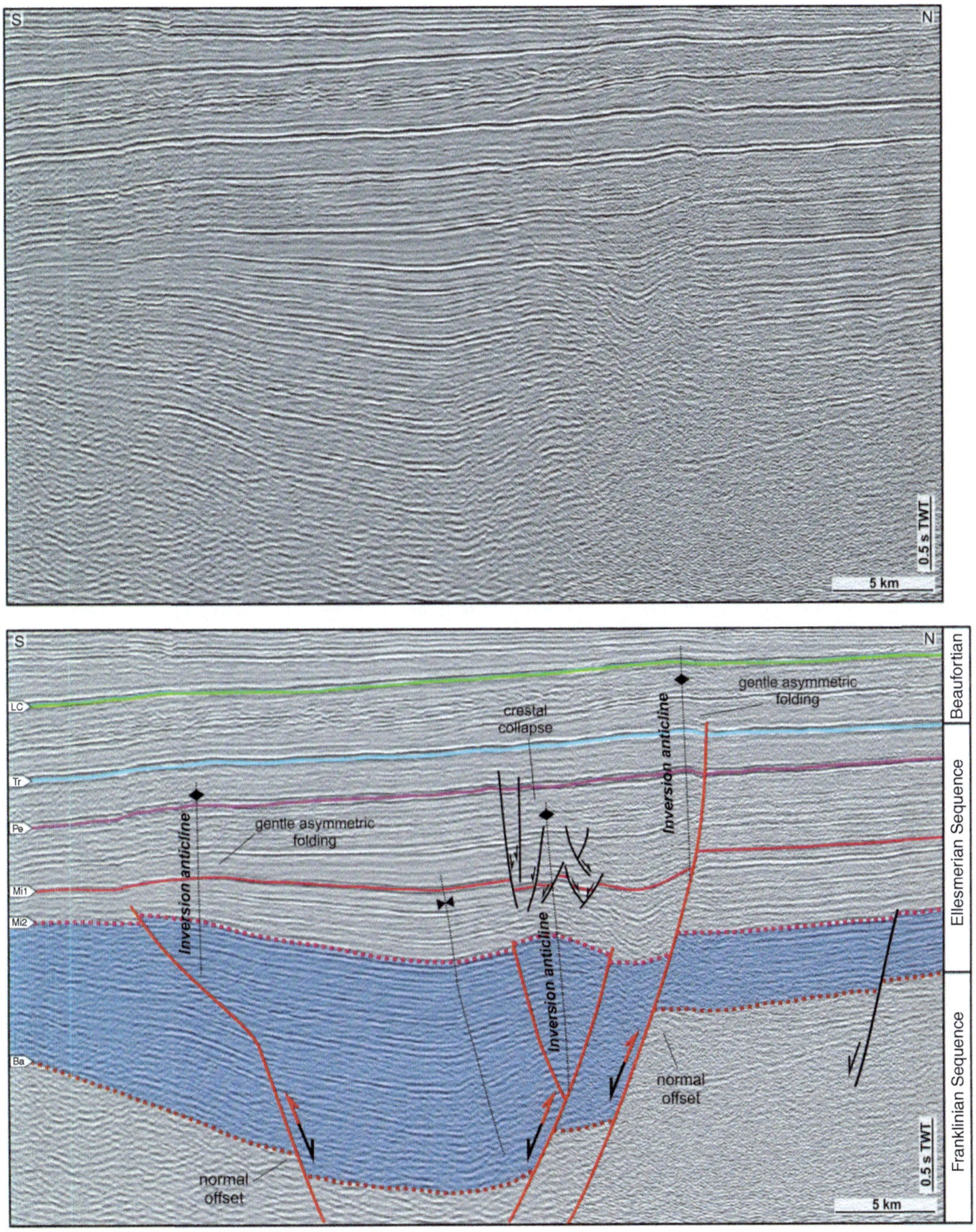

Figure 12.3 Uninterpreted and interpreted time seismic showing section the characteristics of the positive inversion of a minor basin within the Umiat Basin. Some amount of drag related to faults (Mukherjee, 2014) are deciphered. Keys to reflectors: top Lower Cretaceous Unconformity (LC); top Triassic (Tr) – Shublik Fm; top Pennsylvanian (Pe) – Lisburne Group; within Mississippian (Mi1) – top Endicott Group; within Mississippian (Mi2); top Pre-Mississippian Basement (Ba). Seismic data from the USGS reprocessed 2D regional seismic dataset in the NPRA (Miller *et al.*, 2000).

Acknowledgements

The interpretation benefited from stimulating discussions with V. Scisciani. Seismic data are publicly available from the USGS reprocessed 2D regional seismic dataset in the NPRA (Miller *et al.*, 2000).

The Editors, Achyuta Ayan Misra and Soumyajit Mukherjee, are gratefully acknowledged for the editorial handling and for inviting to contribute in this atlas.

References

Bird, K.J. 1981. Petroleum exploration of the North Slope in Alaska, USA. Open-File Report 81-227, United States Geological Survey.

Bird, K.J. 2001. Framework geology, petroleum systems, and play concepts of the National Petroleum Reserve-Alaska. In: Houseknecht, D.W. (ed) *NPRA Core Workshop – Petroleum Plays and Systems in the National Petroleum Reserve*, SEPM Core Workshop 21, Alaska, pp. 5–17.

Cooper, M.A. and Williams, G.D. 1989. *Inversion Tectonics*. Special Publications vol. 44, The Geological Society, London.

Embry, A.F. 1990. Geological and geophysical evidence in support of the hypothesis of anticlockwise rotation of northern Alaska. *Marine Geology*, **93**, 317–329.

Fulk, B.R. 2010. Seismic analysis of the Ikpikpuk-Umiat basin, NPRA, North Slope, Alaska. MSc Thesis, San Diego State University.

Hubbard, R.J., Edrich, S.P. and Rattey, R.P. 1987. Geologic evolution and hydrocarbon habitat of the 'Arctic Alaska microplate.' *Marine and Petroleum Geology*, **4**, 2–34.

Houseknecht, D.W. and Bird, K.J. 2004. Sequence stratigraphy of the Kingak Shale(Jurassic–Lower Cretaceous), National Petroleum Reserve in Alaska. *AAPG Bulletin*, **88**, 279–302.

Kulander, C.S., Grow, J.A., Potter, C.J. *et al.* 2005. Interpreted Regional Seismic Reflection Lines, National Petroleum Reserve in Alaska. Open-File Report 2005-1402, United States Geological Survey.

Mauch, E.A. 1987. Foredeep: Stratigraphic and structural interpretation of selected seismic lines, Ikpikpuk-Umiat Basin, National Petroleum Reserve, Alaska. In: Bally A.W. (ed) *Atlas of Seismic Stratigraphy*, AAPG Studies in Geology vol. 3, American Association of Petroleum Geologists, Tulsa, OK.

Miller, J.J., Agena, W.F., Lee, W.L. *et al.* 2000. Regional seismic lines reprocessed using post-stack processing techniques: National Petroleum Reserve – Alaska (NPRA). Open File Report 00-286, United States Geological Survey.

Mukherjee, S. 2014. Review of flanking structures in meso- and micro-scales. *Geological Magazine*, **151**, 957–974.

Schultz, L.E. 2013. Sequence Stratigraphic and Structural Interpretation, Modeling and Restoration in the National Petroleum Reserve in Alaska (NPRA). BSc Thesis, Washington and Lee University.

Sherwood, K.W., Johnson, P.P., Craig, J.D. *et al.* 2002. Structure and stratigraphy of the Hanna Trough, U.S. Chukchi Shelf, Alaska. *Geological Society of America Special Papers*, **360**, 39–66.

Yamada, Y. and McClay, K.R. 2004. 3-D Analog modelling of inversion thrust structures. In: McClay, K.R. (ed) *Thrust Tectonics and Hydrocarbon Systems*. American Association of Petroleum Geologists Memoir vol. 82, pp. 276–302.

13

Complex Multiphase Inversion Tectonics in the Southern East Shetland Platform, Offshore United Kingdom

Stefano Patruno and William Reid*

Petroleum Geo-Services (PGS), Weybridge, United Kingdom
** Stephano.Patruno@pgs.com*

The sub-Tertiary sedimentary cover is affected by the following tectonic events: (i) Devonian extension, highlighted by few minor faults that do not seem to penetrate the top of the Devonian package (e.g. between faults 6 and 8 in Figures 13.1 and 13.2) (Seranne, 1992; Ziegler, 1992; Marshall and Hewett, 2003; Wilson *et al.*, 2010; Patruno & Reid, 2016, 2017; Patruno, 2017); (ii) Early Carboniferous extension, reflected by normal faults terminating within the lower part of the Carboniferous unit and causing hangingwall expansion of the Carboniferous package (e.g., faults 4, 5, 7 in Figures 13.1 and 13.2) and the regional-wide preservation of relatively thick Carboniferous strata within UKCS Quadrant 15 (Leeder and Boldy, 1990; Whitehead and Pinnock, 1991; Hibbert and Mackertich, 1993; Bruce and Stemmerik, 2003; Kombrink *et al.*, 2010); (iii) Late Carboniferous to Early Permian compression (Variscan orogeny), possibly forming Folds A and B, in the eastern half of the figure (Coward *et al.*, 1989; Seranne, 1992; Corfield *et al.*, 1996; Balson *et al.*, 2002; Zanella and Coward, 2003); (iv) Middle Permian extension, forming the Rotliegend-filled half-graben bounded to the west by Fault 6 and other minibasins in the area (Figure 13.2c) (compare similar structures described by Stemmerik *et al.* (2000) over the Norwegian-Danish Basin); (v) latest Permian to Triassic extension, forming the horst and half-graben system with Permo-Triassic syn-rift wedges around faults 8 and 9 (Hibbert and Mackertich, 1993; Glennie, 1995; Balson *et al.*, 2002; Patruno & Reid, 2017; Goldsmith *et al.*, 2003; Wilson *et al.*, 2010; Whipp *et al.*, 2014); (vi) Aalenian uplift (Underhill and Partington, 1993; Zanella and Coward, 2003; Husmo *et al.*, 2003), recorded by the stratigraphy penetrated by several wells (e.g., uplift in Figure 13.2b and Late Triassic to Aalenian hiatuses

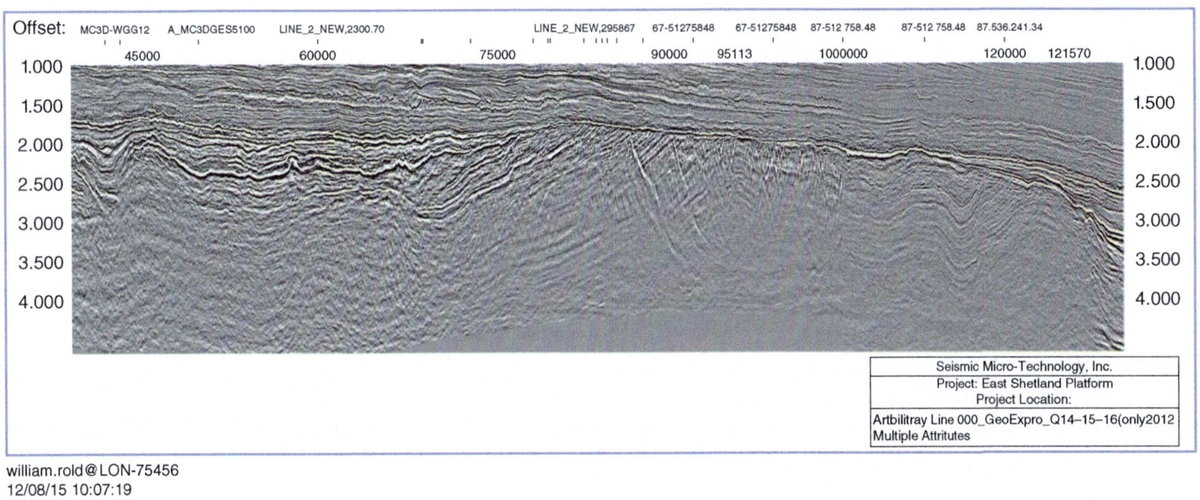

Figure 13.1 Uninterpreted section in the southern East Shetland Platform (UKCS Quadrants 15–16), near the transition between the Platform and the Witch Ground Graben. Courtesy of PGS GeoStreamer®.

Atlas of Structural Geological Interpretation from Seismic Images, First Edition. Edited by Achyuta Ayan Misra and Soumyajit Mukherjee.

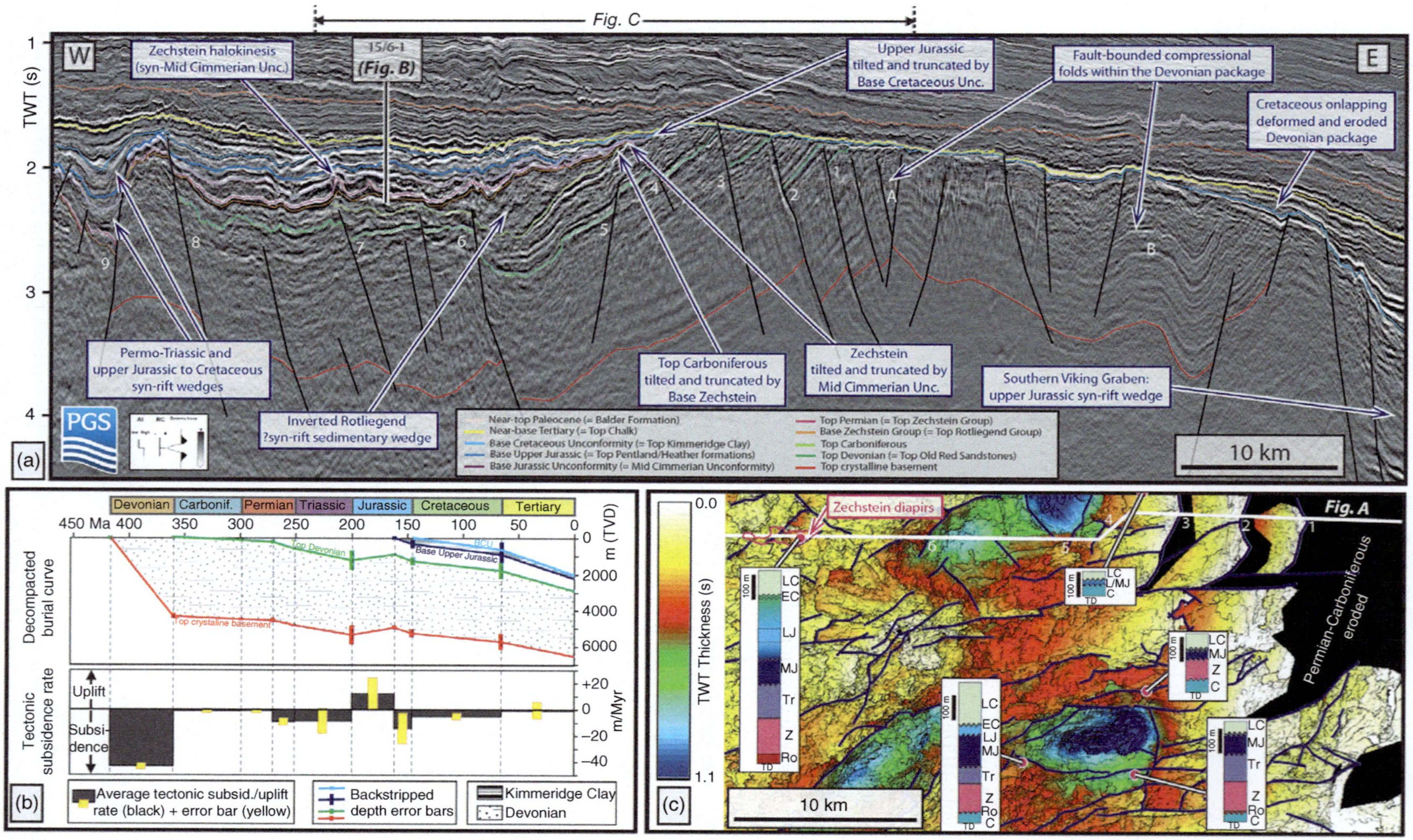

Figure 13.2 Complex multiphase inversion tectonics in the southern East Shetland Platform (UKCS Quadrants 15–16) near the transition between the Platform and the Witch Ground Graben (about 36 km further south) and the South Viking Graben (right-hand side in a). This figure includes the following parts: (a) interpreted seismic cross-section (Courtesy of PGS GeoStreamer® MC3D-Q15-2014 3D seismic survey); (b) total subsidence curve below present-day sea level and tectonic subsidence rate graphs from Well 15/6-1 (backstripping method of Allen and Allen, 2005); (c) TWT thickness map of the Permian to Carboniferous interval (PGS GeoStreamer® MC3D-Q15-2014 3D seismic survey). LC = Late Cretaceous; EC = Early Cretaceous; LJ = Late Jurassic; MJ = Middle Jurassic (post-Aalenian); Tr = Triassic (mostly Early Triassic); Z = Zechstein Group (latest Permian); Ro = Rotliegend Group (Late Permian); C = Carboniferous.

in Figure 13.2c) and leading to the partial inversion of Faults 6 and 7 similar inheritance is widespread in fold-and-thrust belts worldwide (e.g., Calamita *et al.*, 2007); (vii) renewed rifting in the Late Jurassic to earliest Cretaceous (Ziegler, 1992; Fraser *et al.*, 2003), leading to the formation of the main border faults of the South Viking Graben (eastern side of Figure 13.2a), as well as to the formation or reactivation of pre-existing faults over the platform (e.g. Faults 8 and 9) Patruno & Reid, 2016, 2017; Patruno, 2017. For general information about inheritance, consult Misra and Mukherjee (2015).

The 'Base Cretaceous Unconformity' (BCU) represents the end of the main Mesozoic rifting, with the subsequent formation of an angular unconformity over the tilted fault blocks. The 'Mid Cimmerian Unconformity' (MCU) reflects the widespread Aalenian 'Mid North Sea doming' event (Underhill and Partington, 1993; Husmo *et al.*, 2003; Graversen, 2006) leading to the regional erosion of upper Triassic to Aalenian sediments in this region (e.g. wells in Figure 13.2c). Halokinesis of the thin Zechstein succession around well 15/6-1 (Figures 13.2a and 13.2c) can be roughly dated as syn-Aalenian compression, as the diapirs pierce the Triassic succession (pre-halokinesis) and form Middle Jurassic rim syncline (syn- to post-halokinesis). Furthermore, this event leads to the detachment of the continental red beds of the Rotliegend syn-rift wedge (Brown, 1991) from the underlying tabular Carboniferous and its inversion, with the fold above Fault 6 affecting the Rotliegend, Zechstein and Triassic to lower Jurassic sediments (Figure 13.2a). Observing diapirs in the 'Northern Permian Basin' (Ziegler, 1992; Glennie *et al.*, 2003) is a novelty, as the Zechstein evaporites in this basin were previously deemed as too thin and halite-poor to serve as a seal rock or to form diapiric structures (Clark *et al.*, 1998). Finally, a composite unconformity surface is present in the area between Fault 6 and Anticline B, where the BCU, MCU and older Paleozoic unconformities are merging. At least two different phases of regional easterly-tilting affected the sediments over this area. The first phase is likely associated to the formation of Anticline A (Variscan age) and with the hangingwall tilting associated with the activity of Fault 6 (Early Permian age?); the second phase is related to the Aalenian regional uplifting, which had an increasingly greater magnitude towards the proto-Viking Graben, where the focus of the regional thermal doming and erosion was situated (Coward *et al.*, 2003). As a consequence, the Devonian and Carboniferous succession between Fault 6 and Anticline A underwent two tilting events and are now characterized by a greater overall easterly tilt ($\sim 83\,\mathrm{ms\,km^{-1}}$) than the Permo-Triassic over the same area ($\sim 35\,\mathrm{ms\,km^{-1}}$), which was only affected by the Mid Cimmerian tilting. Late Jurassic and Cretaceous strata over this area are subhorizontal and progressively onlap the residual high in the region between Fault 5 and Anticline B (Fladen Ground Spur).

References

Allen, P.A. and Allen, J.R. 2013. *Basin Analysis: Principles and Application to Petroleum Play Assessment*, John Wiley & Sons Ltd, Chichester.

Balson, P., Butcher, A., Holmes, R. *et al.* 2002. North Sea Geology. Strategic Environmental Assessment, SEA2and3. Technical Report 008 Rev1, Geology. British Geological Survey.

Brown, S. 1991. Stratigraphy of the oil and gas reservoirs: UK Continental Shelf. In: Abbotts, I.L. (ed) *United Kingdom Oil and Gas Fields, 25 Years Commemorative Volume*, Geological Society Memoir vol. 14, pp. 9–18.

Bruce, D.R.S. and Stemmerik, L. 2003. Carboniferous. In: Evans, D., Graham, C., Armour, A. and Bathurst, P. (eds) *The Millennium Atlas: Petroleum Geology of the Central and Northern North Sea*, The Geological Society, London, pp. 83–89.

Clark, J.A., Stewart, S.A. and Cartwright, J.A. 1998. Evolution of the NW margin of the North Permian Basin, UK North Sea. *Journal of the Geological Society*, **155**, 663–676.

Corfield, S.M., Gawthorpe, R.L., Gage, M. *et al.* 1996. Inversion tectonics of the Variscan foreland of the British Isles. *Journal of the Geological Society*, **153**, 17–32.

Coward, M.P., Enfield, M.A. and Fischer, M.W. 1989. Devonian basins of Northern Scotland: extension and inversion related to Late Caledonian – Variscan tectonics. In: Cooper, M.A. and Williams, G.D. (eds), *Inversion Tectonics*. Special Publications vol. 44, The Geological Society, London, pp. 275–308.

Coward, M.P., Dewey, J.F., Hempton, M. and Holroyd, J. 2003. Tectonic evolution. In: Evans, D., Graham, C., Armour, A. and Bathurst, P. (eds) *The Millennium Atlas: Petroleum Geology of the Central and Northern North Sea*, The Geological Society, London, pp. 17–33.

Fraser, S., Robinson, A., Johnson, H. *et al.* 2003. Upper Jurassic. In: Evans, D., Graham, C., Armour, A. and Bathurst, P. (eds) *The Millennium Atlas: Petroleum Geology of the Central and Northern North Sea*, The Geological Society, London, pp.157–189.

Glennie, K.W. 1995. Permian and Triassic rifting in northwest Europe. In: Boldy, S.A.R. (ed) *Permian and Triassic Rifting in Northwest Europe*. Special Publications vol. 91, The Geological Society, London, pp. 1–5.

Glennie, K.W., Higham, J. and Stemmerik, L. 2003. The Permian of the Northern North Sea. In: *Millennium Atlas: Petroleum Geology of the Central and Northern North Sea*, The Geological Society, London, pp. 91–103.

Goldsmith, P.J., Hudson, G. and Van Veen, P. 2003. Triassic. In: Evans, D., Graham, C., Armour, A. and Bathurst, P. (eds) *The Millennium Atlas: Petroleum Geology of the Central and Northern North Sea*, The Geological Society, London, pp. 105–127.

Graversen, O. 2006. The Jurassic-Cretaceous North Sea Rift Dome and Associated Basin Evolution. Search and Discovery Article #30040.

Hibbert, M.J. and Mackertich, D.S. 1993. The structural evolution of the eastern end of the Halibut Horst, Block 15/21, Outer Moray Firth, UK North Sea. In: Parker, J.R. (ed) *Petroleum Geology of Northwest Europe*, Proceedings of the 4th Conference, The Geological Society, London, pp. 1179–1188.

Husmo, T., Hamar, G., Høiland, O. *et al.* 2003. Lower and Middle Jurassic. In: Evans, D., Graham, C., Armour, A. and Bathurst, P. (eds) *The Millennium Atlas: Petroleum Geology of the Central and Northern North Sea*, The Geological Society, London, pp. 129–155.

Kombrink, H., Besly, B.M., Collinson, J.D. *et al.* 2010. Carboniferous. In: Doornenbal, J.C. and Stevenson, A.G. (eds) *Petroleum Geological Atlas of the Southern Permian Basin Area*, EAGE Publications b.v., Houten, The Netherlands, pp. 81–99.

Leeder, M.R. and Boldy, S.R. 1990. The Carboniferous of the Outer Moray Firth Basin, quadrants 14 and 15, Central North Sea. *Marine and Petroleum Geology*, **7**, 30–37.

Marshall, J.E.A. and Hewett, A.J. 2003. Devonian. In: Evans, D., Graham, C., Armour, A. and Bathurst, P. (eds) *The Millennium Atlas: Petroleum Geology of the Central and Northern North Sea*, The Geological Society, London, pp. 65–81.

Misra, A.A., Mukherjee, S. 2015. Tectonic Inheritance in Continental Rifts and Passive Margins. Springerbriefs in Earth Sciences. ISBN 978-3-319-20576-2.

Patruno, S., Reid., W. 2016. New plays on the Greater East Shetland Platform (UKCS Quadrants 3, 8–9, 14–16) – part 1: Regional setting and a working petroleum system. First Break (peer-reviewed 'Technical Article' sections), **34**, 33–45.

Patruno, S., Reid., W. 2017. New plays on the Greater East Shetland Platform (UKCS Quadrants 3, 8–9, 14–16) – part 2: Newly reported Permo-Triassic intra-platform basins and their influence on the Devonian-Paleogene prospectivity of the area. First Break (peer-reviewed 'Technical Article' sections), **35**, 59–69.

Patruno, S. 2017. A new look at the geology and prospectivity of a North Sea frontier area with modern seismic: the East Shetland Platform. 79th EAGE Conference and Exhibition 2017, 15 June 2017, At Paris, France (Expo Porte de Versailles), Volume: Oral Presentation Session: Exploration Plays, Prospects and Prospects Evaluation II, Abstract Th C2 09. Extended abstract and oral presentation. Extended abstract available online at: http://www.earthdoc.org/publication/publicationdetails/?publication=88346

Seranne, M. 1992. Devonian extensioinal tectonics versus Carboniferous inversion in the northern Orcadian basin. *Journal of the Geological Society*, **149**, 27–37.

Stemmerik, L., Ineson, J.R. and Mitchell, J.G. 2000. Stratigraphy of the Rotliegend Group in the Danish part of the Northern Permian Basin, North Sea. *Journal of the Geological Society*, **157**, 1127–1136.

Underhill, J.R. and Partington, M.A. 1993. Jurassic thermal doming and deflation in the North Sea: implication of the sequence stratigraphic evidence. In: Parker, J.R. (ed) *Petroleum Geology of Northwest Europe*, Proceedings of the 4th Conference, The Geological Society, London, pp. 337–346.

Whipp, P.S. and Jackson, C.A.-L., Gawthorpe, R.L. *et al.* 2014. Normal fault array evolution above a reactivated rift fabric; a subsurface example from the northern Horda Platform, Norwegian North Sea, *Basin Research*, **26**, 523–549.

Whitehead, M. and Pinnock, S.J. 1991. The Highlander Field, Block 14/20b, UK North Sea. In: Abbotts, I.L. (ed) *United Kingdom Oil and Gas Fields, 25 Years Commemorative Volume*, Geological Society Memoir vol. 14, pp. 323–329.

Wilson, R.W., Holdsworth, R.E., Wild, L.E. *et al.* 2010. Basement-influenced rifting and basin development: a reappraisal of post-Caledonian faulting patterns from the North Coast Transfer Zone, Scotland. In: Law, R.D., Butler, R.W.H., Holdsworth, R.E. *et al.* (eds) *Continental Tectonics and Mountain Building: The Legacy of Peach and Horne*, Special Publications vol. 335, The Geological Society, London, pp. 795–826.

Zanella, E. and Coward, M.P. 2003. Structural framework. In: Evans, D., Graham, C., Armour, A. and Bathurst, P. (eds) *The Millennium Atlas: Petroleum Geology of the Central and Northern North Sea*, The Geological Society, London, pp. 42–59.

Ziegler, P.A. 1992. North Sea rift system. *Tectonophysics*, **208**, 55–75.

14

Impact of Activation of a Syn-Depositional Normal Fault on Sedimentation in Greater Mid North Sea High

Stefano Patruno and William Reid*

Petroleum Geo-Services (PGS), Weybridge, United Kingdom
** Stephano.Patruno@pgs.com*

This example (Figures 14.1, 14.2 and 14.3) shows the activation of a syn-depositional normal fault and its impact on the stratigraphic architecture of a prograding Zechstein-age sulfate/carbonate platform complex in the 'Greater Mid North Sea High' area (modified after Patruno *et al.*, 2017).

Figure 14.1 shows uninterpreted and interpreted seismic cross-sections in parts of UKCS Quadrants 36 and 42 (PGS MC3D-SNS2013M 3D survey), showing the 'normal' stratigraphic architecture of a prograding sulfate/carbonate platform complex (Taylor, 1980, 1998; Tucker, 1991; Patruno *et al.*, in press) in the absence of significant syn-depositional tectonics. Above the fault-bounded Carboniferous folds (Kombrink *et al.*, 2010), the Zechstein cyclothems (Richter-Bernburg, 1955a, 1955b) can be subdivided into a series of basin-fill halite units and prograding carbonate- and sulfate-rich marginal sedimentary wedges (clinoform-bounded units or 'clinothems') (Rich, 1951; Tucker, 1991; Steel and Olsen, 2002; Patruno *et al.*, 2015a, 2015b). In the latter category are found: the Z1 sulfate-prone clinothems (Patruno *et al.*, in press), mostly comprising the Werraanhydrit Formation (between the 'Base Zechstein' and 'Top Werraanhydrit' horizons); the lower Z2 carbonate and sulfate-prone clinothem, comprising Hauptdolomit and Basalanyhdrit Formations (between the 'Top Werraanhydrit' and 'Top Basalanhydrit' horizons); and the thinner lower Z3 carbonate and sulfate-prone clinothem, comprising Plattendolomit and Hauptanhydrit Formations (between the 'Top Stassfurt Halit' and the 'Top Hauptanhydrit' horizons). A thick upper Z2 Stassfurt Halite succession fills the accommodation inherited by the frontal rollover of the pre-existing top Basalanhydrit horizon, at the top of the Z1–Z2 composite sulfate/carbonate platform complex, and separates this from the overlying 'lower Z3' carbonate/sulfate platform. The clinoform trajectories (Helland-Hansen and Hampson, 2009; Patruno *et al.*, 2015c) suggest a very high progradation/aggradation rate between the Z1 clinothem and the 'lower Z2' clinothem, and aggradational trajectories between the 'lower Z2' and the 'lower Z3' clinothems. As a consequence, the frontal rollover of the 'lower Z2' clinothem steps approximately 2.5–3.5 km further basinwards of the frontal Z1 clinothem rollover, whereas the frontal Z3 clinothem rollover does not show any further basin-wards migration, and the entire 'lower Z3' clinothem lies above the area previously occupied by the Z1–Z2 sulfate/carbonate platform complex.

Figure 14.3b shows interpreted seismic cross-sections across parts of UKCS Quadrants 36-37 (PGS GeoStreamer® MC2D-MNSG2009 2D survey), showing significantly different geometry for the lower Z3 clinothem. Particularly, at the eastern end of the line, an anomalously thick lower Z3 clinothem has been interpreted, with at least 7 km of basin-ward advancement from the rollover of the pre-existing Z1-Z2 sulfate/carbonate platform complex and that of the lower Z3 platform edge. This anomalous architecture is interpreted to reflect the reactivation of the normal fault in the middle of the line during the deposition of the lower Z3 clinothem (Patruno *et al.*, in press). This is further explained by sketches showing the evolution of the Zechstein along this line in stages: (i) at the end of the deposition of the Z1 cyclothem (Figure 14.3c); (ii) during the early deposition of the Plattendolomit (lower Z3) (Figure 14.3d); (iii) at the end of the deposition of the 'lower Z3' clinothem (top Hauptanhydrit) (Figure 14.3e); (iv) at the end of the deposition of the Zechstein Group (Figure 14.3f). During the deposition of the lower Zechstein Group, inherited footwall relief associated to a pre-existing inactive normal fault caused the formation of an anomalously thin Z1 clinothem (Figure 14.3c) and Z2 unit (Figure 14.3d). A relatively minor reactivation of this fault during the deposition of the lower Z3 clinothem caused footwall crest uplift (destruction of accommodation) and the rejuvenation of the gradient of the footwall

Atlas of Structural Geological Interpretation from Seismic Images, First Edition. Edited by Achyuta Ayan Misra and Soumyajit Mukherjee.

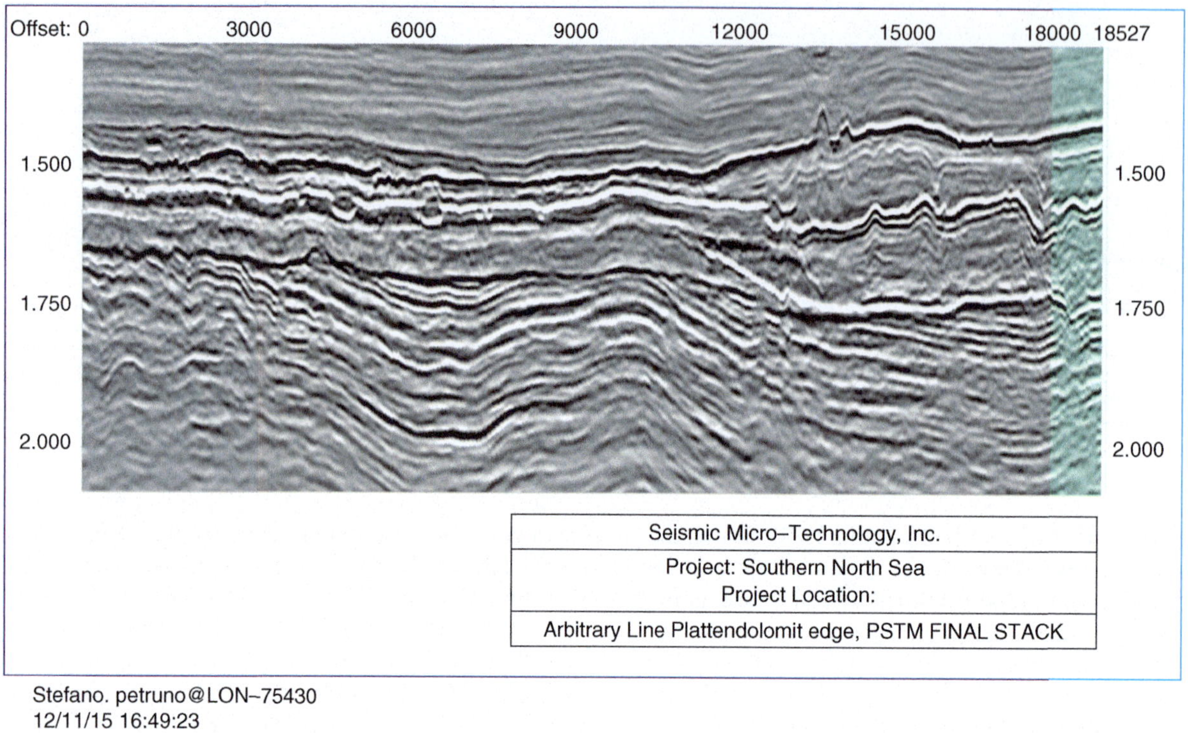

Figure 14.1a Uninterpreted seismic section for the area with normal edge unit thickness.

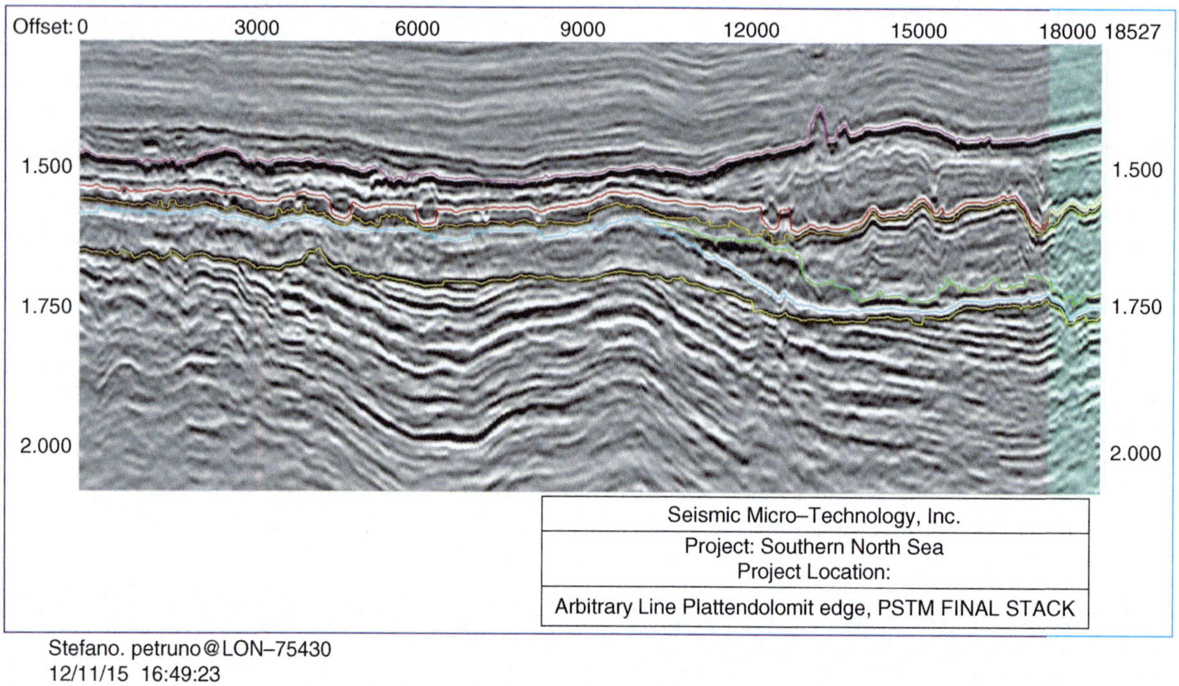

Figure 14.1b Interpreted seismic section for the area with normal edge unit thickness.

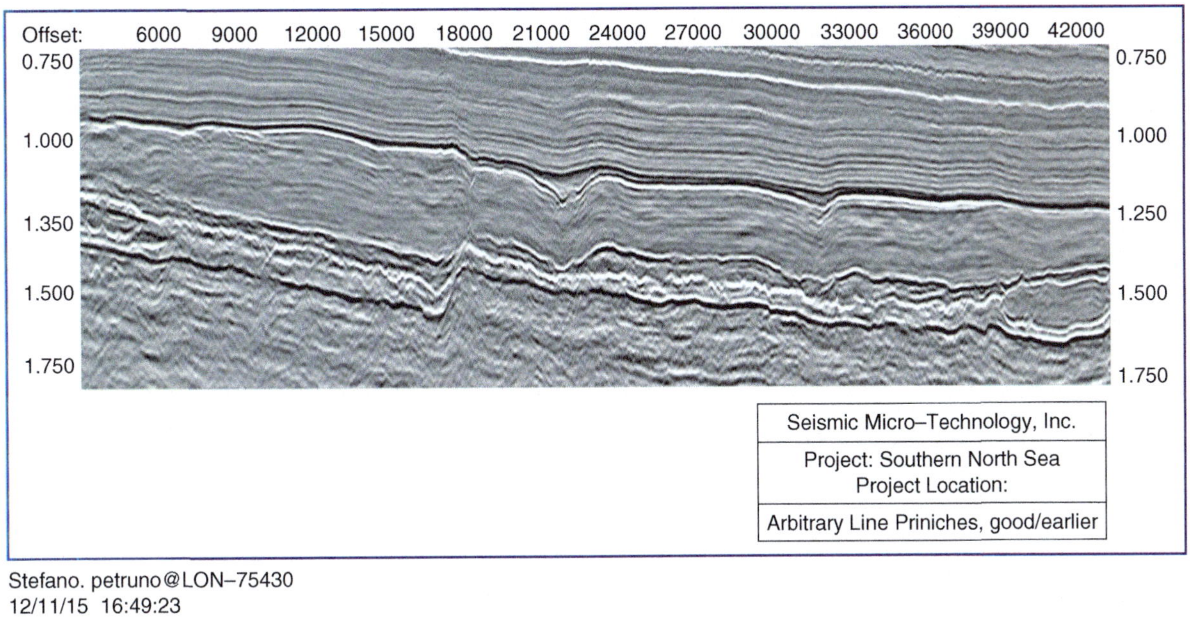

Figure 14.2a Uninterpreted seismic section for the area with edge unit of anomalous thickness.

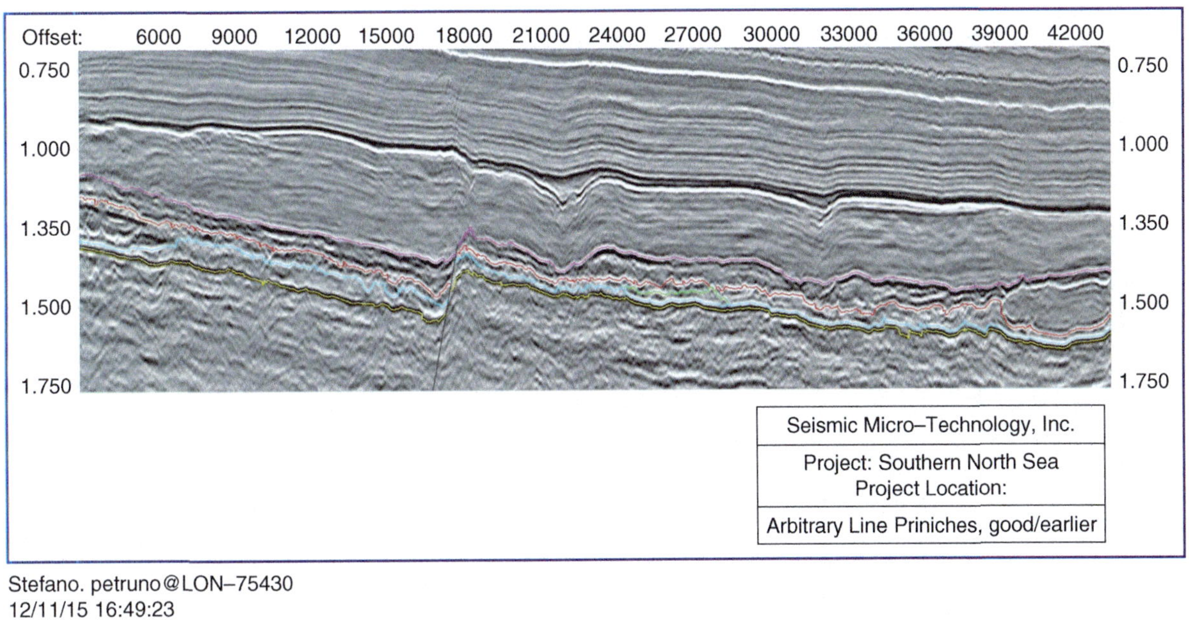

Figure 14.2b Interpreted seismic section for the area with edge unit of anomalous thickness. A normal drag (Mukherjee, 2014) related to faulting is deciphered.

dip slope (creation of accommodation) (Figure 14.3e). This caused the forced progradation of the lower Z3 clinothem down the flank of the footwall dip slope, beyond the limits that are normally observed elsewhere. This also led to the progressive increase in the clinothem thickness as it advanced towards the increasingly deeper-water footwall dip slope substrate. Further low-magnitude rotational events across this faults continued during the deposition of the Z4 package and possibly in the Triassic, leading to the formation of 'classical' syn-rift wedges (Figure 14.3f). Vide Mukherjee and Khonsari (in press) for mechanisms of fault block rotation.

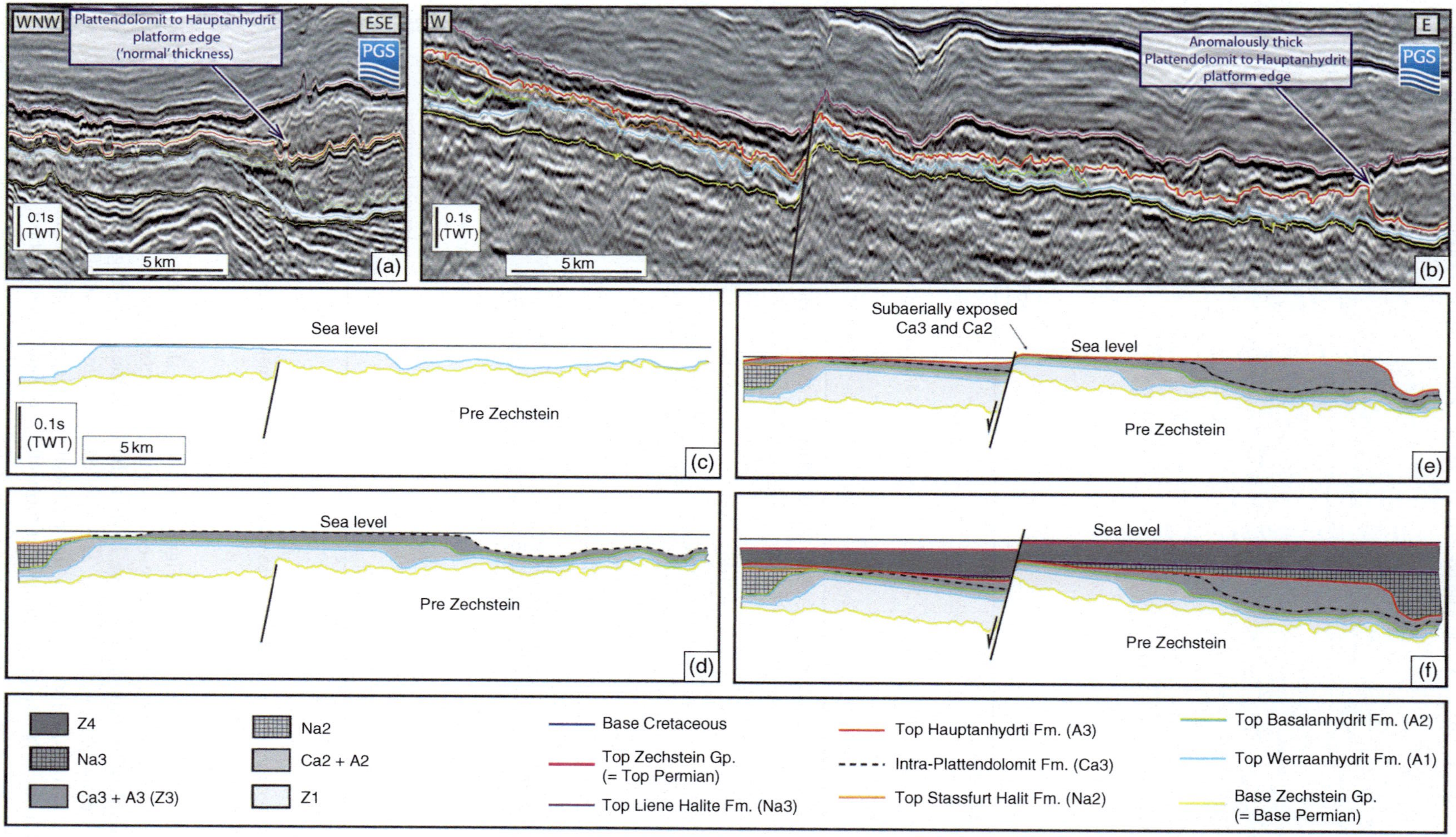

Figure 14.3 Interpreted seismic sections and the different stages of formation of the sedimentary units.

References

Helland-Hansen, W. and Hampson, G.J. 2009. Trajectory analysis: concepts and applications. *Basin Research* **21**, 454–483.

Kombrink, H., Besly, B.M., Collinson, J.D. *et al.* 2010. Carboniferous. In: Doornenbal, J.C. and Stevenson, A.G. (eds) *Petroleum Geological Atlas of the Southern Permian Basin Area*, EAGE Publications bv, Houten, The Netherlands, pp. 81–99.

Mukherjee, S. 2014. Review of flanking structures in meso- and micro-scales. *Geological Magazine*, **151**, 957–974.

Mukherjee, S., Khonsari. M.M. (in press). Brittle rotational faults and the associated shear heating. *Marine and Petroleum Geology*. DOI: https://doi.org/10.1016/j.marpetgeo.2017.09.003

Patruno, S., Hampson, G.J. and Jackson, C.A.-L. 2015a. Quantitative characterisation of deltaic and subaqueous clinoforms. *Earth-Science Reviews*, **142**, 79–119.

Patruno, S., Hampson, G.J., Jackson, C.A.-L. and Dreyer, T. 2015b. Clinoform geometry, geomorphology, facies character and stratigraphic architecture of an ancient sand-prone subaqueous delta: Upper Jurassic Sognefjord Formation, Troll Field, Offshore Norway. *Sedimentology*, **62**(1), 350–388.

Patruno, S., Hampson, G.J., Jackson, C.A.-L. and Whipp, P.S. 2015c. Quantitative progradation dynamics and stratigraphic architecture of ancient shallow-marine clinoform sets: a new method and its application to the Upper Jurassic Sognefjord Formation, Troll Field, offshore Norway. *Basin Research*, **27**, 412–452.

Patruno, S., Reid, W., Jackson, C.A. and Davies, C. in press. New insights into the unexploited reservoir potential of the Mid North Sea High (UKCS quadrants 35–38 and 41–43): a newly described intra-Zechstein sulphate–carbonate platform complex. In: Petroleum Geology Conference series Vol. 8, The Geological Society, London. https://doi.org/10.1144/PGC8.9.

Rich, J.L. 1951. Three critical environments of deposition and criteria for recognition of rocks deposited in each of them. *GSA Bulletin*, **62**, 1–20.

Richter-Bernurg, G. 1955a. Über salinare Sedimentation. *Zeitschrift der Deutschen Geologischen Gesellschaft*, **105**, 593–645.

Richter-Bernurg, G. 1955b. Stratigraphische Gliederung des deutschen Zechsteins. *Zeitschrift der Deutschen Geologischen Gesellschaft*, **105**, 843–854.

Steel, R.J. and Olsen, T. 2002. Clinoforms, clinoform trajectory and deepwater sands. In: Armentrout, J.M. and Rosen, N.C. (eds) *Sequence Stratigraphic Models for Exploration and Production: Evolving Methodology, Emerging Models and Application Histories*. GCS-SEPM Special Publication, Houston, TX, pp. 367–381.

Taylor, J.C.M. 1980. Origin of the Werranhydrit in the U.K. Southern North Sea – a reappraisal. In: Fuchtbauer, H. and Peryt, T. (eds.) *The Zechstein Basin with Emphasis on Carbonate Sequences*. Contributions to Sedimentology No. 9, pp. 91–113.

Taylor, J.C.M. 1998. Upper Permian – Zechstein. In: Glennie, K.W. (ed) *Petroleum Geology of the North Sea: Basic Concepts and Recent Advances*, 4th edn, Blackwell Science Ltd, Oxford, pp. 174–211.

Tucker, M.E. 1991. Sequence stratigraphy of carbonate-evaporite basins: models and application to the Upper Permian (Zechstein) of northeast England and adjoining North Sea. *Journal of the Geological Society*, **148**, 1019–1036.

15

Imaging Normal Faults Using Dual Sensor Streamers

Stefano Patruno and William Reid*

Petroleum Geo-Services (PGS), Weybridge, United Kingdom
** Stephano.Patruno@pgs.com*

This example shows improvement of fault imaging due to recording and imaging of low to ultra-low frequencies (Figures 15.1 and 15.2). As low frequencies are less strongly attenuated with increasing depth compared to high frequencies, a 'deep' subsurface image would require the recording and preservation of low to ultra-low frequencies (2–7 Hz). The wells shown in Figure 15.3a were drilled based on seismic data lacking low frequencies, which resulted in a very poor sub-Tertiary imaging. This is, however, not an unusual case for 1980s and 1990s seismic data. Figure 15.3b illustrates that the 6 m ghost function attenuates the low frequencies much more than the 15 m ghost function; so towing deeper enables the recording of more low frequency information. An obvious solution to simultaneously reduce sea-surface generated noise and enrich the low frequency content would have been, therefore, to tow deeper (e.g., 15 m instead of 6 m) (Figure 15.3b).

Increasing the tow depth of hydrophone-only acquisition systems from 6 to 15 m, however, moves the notches created by the sea-surface ghost reflections to mid-range frequencies (e.g., from 125 Hz to 100 and 50 Hz) (Figure 15.3b). Notching occurs when the upgoing wavefield (i.e. reflected from the geological reflectors) superimposes and destructively interferes at certain frequencies with the down-going wavefield (i.e. reflected from the air-sea interface) (Figures 15.3b and 15.3c) (Cieslewicz and Lawton, 1998). As a consequence, in order to preserve the mid-frequency range, which is important for imaging 'traditional' target intervals in the subsurface (e.g., Palaeogene and Jurassic for the North Sea), the compromise was to tow the streamers shallow, resulting in higher noise levels and poor imaging of the deeper reflectors (e.g., Early Mesozoic to Paleozoic in Figure 15.3a).

Since 2007, the marine seismic industry has designed a number of solutions to this problem, giving rise to broad-bandwidth de-ghosted seismic data. The first effective acquisition-based solution to the ghost notch problem was developed by PGS and involves dual-sensor streamers (GeoStreamer®) (Day *et al.*, 2013). This technology is briefly described here.

In contrast to conventional, hydrophone-only acquisition systems, GeoStreamer® is equipped with a hydrophone and a vertical velocity sensor, which detects the time-delayed receiver ghost reflection with the opposite polarity to the hydrophone (Figure 15.3c) (Day *et al.*, 2013). This complementary property enables the removal of the receiver ghost by combining the recordings of the two different sensors and performing wavefield separation into up- and downgoing wavefields (Figure 15.3c). The ability to remove the ghost accurately allows the streamers to be towed deeper (usually 15–25 m), further away from the noise generated by waves of the ocean's surface. The signal-to-noise ratio is improved by a deeper tow, especially at the low frequency end of the seismic spectrum. Improved low frequency content enhances the quality and accuracy of seismic inversion and reservoir characterization. Figure 15.3d shows the theoretical amplitude spectra for a de-ghosted conventional and dual-sensor streamer data. The extended bandwidth of the GeoStreamer data has a significant effect on the low end of the amplitude spectrum, reducing the dependencies on well-derived data (Day *et al.*, 2013).

Atlas of Structural Geological Interpretation from Seismic Images, First Edition. Edited by Achyuta Ayan Misra and Soumyajit Mukherjee.

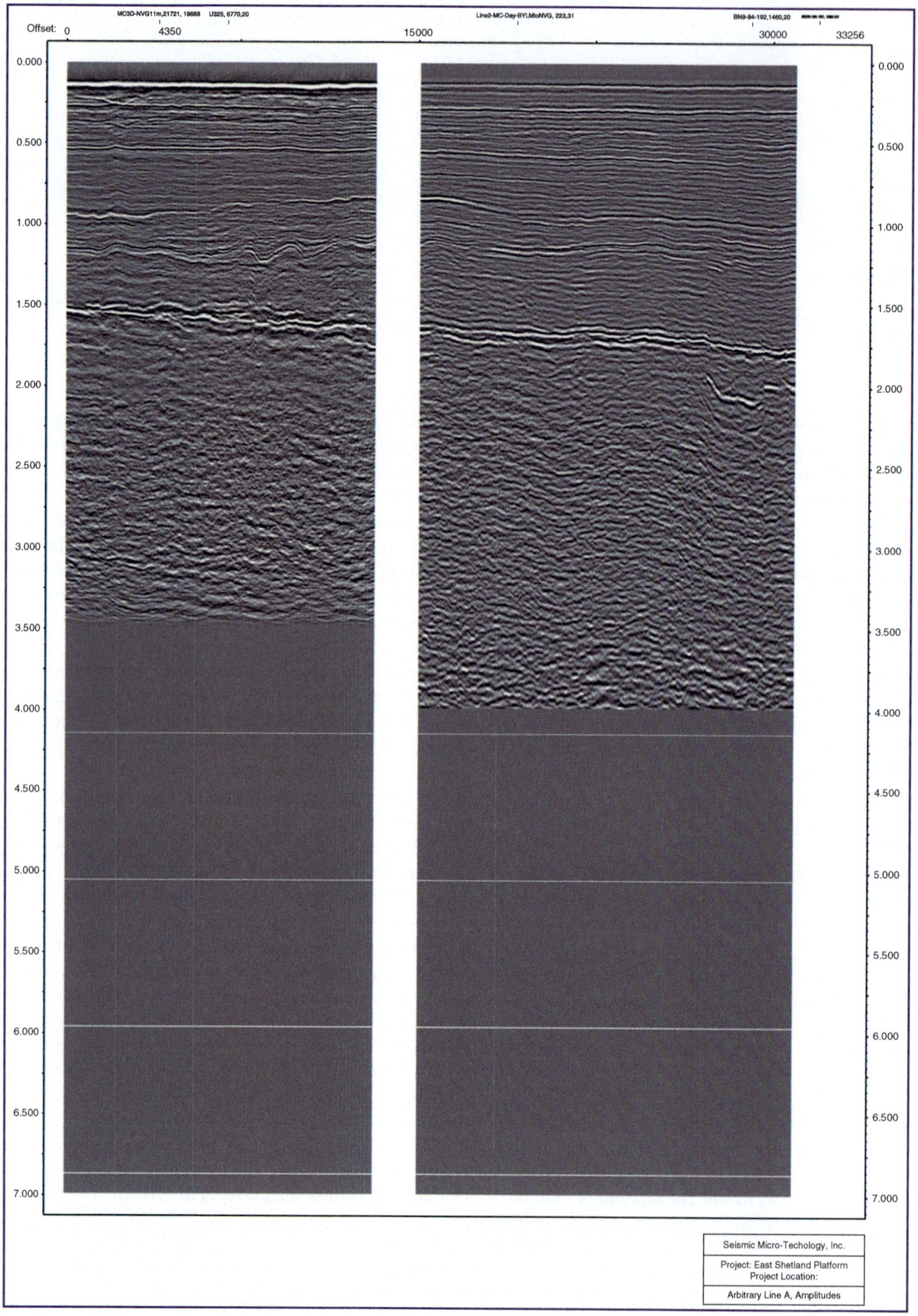

Figure 15.1 Uninterpreted reflection seismic section acquired in 1986 with a conventional band-limited seismic line with a shallow tow.

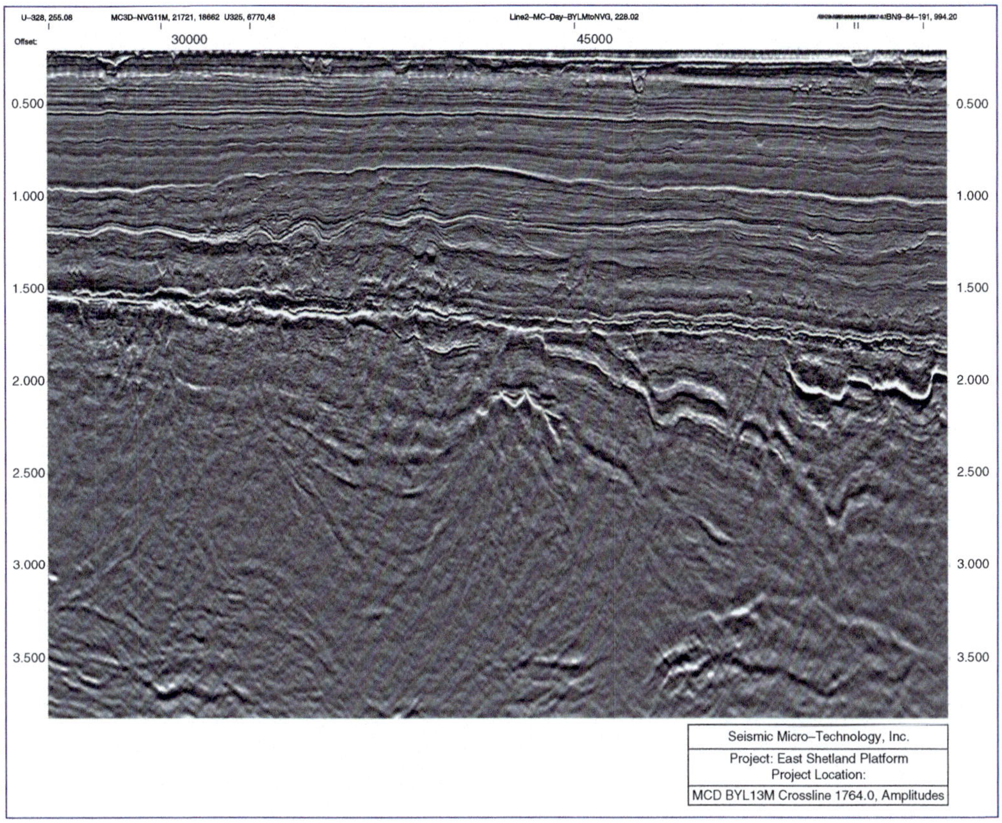

Figure 15.2 Uninterpreted reflection seismic section of the same area as in Figure 15.1 acquired in 2013 with a broadband configuration.

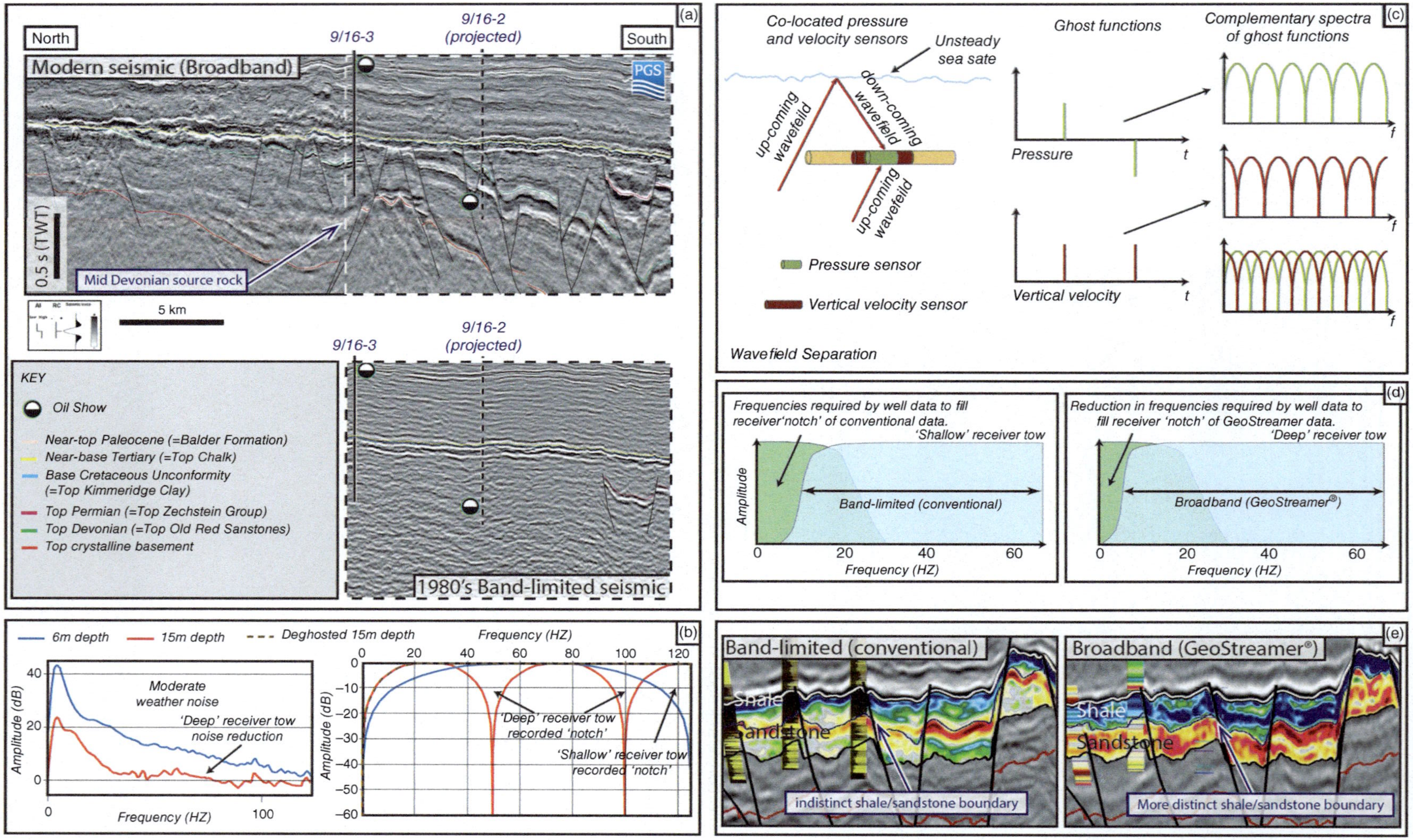

Figure 15.3 (a) Interpreted section for the conventional band-limited seismic line with a shallow tow (1986 acquisition; Figure 15.1) compared with the broadband (2013) dual-sensor streamer with a deep tow (Geostreamer®) (Parkes and Hegna, 2011a, 2011b; Day *et al*., 2013; Figure 15.2). The eastern half of the GeoStreamer line (highlighted by a white dashed polygon) has been extracted along the same north-to-south transect as the conventional line, on the northern East Shetland Platform (Central North Sea). A significant improvement in the definition and clarity of reflectors and geological structures can be observed. In particular, the sub-Tertiary reflectors visible in the GeoStreamer line are absent in the conventional seismic and the Tertiary reflectivity is less well-defined. A Devonian horst and syn-rift wedge package are now visible, making sense of old wells, drilled in 1987 (9/16-2) and 1991 (9/16-3) targeting Paleocene prospects. These wells respectively reported oil shows in the Devonian reservoir (9/16-2) and Devonian source rock (9/16-3) (Duncan and Buxton, 1995), but are clearly missing the top of the horst structure. A new deep potential play is therefore opened up in this frontier exploration region as a result of the added clarity from modern seismic acquisition and processing. (b–e) Figures show the challenges faced by conventional marine seismic data and the technological solution developed in recent years, resulting in broad-bandwidth de-ghosted seismic data such as those sections shown in (a). Adapted from Patruno and Reid (2017). Courtesy of PGS GeoStreamer® MC3D-Q16-2013 3D survey.

Acknowledgements

The authors gratefully acknowledge Eivind Fromyr and Volker Dirks (PGS) for the insightful discussions and the very helpful suggestions. The quality of Figure 15.2 and the entire text has been highly improved thanks to their support.

References

Cieslewicz, D. and Lawton, D.C. 1998. Receiver notching in a linear V(z) near-surface medium. CREWES Research Report, vol. 10, pp. 3-1–3-17.

Day, A., Klüver, T., Söllner, W. *et al.* 2013. Wavefield-separation methods for dual-sensor towed-streamer data. *Geophysics*, **78**, WA55–WA70.

Duncan, W.I. and Buxton, W.K. 1995. New evidence for evaporitic Middle Devonian lacustrine sediments with hydrocarbon source potential on the East Shetland Platform, North Sea. *Journal of the Geological Society*, **152**, 251–258.

Parkes, G. and Hegna, S. 2011a. An acquisition system that extracts the earth response from seismic data. *First Break*, **29**, 81–87.

Parkes, G. and Hegna S. 2011b. A marine seismic acquisition system that provides a full 'ghost-free' solution. Expanded Abstract, 81st SEG Annual International Meeting, pp. 37–41.

Patruno, S. and Reid., W. 2016. New plays on the Greater East Shetland Platform (UKCS Quadrants 3, 8-9, 14-16) – part 1: Regional setting and a working petroleum system. *First Break*, **34**, 33–45.

16

Understanding Sinkholes Using Shear Wave Reflection Seismic Imaging in N Germany

Charlotte M. Krawczyk[1]* *and U. Polom*[2]

[1] *GFZ German Research Centre for Geosciences, Potsdam, Germany; Technische Universität, Berlin, Germany*
[2] *Leibniz Institute for Applied Geophysics, Hannover, Germany*
* *lotte@gfz-potsdam.de*

Sinkholes are challenging hazards in many countries, especially their structural characterization and monitoring (Krawczyk and Dahm, 2011). In northern Germany, the densely populated, residential quarter Hamburg-Flottbek is a sinkhole-prone area that also suffers repeated microseismicity (Dahm *et al.*, 2010). To overcome the problems urban geophysics is facing with respect to limitations in geophysical imaging, shear wave reflection seismic imaging, for instance, has been put forward in the last decade (Krawczyk *et al.*, 2013).

The surveyed area is located in the Southern Permian Basin in the transition zone between the Caledonian Deformation Front to the north and the Elbe Line to the south (Krawczyk *et al.*, 2008, and references therein). Below the Mesozoic Basin sediments, thick Rotliegend and Zechstein salt moved into salt walls, diapirs and pillows during post-Permian times. The reflection seismic profile shown crosses the western flank of the Othmarschen-Langenfelde diapir.

The shear wave reflection seismic sections presented are composed as stacked time sections down to 1200 ms TWT (two-way travel time) (Figure 16.1a), a 100-m deep, migrated depth section that encompasses the upper 600 ms TWT of the time section that is overlain by shear wave velocities (colour-coded) (Figure 16.1b) and the interpreted depth section (Figure 16.1c). More details and processing information can be found elsewhere (Krawczyk *et al.*, 2012).

The sections show partly continuous and strong reflectors. The uppermost reflective events have small but largely continuous segments of high frequency amplitudes down to about 150 ms TWT or about 10 m depth. Between 20 and 40 m depth, a sedimentary unit with incoherent, high frequency reflectivity and a rather transparent appearance occurs (100–200 m distance, 200–300 ms TWT). This is followed at depth by strong, lower frequent reflectors of good continuity on the western part of the profile (175–260 m distance, 400–500 ms TWT). Rather unclear reflective elements are observed between 200 and 500 ms (25–60 m depth).

The sedimentary layers are correlated with a drill hole projected from about 100 m distance. The base of the artificial infill of the area has been interpreted using the uppermost coherent reflector (about 2–4 m depth). Below, almost flat-lying sandy layers are found down to around 10 m depth (green line). The top of a basin structure is interpreted at 15 m at the borders to 30 m depth in the section centre (pink line). A faulted surface between 55 and 80 m depth (orange lines) deepens towards the profiles centre by en echelon normal faults. The western side of this structure is much clearer imaged than its eastern flank, perhaps due to acquisition geometry or surface-related energy loss.

In general, the shear wave interval velocities increase towards depth (150–650 m s^{-1}) and the uppermost 50 m show values between 150 and 300 m s^{-1}, typical for unconsolidated silt, sand and gravel. The low velocity zone (150–200 m s^{-1}, 180–260 m distance, 15 m depth) cannot be explained by the borehole, but peat lenses are

Atlas of Structural Geological Interpretation from Seismic Images, First Edition. Edited by Achyuta Ayan Misra and Soumyajit Mukherjee.

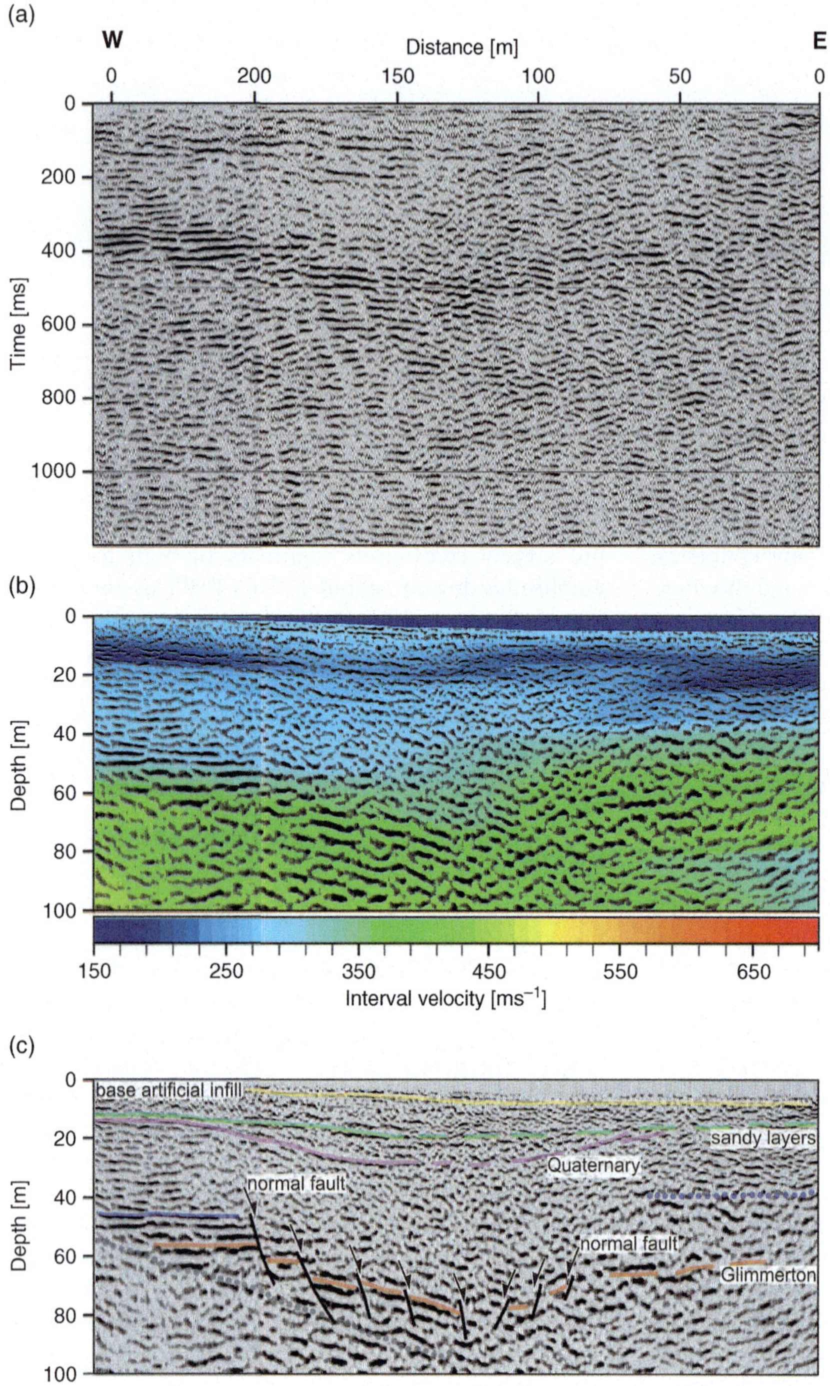

Figure 16.1 (a) Uninterpeted shear wave reflection profile; (b) shear wave velocity profile; (c) interpreted shear wave reflection profile showing normal faults corresponding to the sinkhole. *Source:* Krawczyk *et al*., 2012. Reproduced with permission of Elsevier.

found at 2–4 m depth, so that fine intercalations in the sandy sequences seem reasonable.

The highlights of this survey are:

- shear wave reflection seismics enables urban subsurface imaging with high resolution (here: 1 m vertical);
- small-scale structures in the sediments are resolved down to about 100 m depth;
- shear wave velocity and reflection character as combined attribute may be used for subrosion mapping;
- the collapse depth at 80 m updates the local salt model (Krawczyk *et al*., 2012).

References

Dahm, T., Kühn, D., Ohrnberger, M. *et al.* 2010. Combining geophysical data sets to study the dynamics of shallow evaporates in urban environments. *Geophysical Journal International*, **181**, 154–172. http://dx.doi.org/10.1111/j.1365-246X.2010.04521.x.

Krawczyk, C.M. and Dahm, T. 2011. Charakterisierung und Überwachung von Salz-bezogenen Erdfällen in urbanen Gebieten. http://docplayer.org/5000684-Charakterisierung-und-ueberwachung-von-salz-bezogenen-erdfaellen-in-urbanen-gebieten.html; last accessed 30 September 2017.

Krawczyk, C.M., McCann, T., Cocks, L.R.M. *et al.* 2008. Caledonian Tectonics. In: McCann, T. (ed) *The Geology of Central Europe*, vol. **1**, The Geological Society, London, pp. 301–381; ISBN 978-1-86239-245-8.

Krawczyk, C.M., Polom, U., Trabs, S. and Dahm, T. 2012. Sinkholes in the city of Hamburg – New urban shear wave reflection seismic system enables high-resolution imaging of subrosion structures. *Journal of Applied Geophysics*, **78**, 133–143. doi: 10.1016/j.jappgeo.2011.02.003.

Krawczyk, C.M., Polom, U. and Beilecke, T. 2013. Shear wave reflection seismics as a valuable tool for near-surface urban applications. *The Leading Edge*, **32**(3), 256–263. http://dx.doi.org/10.1190/tle32030256.1.

17

Post-Earthquake Subsurface Structure for Reconstruction Planning Using Shear Wave Reflection Seismic Imaging in Banda Aceh, Indonesia

Charlotte M. Krawczyk[1]* *and U. Polom*[2]

[1] *GFZ German Research Centre for Geosciences, Potsdam, Germany; Technische Universität, Berlin, Germany*
[2] *Leibniz Institute for Applied Geophysics, Hannover, Germany*
* *lotte@gfz-potsdam.de*

After the Great Sumatra Earthquake in 2004 and its devastating consequences, the region of Banda Aceh, Indonesia, was subject to extended geoscientific research. Our investigations were intended to support classification of the alluvial deposits at the earthquake site, ahead of reconstruction planning, by providing both subsurface structure and shear wave velocities (Arsyad and Polom, 2006; Arsyad *et al.*, 2007; Polom *et al.*, 2008).

In addition to the majority of river-transported alluvial deposits, parts of the shallow alluvial fan deposits in the region, bound in the SW by the Great Sumatra Fault, came from offshore and are characterized by clays, silt, sands up to coral boulders, or pyroclastic deposits at larger depths and further towards the NE (Paris *et al.*, 2010). Post-tsunami groundwater exploration drillings yielded stacks of organic clay, silt, sand and gravel down to a depth of 220 m without any hints of basement contact.

The map (Figure 17.1) of the Banda Aceh region indicates the locations chosen for shear wave reflection seismic profiling and yields corresponding velocities (colour-coded VSav30 – average shear wave velocity in the upper 30 m). Critical soil conditions were detected for areas near the coastline and between the Krueng Aceh River and the dextral strike-slip Great Sumatra Fault (red to orange markers). Further, critical areas are near the riverbed in the south-east part of the basin, which has not been expected before. The only stable region seems to be located close to the military airport (green marker), which lies in the transition to pyroclastic deposits (Polom *et al.*, 2008). Five examples across the area are shown here, demonstrating the variability of structures and shear wave interval velocities.

17.1 Figure 17.2a

At site SMK Negeri, horizontal layering is observed from surface down to 100 m depth. Reflection amplitudes and continuity are high, with the strongest events observed between 40 and 50 m depth, sealing presumably an unconsolidated clay zone above thin layers of methane gas and an aquifer (Arsyad and Polom, 2006). In the upper ten metres, weak amplitudes may mark the incision of a river bed, subsequently filled up; a similar structure occurs further to the NW at 20–30 m depth.

The average shear wave velocity in the upper 30 m is 167 ms^{-1}, due to the very soft soil in the area. Translated to the international building code (IBC), this class belongs to category F, which means that the undrained shear strength is below 23.9 kPa, which requires a site-specific inspection before any reconstruction.

17.2 Figure 17.2b

The Babah Jurong site shows strongest, almost horizontal, reflectivity between 50 and 70 m depth. Here, three bright reflectors can be traced across the entire section, delineating presumably clay plus pyroclastic deposits (Arsyad and Polom, 2006). Re-filled riverbed terraces are indicated in the upper 15–35 m in the centre of the section, overlying a lens of horizontally layered very low velocity silts and clays (60–95 m distance, 25–50 m depth, 200 ms^{-1}). A lower frequency but still very coherent signal is envisaged in the lower part of the section (80–120 m depth), where consolidated clay, silt and sand are met (Arsyad and Polom, 2006).

Atlas of Structural Geological Interpretation from Seismic Images, First Edition. Edited by Achyuta Ayan Misra and Soumyajit Mukherjee.

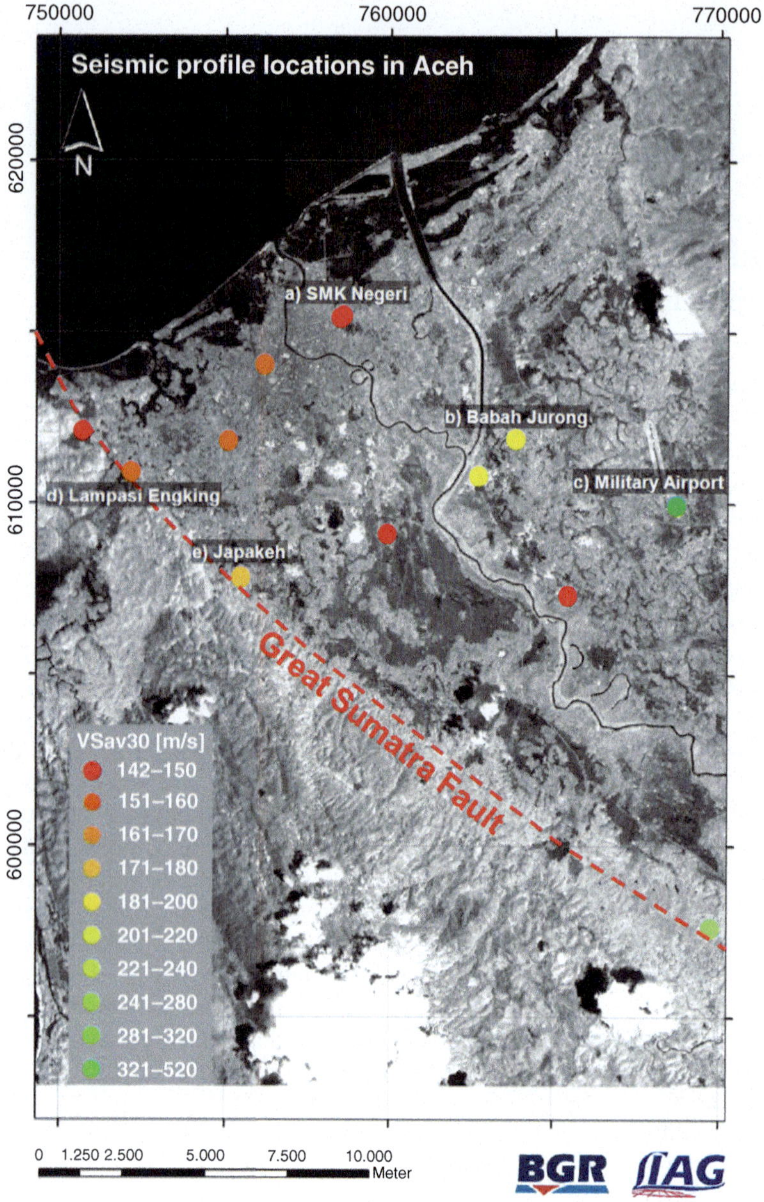

Figure 17.1 Map of the Banda Aceh area showing the locations of the shear wave seismic reflection profiles and the dextral strike-slip Great Sumatra Fault. Colour-coded shear wave velocities are also indicated for the survey locations.

The average shear wave velocity in the upper 30 m is 319 ms^{-1} here. Translated to the international building code, this class belongs to category D, which means that the undrained shear strength is between 50 and 100 kPa, marking stiff soil in the area.

17.3 Figure 17.2c

The profile at the military airport has a very strong amplitude and lower frequency signal in its central part, with diminishing amplitudes and coherency further down. The NE-most 20 m of the profile suffer incoherent and disturbed amplitudes due to strong ground roll and surface waves. To a lesser degree, this is also the case at the SW-end of the line. The bright amplitudes are typical for pyroclastic deposits, identified between 20 and 60 m depth (Arsyad *et al.*, 2007). Unconsolidated clay and silt follow further below identified from weak reflections and in a slightly domed structure (25–55 m distance, 60–90 m depth). Less continuous but in parts bright amplitudes characterize the depths below 100 m, also showing bedrock velocities of more than 700 ms^{-1}, presumably caused by volcanic deposits.

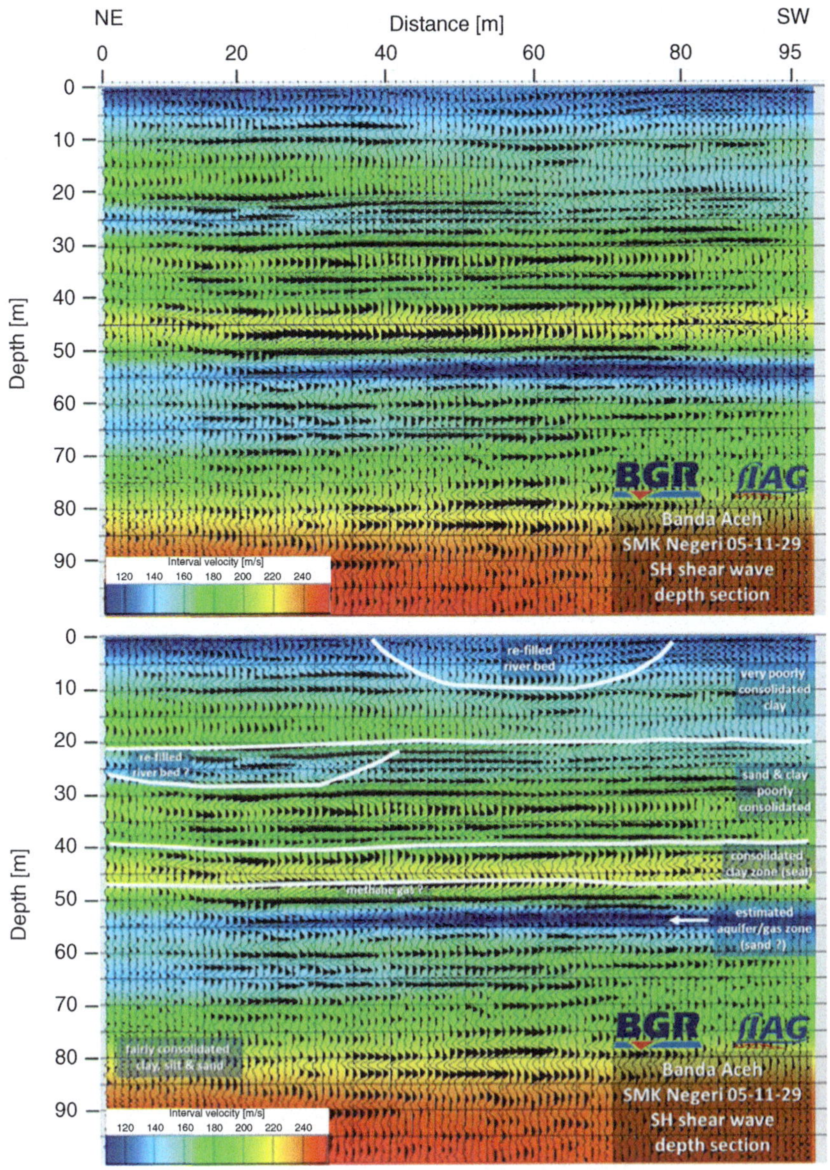

Figure 17.2a Uninterpreted (top) and interpreted (bottom) shear wave profile. Modified from Polom *et al.* 2008.

This part of the investigation area is the only one where the average shear wave velocity in the upper 30 m is higher, with a value of 497 ms^{-1}. Translated to the international building code, this class belongs to category C, which means that the undrained shear strength is higher than 100 kPa, thus consisting of very stiff soil and soft rock in the area.

17.4 Figure 17.2d

Lampasi Engking is immediately located at the Great Sumatra Fault, which is considered as dextral strike-slip fault (Noda, 2013). From this location arises the strong backscattering that dominates the southwestern lower corner of the seismic section. The best traceable and widely continuous reflectors of strong amplitude run from 5 to 15 m depth and bound a sandy aquifer between clay lithologies. All layers deepen NE-wards, with some faint indications for folding in the lower clay–silt–marl sequence (24–40 m distance, 20–40 m depth). The shear wave interval velocities gradually increase downwards, with low values between 100 and 300 ms^{-1}.

The average shear wave velocity in the upper 30 m is 195 ms^{-1} only. Translated to the international building code, this class belongs to category D, which means that the undrained shear strength is between 50 and 100 kPa, marking stiff soil in the area.

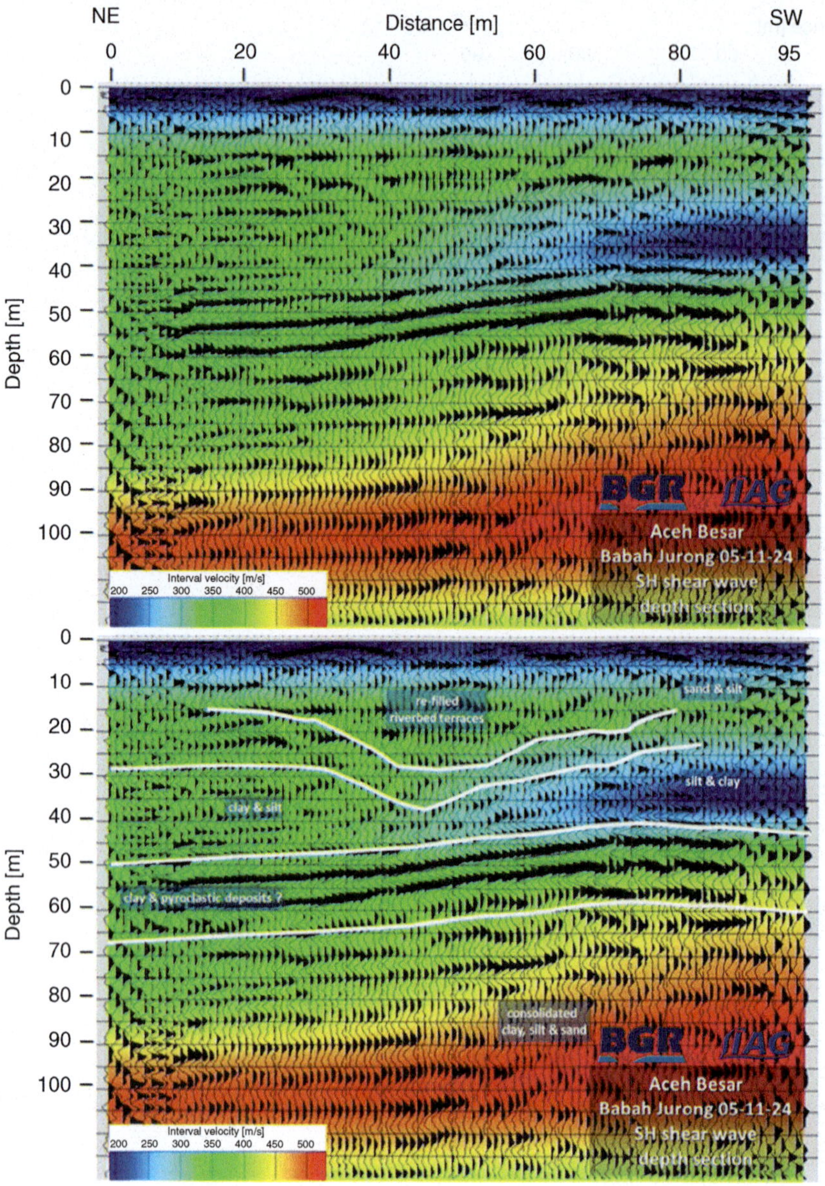

Figure 17.2b Uninterpreted (above) and interpreted shear wave profile. Modified from Polom *et al.* 2008.

17.5 Figure 17.2e

Site Japakeh is short distance from the Great Sumatra Fault, the large dextral strike-slip fault in the region. The N–S trending profile consists of strong, subhorizontal reflectors in the upper 35 m that are imaged with very fine structuring and a high frequency content, especially at the northern end. Further below, the folding of deeper strata dominates the image in the centre of the profile marked by high amplitude reflectors (20–80 m distance, 50–80 m depth). East of the fold, horizontal rather incoherent but strong-amplitude reflectors delimit the folded, mainly sandy package with an area of high velocities (60–95 m distance, 60–100 m depth, velocities higher than 600 ms^{-1}). The lower half of the profile is governed by consolidated clay, sand and silt, but appears almost unstructured.

The average shear wave velocity in the upper 30 m is 247 ms^{-1} only. Translated to the international building code, this class belongs to category D, which means that the undrained shear strength is between 50 and 100 kPa, marking stiff soil in the area.

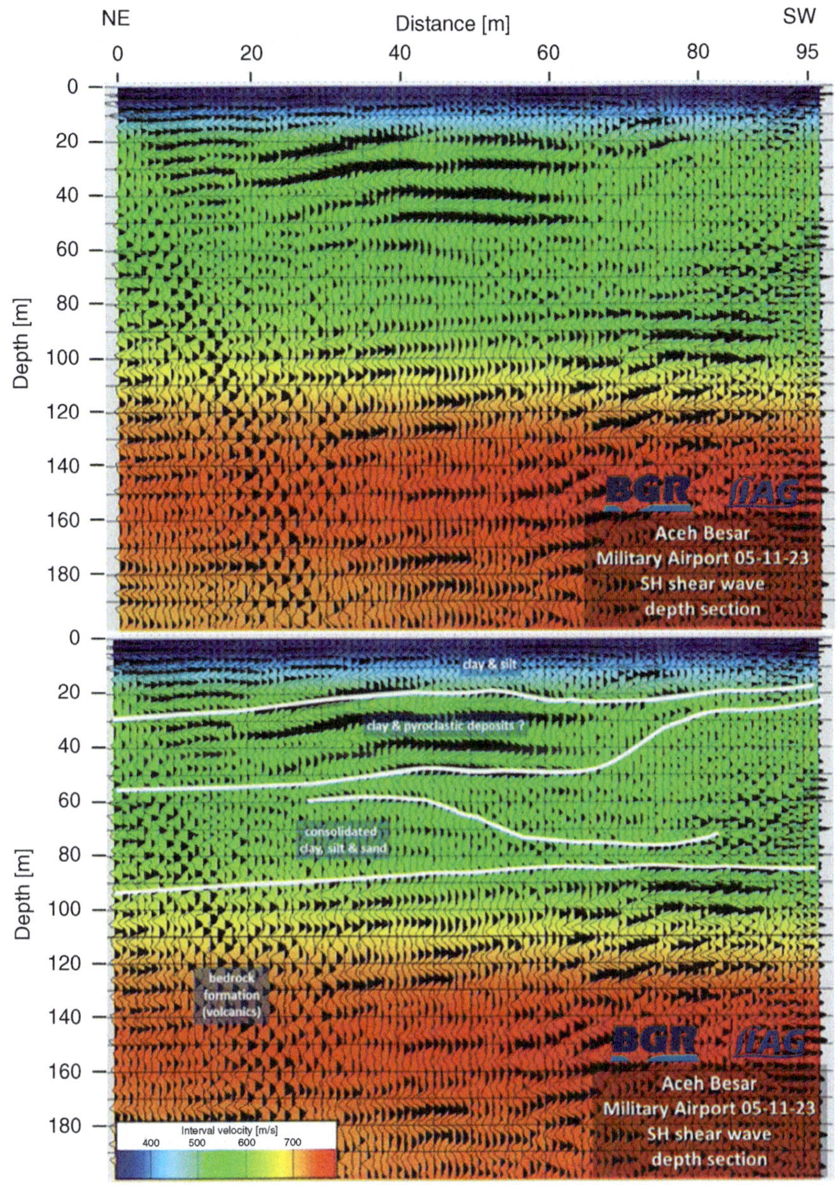

Figure 17.2c Uninterpreted (above) and interpreted shear wave profile. Modified from Polom *et al*. 2008.

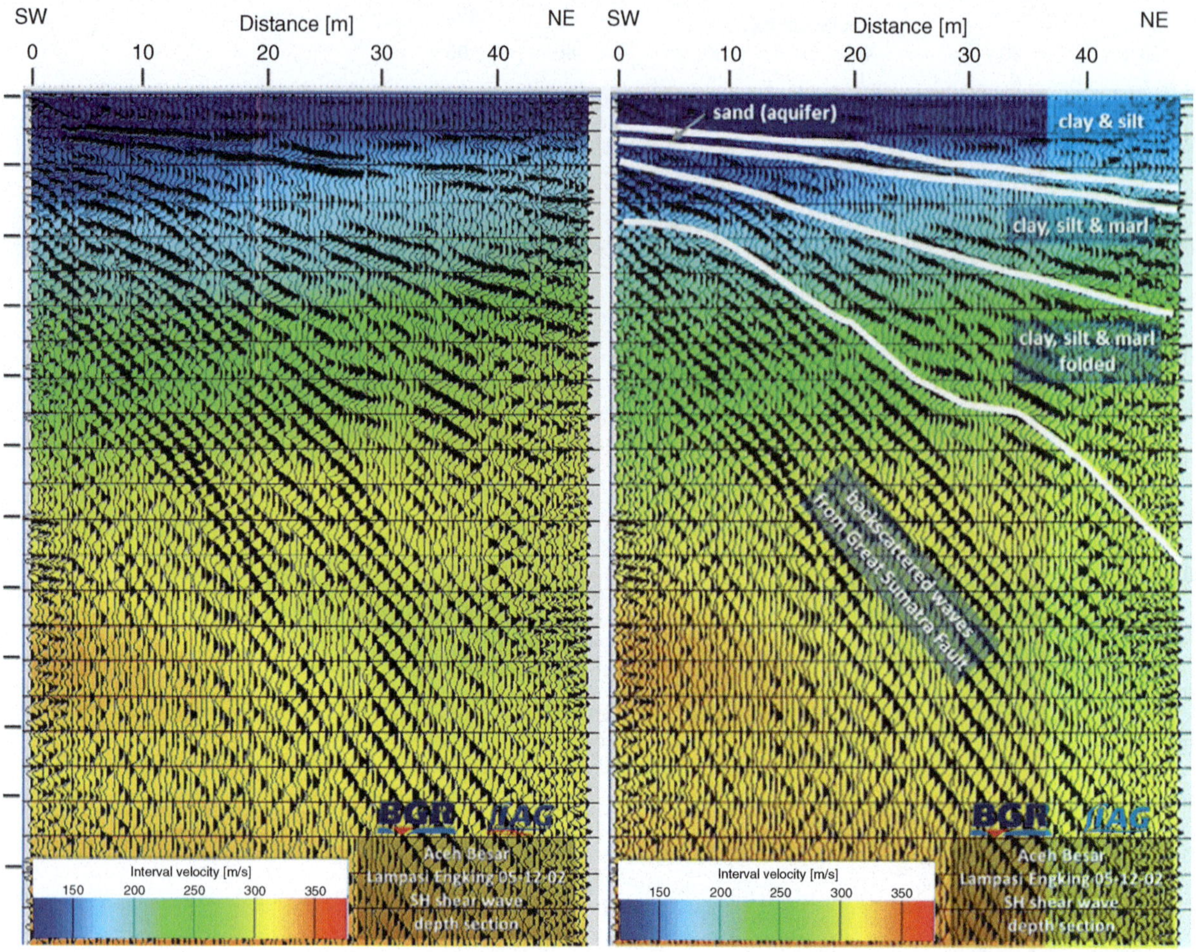

Figure 17.2d Uninterpreted (above) and interpreted shear wave profile. Note the backscattered energy from the Great Sumatra Fault. Modified from Polom *et al.* 2008.

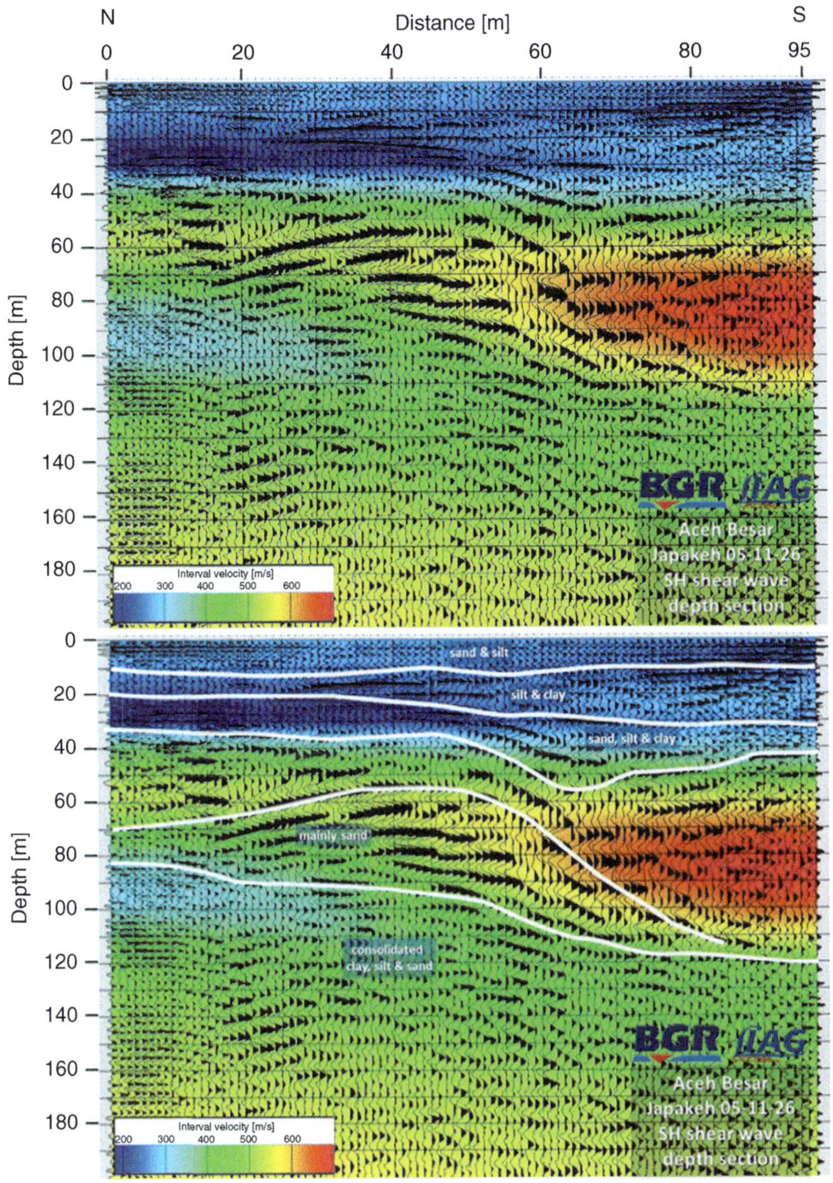

Figure 17.2e Uninterpreted (above) and interpreted shear wave profile. Modified from Polom *et al.* 2008.

References

Arsyad, I. and Polom, U. 2006. Shallow shear wave seismic applications in the Krueng Aceh river delta. *68th EAGE Conference and Exhibition incorporating SPE EUROPEC 2006*. doi: 10.3997/2214-4609.201402212 (www.earthdoc.org).

Arsyad, I., Polom, U. and Wiyono, W. 2007. Shallow Reflection Seismic Shear Wave Velocity Analysis on Paved Soils Using a Land Streamer Unit. *Near Surface 2007 – 13th EAGE European Meeting of Environmental and Engineering Geophysics*. doi: 10.3997/2214-4609.20146637 (www.earthdoc.org).

Noda, A. 2013. Strike-slip basin – its configuration and sedimentary facies. In: Itoh Y. (ed) *Mechanism of Sedimentary Basin Formation – Multidisciplinary Approach on Active Plate Margins*, InTech, pp. 27–57. doi: 10.5772/56593.

Paris, R., Fournier, J., Poizot, E. *et al.* 2010. Boulder and fine sediment transport and deposition by the 2004 tsunami in Lhok Nga (western Banda Aceh, Sumatra, Indonesia): A coupled offshore–onshore model. *Marine Geology*, **268**(3–4), 43–54. doi: 10.1016/j.margeo.2009.10.011.

Polom, U., Arsyad, I. and Kümpel, H.-J., 2008. Shallow shear wave reflection seismics in the tsunami struck Krueng Aceh river basin, Sumatra. *Advances in Geosciences*, **14**, 135–140. doi: 10.5194/adgeo-14-135-2008.

18

Shallow Detachment Along a Transform Margin in Cauvery Basin, India

Achyuta Ayan Misra and Swagato Dasgupta*

Reliance Industries Ltd, Mumbai, India
**Achyuta.Misra@ril.com; achyutaayan@gmail.com*

The Cauvery Basin is the southernmost basin along the eastern continental margin of India (Mazumdar *et al.*, 2018). It is a transform/sheared margin with a two-stage rift history forming the Coromandal strike-slip fault (Nemčok *et al.*, 2013; Sinha *et al.*, 2015). A seismic section from the Cauvery Basin is presented showing a series of rift fault blocks with a shallow crustal detachment (Figure 18.1). The detachment appears at approximately 9 s. The rift is multiphase, as mentioned earlier, so the synrift and postrift sequences are difficult to separate from seismic data alone. In such a case, well data are used to distinguish the synrift and postrift. There is a large rifted main graben, which is comprised of numerous smaller faults (Figures 18.2a and 18.2b). The large fault propagates deeper and due to changes in basement rheology it acts like a detachment for the series of faults appearing above. The extension is so large that normal fault or extensional duplexes (see Figure 18.10 of Fossen, 2016) form along the fault plane (Figures 18.2c and 18.2d). The smaller normal faults detach on the shallow crustal detachment formed by the above mentioned large fault plane (Figures 18.2e and 18.2f). The faults are identified by the offset of the displaced markers, that is the basement top, fault plane reflections and growth of sedimentary sequences towards the large fault plane. There is prolific growth in the peak and late synrift stages (Chapter 4) (Figures 18.2c and 18.2d). The synrift top or breakup unconformity can be identified in the NW of this section from the downlaps of the postrift sediments. However, in the SE, near the large graben bounding fault, there is a possible growth pack above the presently interpreted synrift top. This is possibly a part of the thermal sag. Well data are required to resolve such problems. Adjacent well data indicate that synrift top at the marked horizon, which is the base of postrift. So the possible growth pack is part of the thermal sag. Pronounced footwall erosion can also be interpreted on the graben bounding fault (Figure 18.2d).

Atlas of Structural Geological Interpretation from Seismic Images, First Edition. Edited by Achyuta Ayan Misra and Soumyajit Mukherjee.

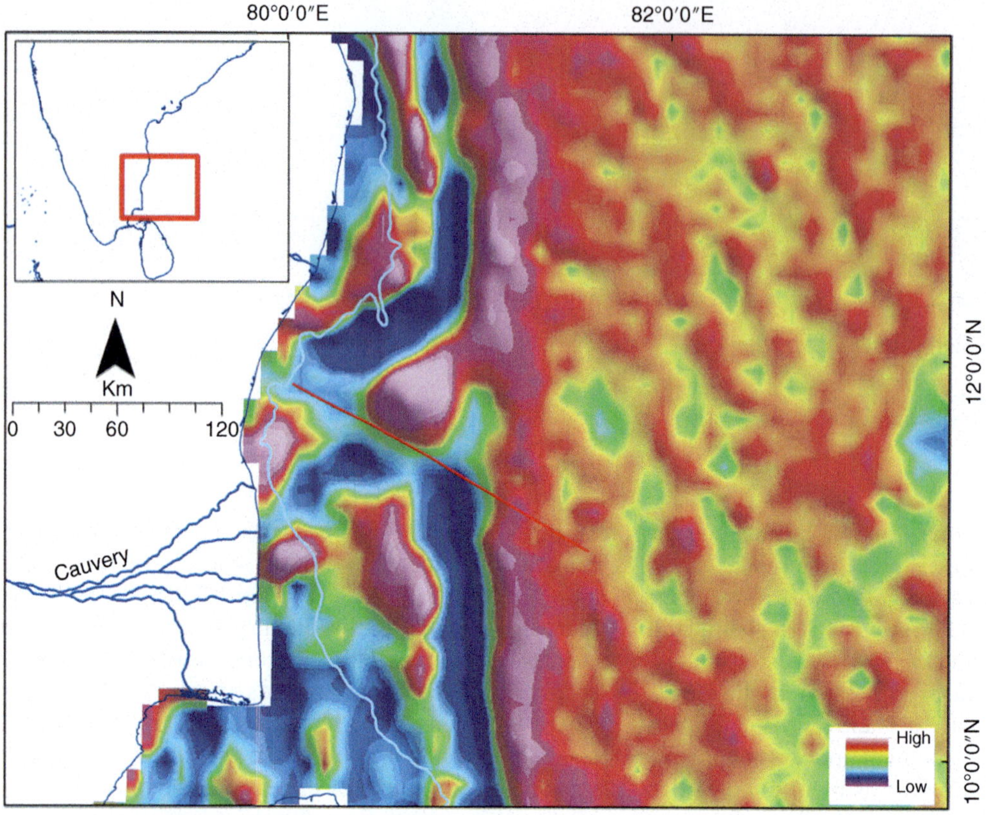

Figure 18.1 Map of the Cauvery Basin along the ECMI overlaid on hill-shaded, upward-continued (10 km) Bouguer gravity anomaly data (see the colour scale). Blue line: Shelf-slope break. Red line: position of seismic section. Note the straight approximate NS trend of the Coromandal strike-slip fault.

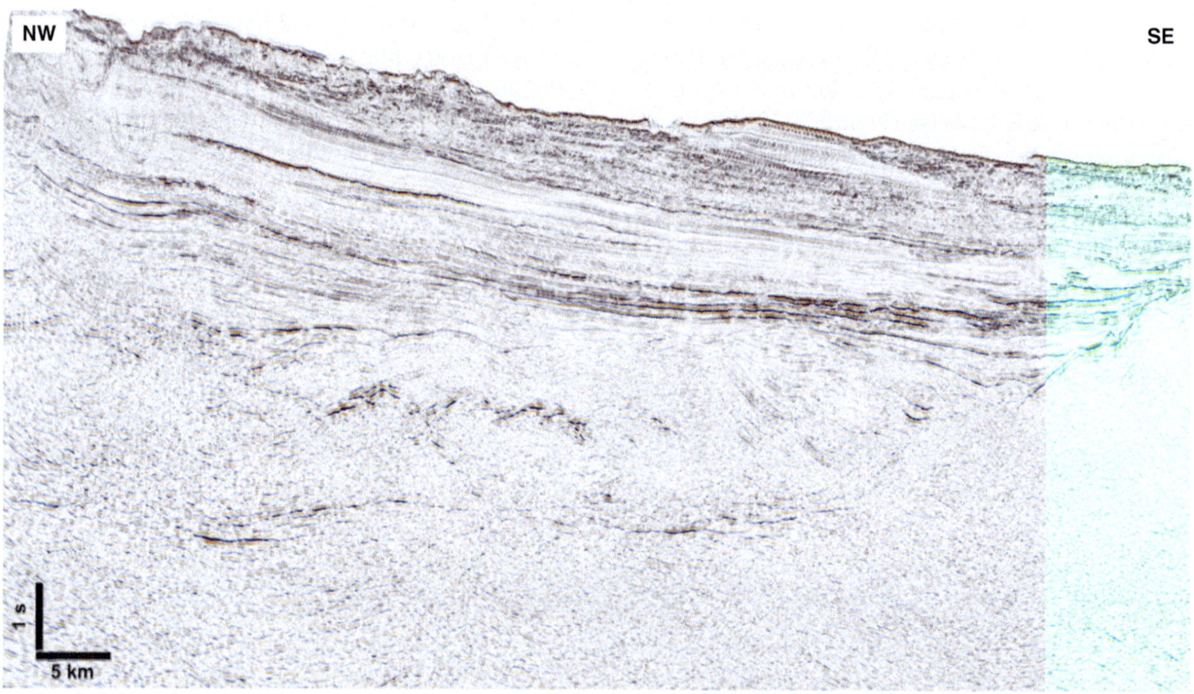

Figure 18.2a Uninterpreted seismic section in the Cauvery Basin. See Figure 18.1 for location. Seismic data courtesy Reliance Industries Ltd. Reproduced with permission from the Directorate General of Hydrocarbons (DGH), India.

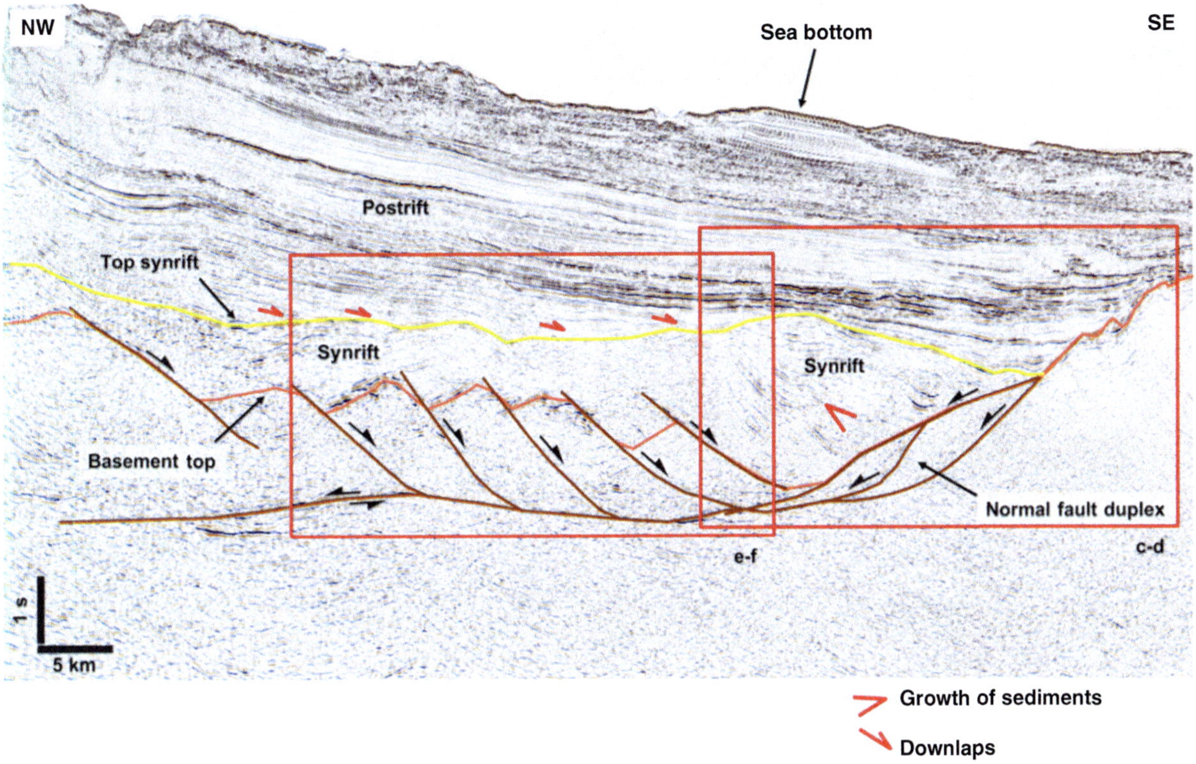

Figure 18.2b Interpretation of seismic section shown in Figure 18.2a. Note the faults were identified on the basis of the offset of the basement top, fault plane reflections and growth packages (see Chapter 2.3). Red rectangles: details of the section in Figures 18.2c–18.2f). Seismic data courtesy Reliance Industries Ltd. Reproduced with permission from the Directorate General of Hydrocarbons (DGH), India.

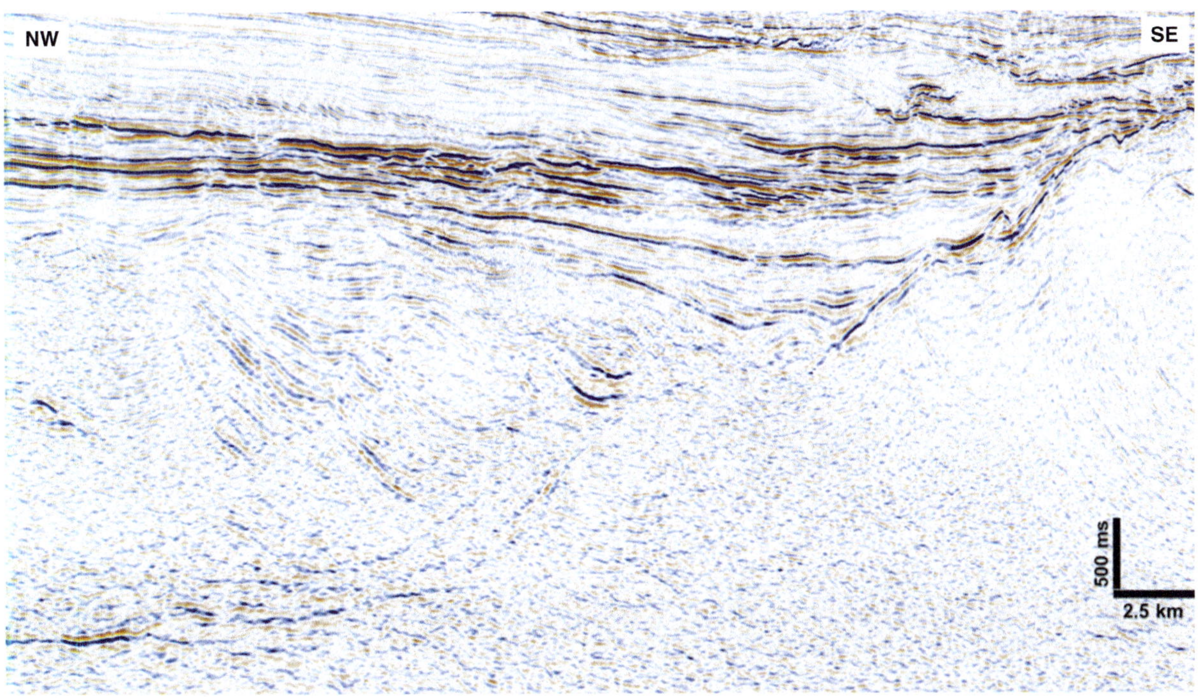

Figure 18.2c Uninterpreted seismic section showing the graben bounding fault (see Figure 18.2b for location). Seismic data courtesy Reliance Industries Ltd. Reproduced with permission from the Directorate General of Hydrocarbons (DGH), India.

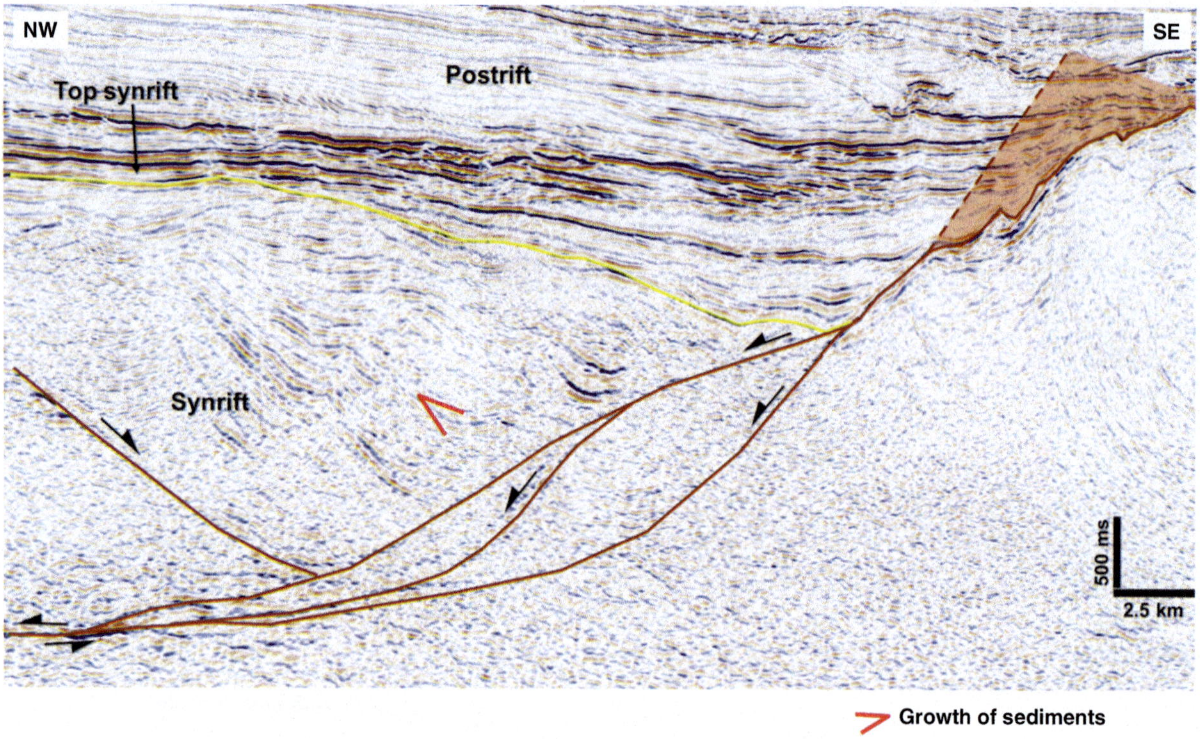

Figure 18.2d Interpretation of seismic section shown in Figure 18.2c. Note the growth of the synrift sediments in the peak synrift stage. Footwall erosion along the graben bounding fault can be interpreted (brown polygon). Seismic data courtesy Reliance Industries Ltd. Reproduced with permission from the Directorate General of Hydrocarbons (DGH), India.

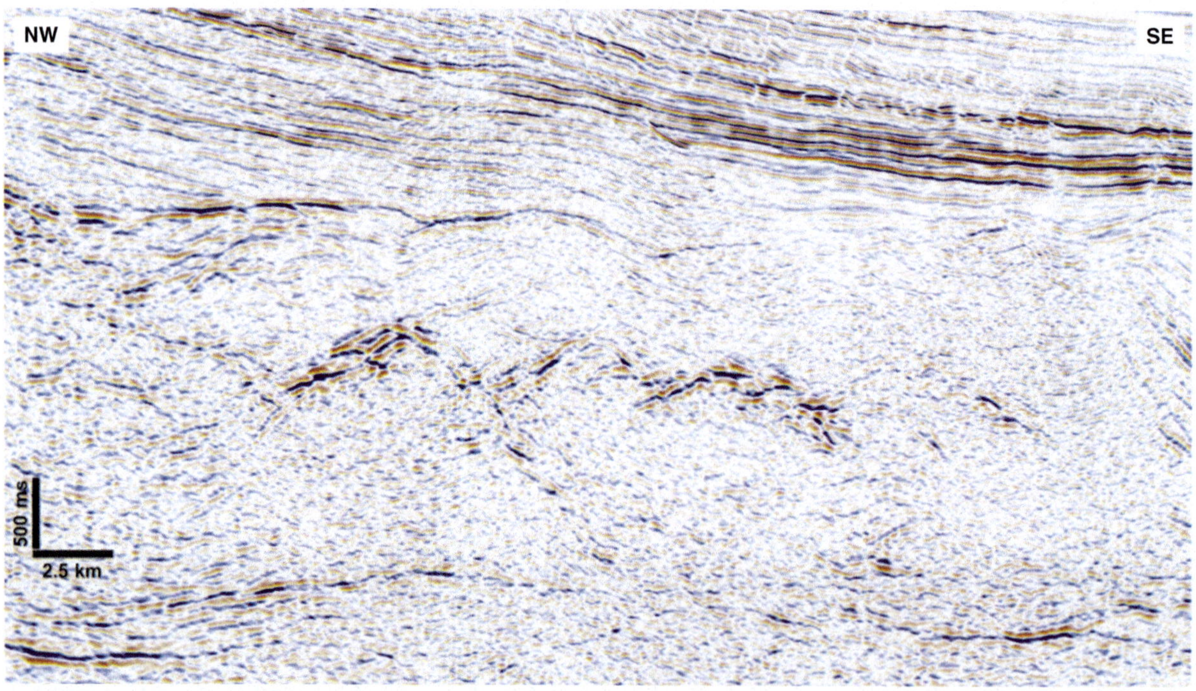

Figure 18.2e Uninterpreted seismic section showing the smaller faults and the shallow crustal detachment. Seismic data courtesy Reliance Industries Ltd. Reproduced with permission from the Directorate General of Hydrocarbons (DGH), India.

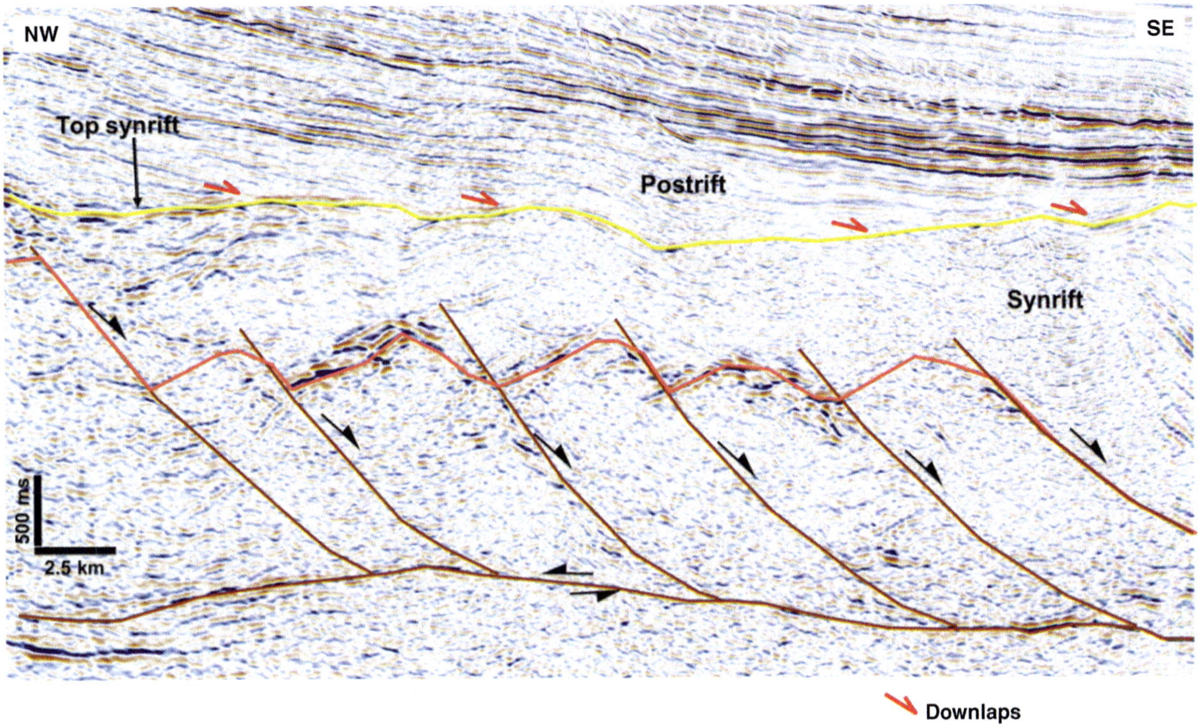

Figure 18.2f Interpretation of seismic section shown in Figure 18.2e. Note that prominent growth is not visible in this seismic section for the synrift sequence. However, the synrift section has been correlated from the SE, where growth is visible. The downlaps denote the start of the postrift sequence. Seismic data courtesy Reliance Industries Ltd. Reproduced with permission from the Directorate General of Hydrocarbons (DGH), India.

References

Fossen, H. 2016. *Structural Geology*. Cambridge University Press. Second Edition.

Mazumdar, S., Tep, B., Pangtey, K.K.S. *et al.* 2018. Basement tectonics and shear zones in Cauvery basin: Implications in hydrocarbon exploration. In: Mukherjee, S. (Ed) Tectonics and Structural Geology: Indian Context. Springer International Publishing AG, Cham. Springer.

Nemčok, M., Sinha, S.T., Stuart, C.J. *et al.* 2013. East Indian margin evolution and crustal architecture: integration of deep reflection seismic interpretation and gravity modelling. In: Mohriak, W.U., Danforth, A., Post, P.J. *et al.* (eds) *Conjugate Divergent Margins*. Special Publications vol. 369, The Geological Society, London, pp. 477–496.

Sinha, S.T., Nemčok, M., Choudhuri, M. *et al.* 2015. The role of break-up localization in microcontinent separation along a strike-slip margin: the East India–Elan Bank case study. In: Nemčok, M., Rybár, S., Sinha, S.T. *et al.* (eds) *Transform Margins: Development, Controls and Petroleum Systems*. Special Publications vol. 431, The Geological Society, London, pp. 95–124.

19

Normal Faulting Related to Differential Compaction on the 85°E Ridge in the Bay of Bengal, India

Achyuta Ayan Misra

Reliance Industries Ltd, Mumbai, India
Achyuta.Misra@ril.com; achyutaayan@gmail.com

The 85°E Ridge is a trace of the Kerguellen Hotspot in the Bay of Bengal, offshore of the east coast of India (Sreejith *et al.*, 2011; Nemčok *et al.*, 2013; Choudhuri *et al.*, 2014). Presently it is buried under 2–3 km of sediments (Choudhuri *et al.*, 2014). It has an elevation of 3–4 km from the surrounding basement. A seismic section acquired by ion-GX Technology where the 85°E Ridge is imaged is shown here (Figure 19.1). Details of the acquisition parameters can be found elsewhere (Choudhuri *et al.*, 2014). The sediments covering the 85°E ridge drape it. There are faults that are present in the overburden and link at the flanks of the ridge (Figures 19.2a and 19.2b). These faults form due to differential or inhomogeneous compaction across the

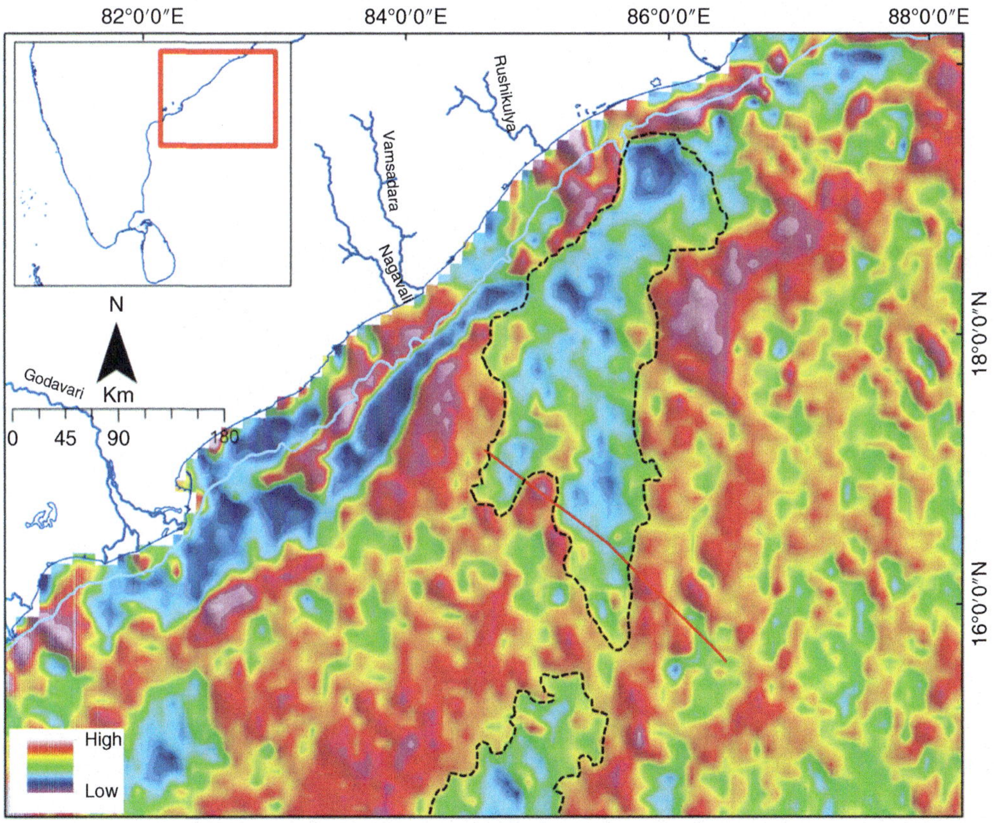

Figure 19.1 Map of the Bay of Bengal overlaid on hill-shaded, upward-continued (10 km) Bouguer gravity anomaly data (see the colour scale). Blue line: Shelf-slope break. Red line: position of seismic section. Dotted black line: outline of the 85°E ridge. Seismic data courtesy Reliance Industries Ltd.

Atlas of Structural Geological Interpretation from Seismic Images, First Edition. Edited by Achyuta Ayan Misra and Soumyajit Mukherjee.

ridge. On top of the ridge there are approximately 3 km thick sediments, while away from the flanks the sediment thickness is approximately 6 km. The compaction of an approximately 6 km thick sedimentary column will be much larger than that of an approximately 3 km thick one. This large difference in compaction cannot be accommodated by draping alone and the sediments are faulted as a result (Figures 19.2c–19.2f). The throws of these faults exceed 100 msec and decrease in the shallower stratigraphies, becoming zero where the effect of the differential compaction is negligible. Normal drags (Mukherjee, 2014) along the faults are present on both sides. They are most possibly related to the bending of the fault plane on top of the ridge (Figures 19.2c–19.2f).

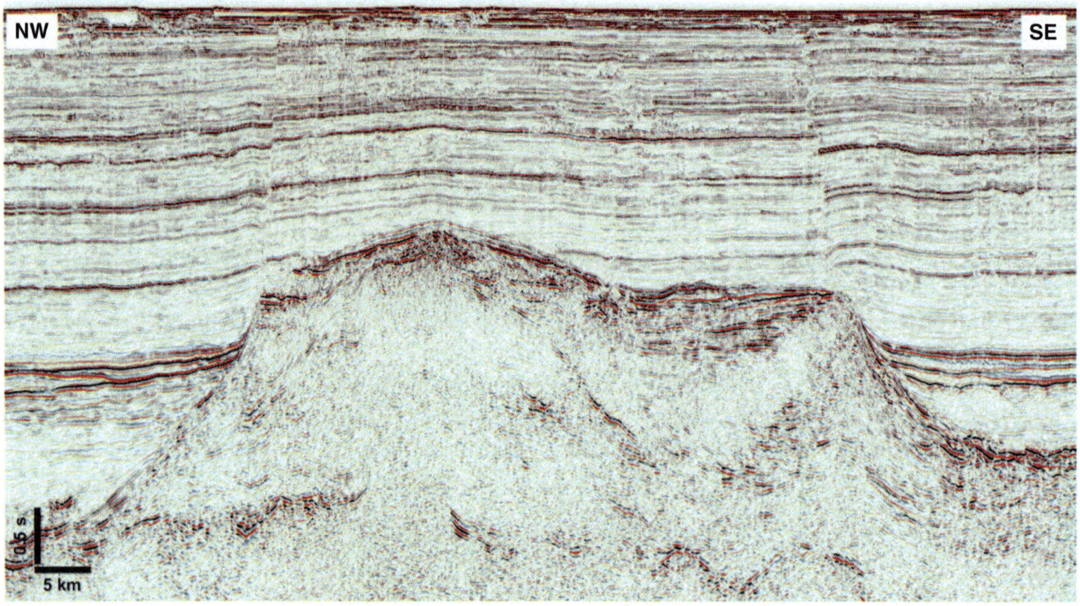

Figure 19.2a Uninterpreted seismic section from the Bay of Bengal showing the 85°E ridge. See Figure 19.1 for location. Seismic data courtesy Reliance Industries Ltd. Reproduced with permission from the Directorate General of Hydrocarbons (DGH), India.

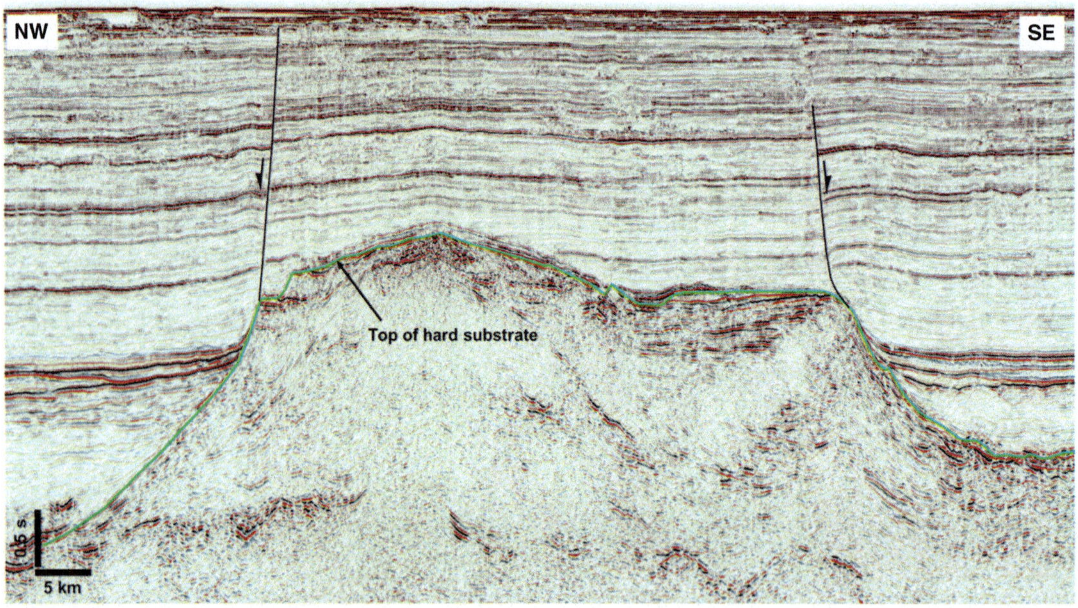

Figure 19.2b Interpretation of seismic section shown in Figure 19.2a. Note the differential compaction related faults that link to the flanks of the ridge. The reverse drags are also clearly observed. Seismic data courtesy Reliance Industries Ltd. Reproduced with permission from the Directorate General of Hydrocarbons (DGH), India.

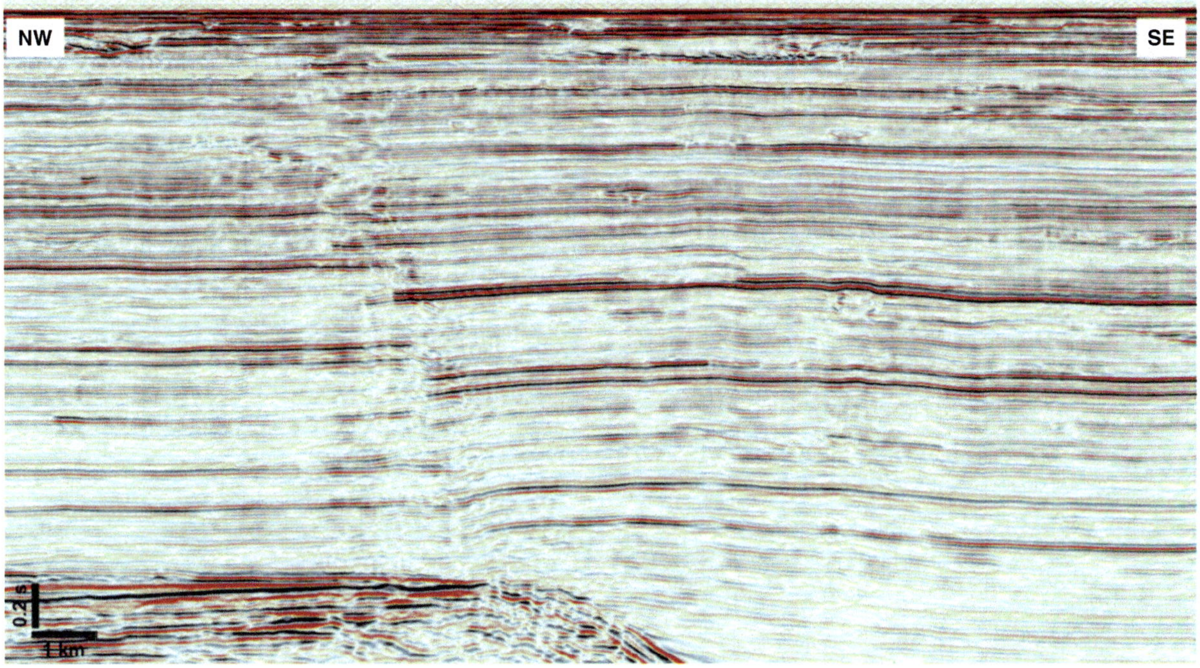

Figure 19.2c Uninterpreted seismic section showing the east flank of the 85°E ridge. Seismic data courtesy Reliance Industries Ltd. Reproduced with permission from the Directorate General of Hydrocarbons (DGH), India.

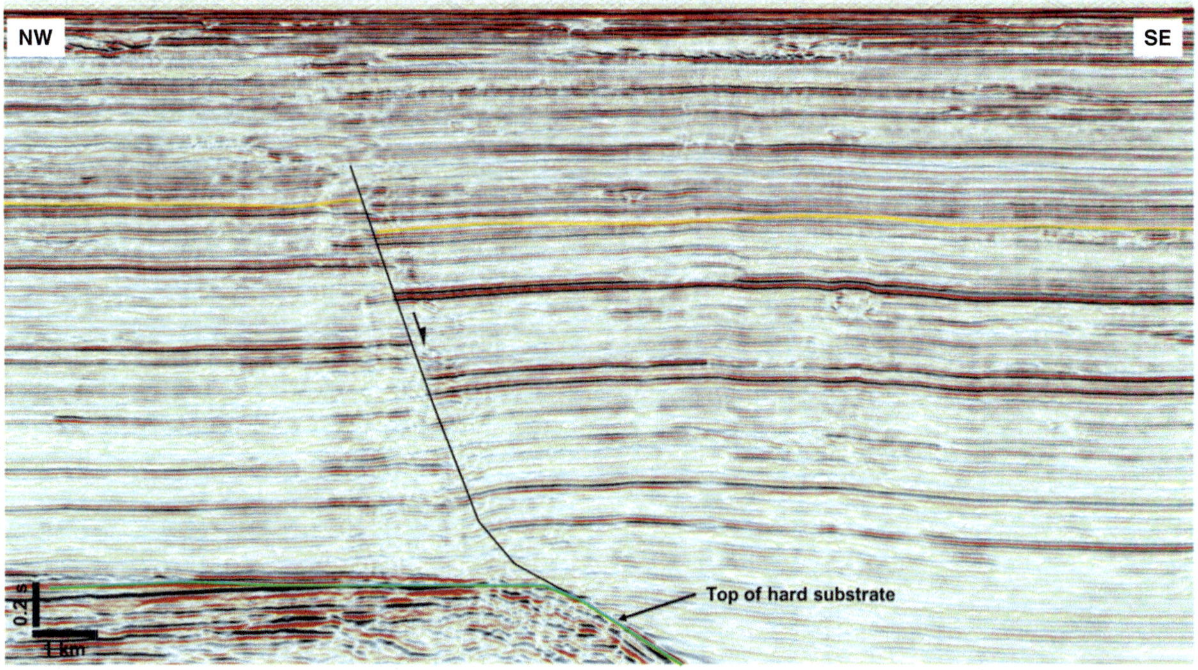

Figure 19.2d Interpretation of seismic section shown in Figure 19.2c. The ochre horizon can be easily correlated. Around 200 ms above that the faults is nonexistent or below seismic resolution. The reverse drags are difficult to identify due to the lower vertical exaggeration in this section. Seismic data courtesy Reliance Industries Ltd. Reproduced with permission from the Directorate General of Hydrocarbons (DGH), India.

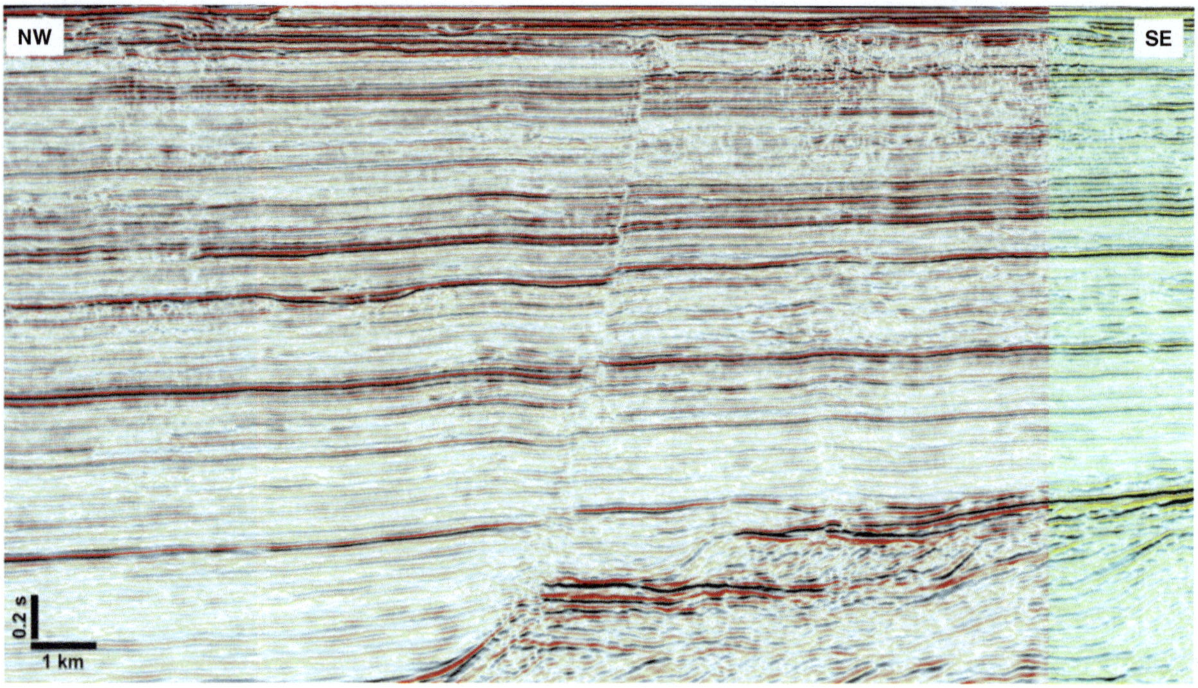

Figure 19.2e Uninterpreted seismic section showing the west flank of the 85°E ridge. Seismic data courtesy Reliance Industries Ltd. Reproduced with permission from the Directorate General of Hydrocarbons (DGH), India.

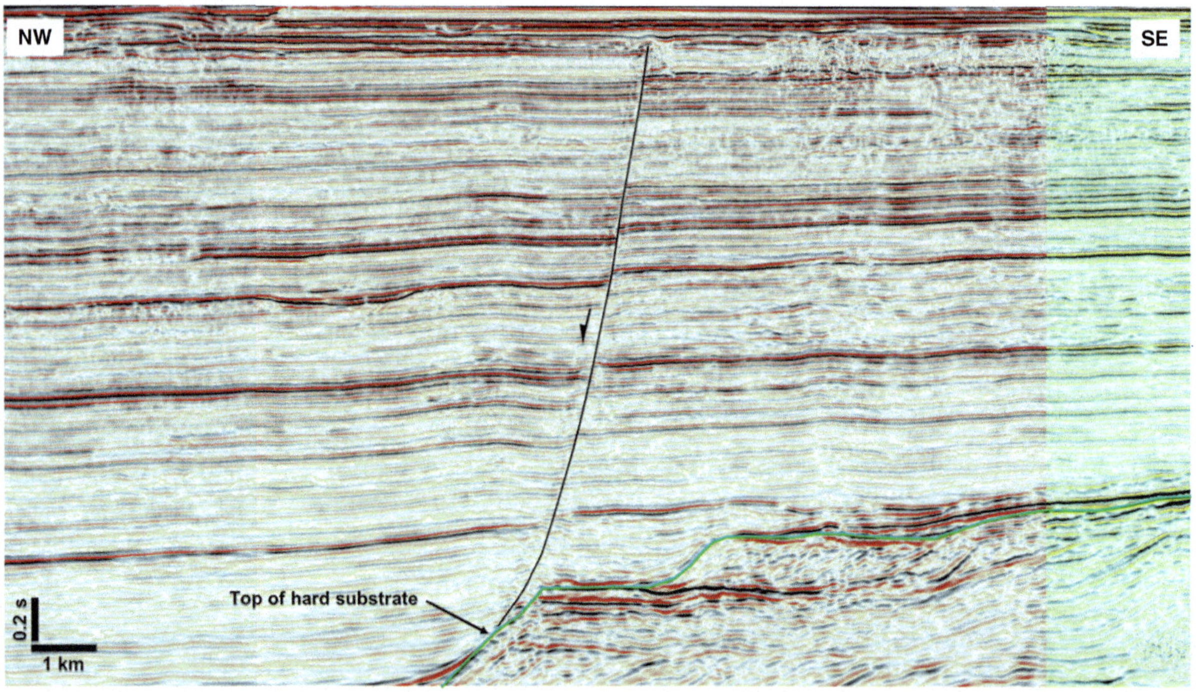

Figure 19.2f Interpretation of seismic section shown in Figure 19.2c. The reverse drags here are slightly more pronounced than in the east flank. Also, this fault continues to a shallower level than its eatern counterpart. Seismic data courtesy Reliance Industries Ltd. Reproduced with permission from the Directorate General of Hydrocarbons (DGH), India.

References

Choudhuri, M., Nemèok, M., Stuart, C. *et al.* 2014. 85°E Ridge, India – constraints on its development and architecture. *Journal of the Geological Society of India*, **84**(5), 513–530.

Mukherjee, S. 2014. Review of flanking structures in meso- and micro-scales. *Geological Magazine*, **151**, 957–974.

Nemčok, M., Sinha, S.T., Stuart, C.J. *et al.* 2013. *East Indian Margin Evolution and Crustal Architecture: Integration of Deep Reflection Seismic Interpretation and Gravity Modelling.* Special Publications vol. 369, The Geological Society, London, pp. 477–496.

Sreejith, K.M., Radhakrishna, M., Krishna, K.S. and Majumdar, T.J. 2011. Development of the negative gravity anomaly of the 85°E Ridge, northeastern Indian Ocean – A process oriented modelling approach. *Journal of Earth System Science*, **120**(4), pp. 605–615.

20

Polygonal Fault System in Late Cretaceous Sediments of Cauvery Deepwater Basin, India

Swagato Dasgupta and Achyuta Ayan Misra*

Reliance Industries Ltd, Mumbai, India
** Swagato.Dasgupta@ril.com, swagato.dg@gmail.com*

Polygonal fault systems are of small-scale, closely spaced, multidirectional fault networks with throws ranging 10 to 100 m (Cartwright and Lonergan, 1996; Watterson *et al.*, 2000; Biswas and Mukherjee, 2014). The individual faults have relatively uniform dips, normally >40–45° (Watterson *et al.*, 2000). The strike of these faults varies widely, thereby giving a polygonal appearance in plan view. In cross-section, they may appear as conjugate set of faults (Watterson *et al.*, 2000) in many cases, mainly due to wide variation in their strike. The network of polygonal faults normally occurs as a layer bound of deformational events, that is its top and bottom layers are undeformed. These sets of normal faults are neither related to extension nor to any gravity-driven slip (Watterson *et al.*, 2000; Cartwright *et al.*, 2003). Rather, they are closely associated with sediment compaction driven dewatering systems and to gravitational instability, due to the density variation of the lithological units. However, they are distinctly different from diapirism, which is guided by gravitational instability (Mukherjee *et al.*, 2010; Mukherjee, 2011; Mukherjee and Mulchrone, 2012). The network of polygonal faults act as fluid escape feature, through which fluids from lower lithological units diffuse out to the upper surface. These polygonal faults in general occur in finer clastics and are distributed randomly along horizontal section. Such polygonal fault systems are present in different passive margins like the North Sea Basin (Ostanin *et al.*, 2012), the Atlantic margin of North America (Hansen *et al.*, 2004), East and West African offshore, Australian offshore, etc.

Polygonal faulting is also observed in the Indian E coast. In this example, a polygonal fault system is shown in the Late Cretaceous finer sediments as identified in the deepwater Cauvery Basin. A set of polygonal faulting is observed within the Santonian to Early Campanian finer clastics at a depth range of 1500 – 2200 m below sea bottom (Figures 20.1 and 20.2). These faults have variable trends and 55 – 70° dip. The throw along these faults ranges 10 – 75 m. A root mean square (RMS) amplitude map generated along the lower part of this polygonal fault system, from the Santonian interval, depicts a typical polygonal geometry (Figure 20.3). The fault system has also affected, to some extent, a channel dominated coarser clastic fairway in the southwest part of the map (Figure 20.3). Some sediment decompaction-related faulting is also seen in the Turonian section over the Early Cretaceous structural high (Figure 20.2). The polygonal fault system as well as the inhomogenous decompaction related faults, as observed in the lower section, act as fluid escape structures. Fluids can migrate to shallower sections through such a fault network.

Atlas of Structural Geological Interpretation from Seismic Images, First Edition. Edited by Achyuta Ayan Misra and Soumyajit Mukherjee.

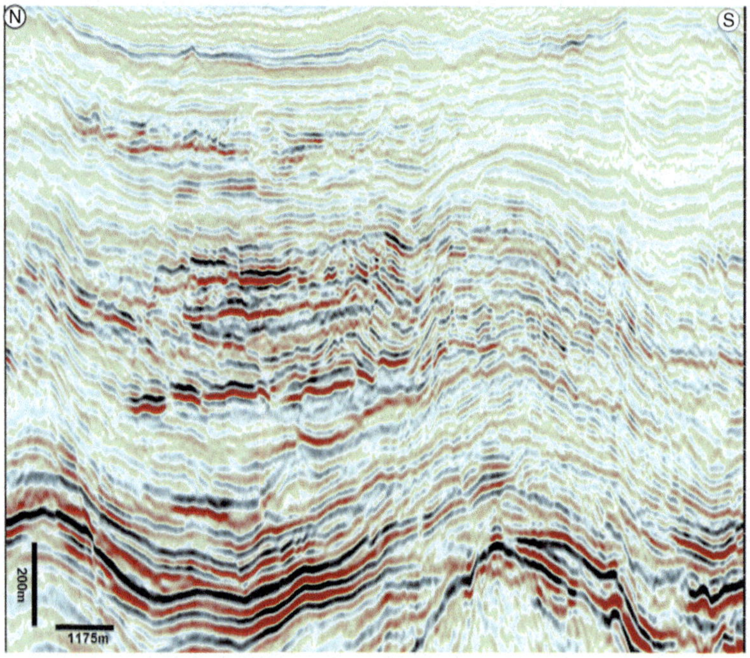

Figure 20.1 N–S seismic section through the Late Cretaceous interval of deepwater Cauvery Basin. Location concealed due to data confidentiality. The location of the Cauvery Basin is shown in Chapter 18. Seismic data courtesy Reliance Industries Ltd. Reproduced with permission from the Directorate General of Hydrocarbons (DGH), India.

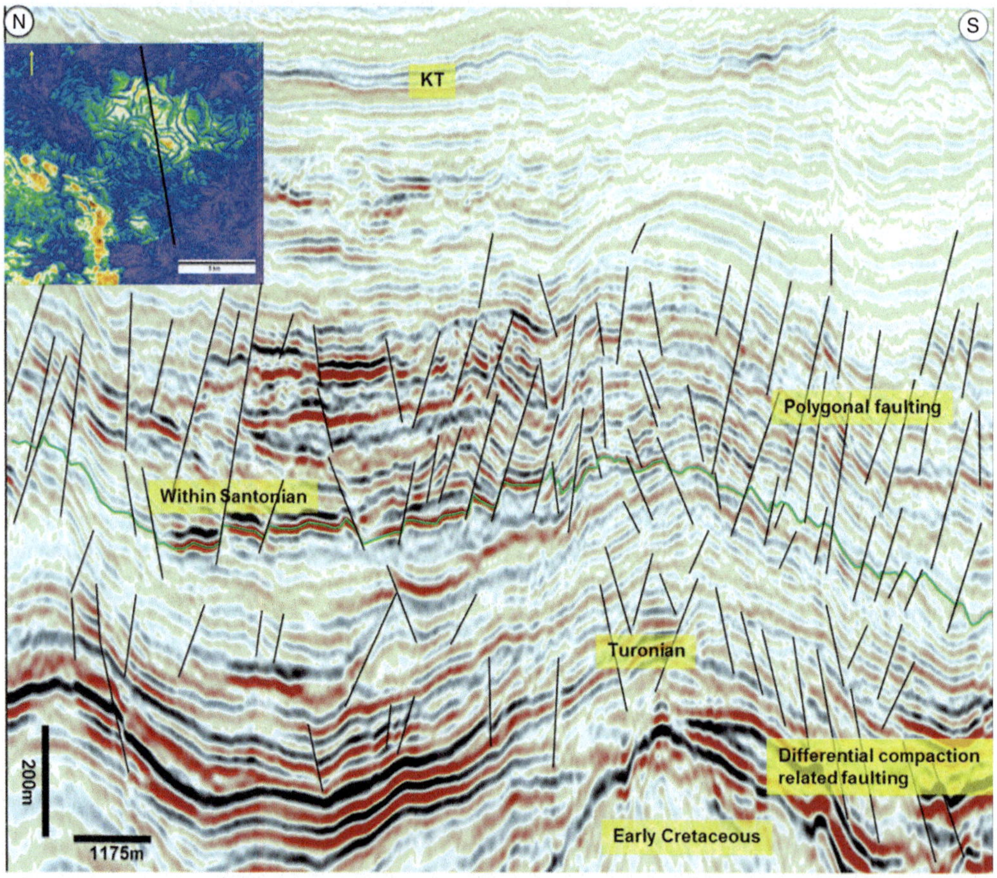

Figure 20.2 Interpreted seismic section of Figure 20.1 depicting a polygonal fault system in Late Cretaceous sediments and differential compaction related faulting in Turonian interval. Note that this section has an approximately 5:1 vertical exaggeration, therefore does not depict the actual drag of layers (Mukherjee, 2014) near the fault planes. Seismic data courtesy Reliance Industries Ltd. Reproduced with permission from the Directorate General of Hydrocarbons (DGH), India.

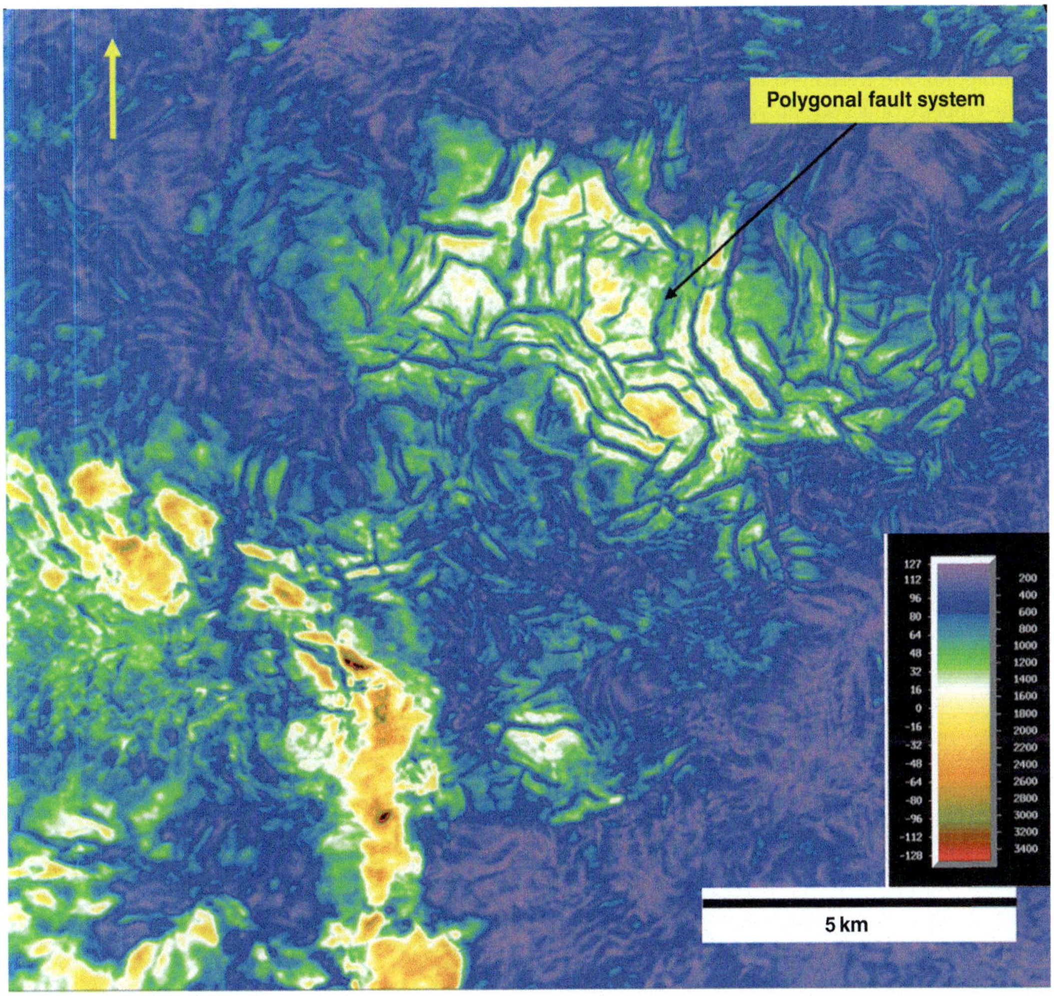

Figure 20.3 RMS amplitude map generated along green horizon of Figure 20.2, depicting distinct polygonal fault pattern. Seismic data courtesy Reliance Industries Ltd. Reproduced with permission from the Directorate General of Hydrocarbons (DGH), India.

References

Biswas, T. and Mukherjee, S. 2014. A review of polygonal fault systems and their hydrocarbon prospects. Tectonic Studies Group, 06–08 January, Cardiff University, UK.

Cartwright, J.A. and Lonergan, L. 1996. Volumetric contraction during the compaction of mudrocks: a mechanism for the development of regional-scale polygonal fault systems. *Basin Research*, **8**, 183–193.

Cartwright, J.A., James, D.M.D. and Bolton, A.J. 2003. The genesis of polygonal fault systems: a review. In: Van Rensbergen, P., Hillis, R. and Morley, C. (eds) *Subsurface Sediment Mobilisation*. Special Publications vol. 216, The Geological Society, London, pp. 223–242.

Hansen, D.M., Shimeld, J.W., Williamson, M.A. and Anderson, H.L. 2004. Development of a major polygonal fault system in Upper Cretaceous chalk and Cenozoic mudrocks of the Sable Subbasin, Canadian Atlantic margin. *Marine and Petroleum Geology*, **21**, 1205–1219.

Mukherjee, S. 2011. Estimating the Viscosity of Rock Bodies- a Comparison Between the Hormuz- and the Namakdan Salt Domes in the Persian Gulf, and the Tso Morari Gneiss Dome in the Himalaya. *The Indian Journal of Geophysical Union*, **15**, 161–170.

Mukherjee, S. 2014. Review of flanking structures in meso- and micro-scales. *Geological Magazine*, **151**, 957–974.

Mukherjee, S., Talbot, C. J. and Koyi, H. A. 2010. Viscosity estimates of salt in the Hormuz and Namakdan salt diapirs, Persian Gulf . *Geological Magazine*, **147**, 497–507.

Mukherjee, S. and Mulchrone, K. 2012. Estimating the viscosity and Prandtl number of the Tso Morari Gneiss

Dome, western Indian Himalaya. *International Journal of Earth Sciences*, **101**, 1929–1947.

Ostanin, I., Anka, Z., Primio, Rd. and Bernal, A. 2012. Identification of a large Upper Cretaceous polygonal fault network in the Hammerfest basin: Implications on the reactivation of regional faulting and gas leakage dynamics, SW Barents Sea. *Marine Geology*, **332–334**, 109–125.

Watterson, J., Walsh, J., Nicol, A. *et al.* 2000. Geometry and origin of a polygonal fault system. *Journal of the Geological Society*, **157**, 151–162.

21

Transfer Zone Geometry in the Offshore Cauvery Basin, India

Swagato Dasgupta and Abhimanyu Maitra*

Reliance Industries Ltd, Mumbai, India
** Swagato.Dasgupta@ril.com, swagato.dg@gmail.com*

The Cauvery Basin, a pericratonic rift basin (Sastri *et al.*, 1981; Biswas *et al.*, 1993; Chaudhuri *et al.*, 2010), evolved during Jurassic to Early Cretaceous time as the eastern Gondwanaland rifted. The basin is located in the southern part of the Indian east coast, covering ~62 000 km^2, of which ~25 000 km^2 is on land and the remaining in offshore. The general trend of the major rift fault is NE–SW, resembling that of the Krishna–Godavari (K–G) Basin in the north. Several studies have been carried out on the rift segments of the east coast of India, especially in the K–G and Cauvery Basins (Veevers and Tewari, 1995; Gaina *et al.*, 2007; Watkinson *et al.*, 2007; Veevers, 2009, 2012; Lal *et al.*, 2009; Sinha *et al.*, 2010; Nemčok *et al.*, 2013), in relation to the break-up of East Gondwana into India, Antarctica and Africa.

The rift architecture varies along the strike of the rift-related faults. Such variation is a consequence of the adjustment of half grabens related to variation in slip of the normal faults. Transfer zones and faults, relay ramps and/or transverse faults accommodate these changes (Gawthorpe and Hurst, 1993; Faulds and Varga, 1998). They divide the rift basin into a number of sub-basins. Researchers have studied these zones in rift basins such as the North Sea, Gulf of Suez and East Africa rift basin, thereby coming-up with different morpho-tectonic terminologies (Gibbs, 1984; Rosendahl, 1987; Morley *et al.*, 1990) that are widely accepted and used (Misra *et al.*, 2014, 2015; Misra and Mukherjee, 2015; Dasgupta and Mukherjee, 2017). Transfer zones conserve the amount of strain between two overlapping normal faults and, thus, accommodate not only slip but also elevation variances of the adjoining foot wall blocks (Morley *et al.*, 1990). The elevation difference is accommodated by a set of relay ramp faults. Pre-existing basement structures, i.e. weak zones in basement, play an important role in formation of transfer zones (Morley, 1999a; Morley *et al.*, 2004; Dasgupta and Mukherjee, 2017). In a typical rift system, the rate of occurrence and dip direction of secondary faults might change on successive dip lines, forming different types of transfer zones (Morley *et al.*, 1990; Morley, 1999).

The change in displacement along the Cauvery basin bounding faults, adjacent to Portonovo High, are accommodated by a set of relay ramps linked to a number of secondary fault systems. These faults trend NE, similar to that of the basin bounding faults of the Karaikal highs. These secondary fault tips overlap mutually, thereby forming transfer zones across which the slip of the basin bounding faults die out. Most of these transfer zones (Figures 21.1 and 21.2) are of synthetic overlapping in nature, except for a few that are conjugate overlapping to collateral category. These relay ramps are one of the major sediment entry points to the sub-basin. The conjugate (divergent) collateral transfer zone creates intrabasinal highs (Figure 21.2), as observed in the SE part of the inset fault map of Figure 21.2. Few faults related to differential compaction are also observed over such intrabasinal highs in the post-rift sediments (Figure 21.2).

In areas where slip along the main basin bounding fault decreases, a set of secondary relay faults evolves along with associated transfer zones to accommodate the displacement related strain (Figures 21.3 to 21.6). These secondary faults are usually synthetic to the basin bounding fault (Figures 21.4 and 21.6). Some collateral synthetic faults could also be expected in the proximal part (Figures 21.3 and 21.4). In such areas, the dip of the synthetic relay faults gradually decreases towards the distal part, that is becomes more listric in nature (Figures 21.5 and 21.6). Due to the change in displacement along these areas, the syn-rift sediment growth varies laterally as well as in vertical sections along these synthetic relay faults.

Atlas of Structural Geological Interpretation from Seismic Images, First Edition. Edited by Achyuta Ayan Misra and Soumyajit Mukherjee.

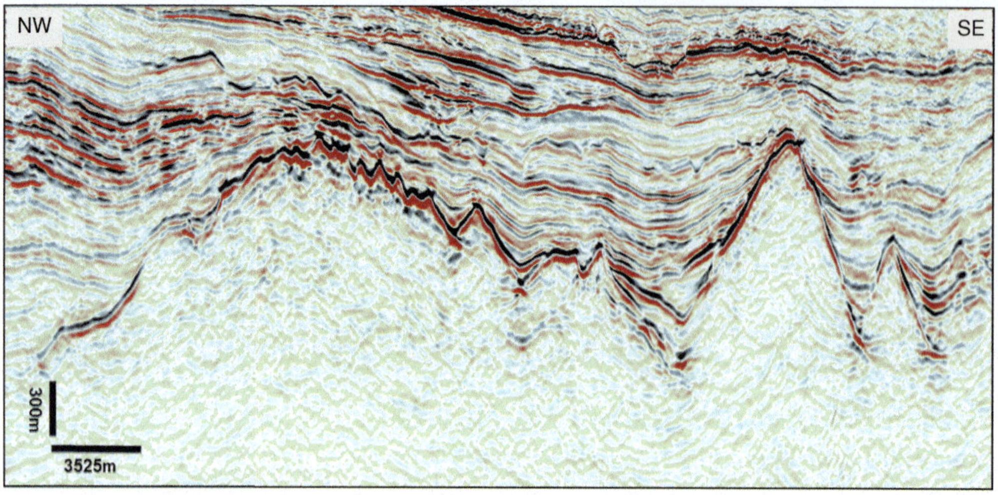

Figure 21.1 NW–SE seismic section across the Portonovo high and adjacent intrabasinal high in the offshore Cauvery Basin. Seismic data courtesy Reliance Industries Ltd. Reproduced with permission from the Directorate General of Hydrocarbons (DGH), India.

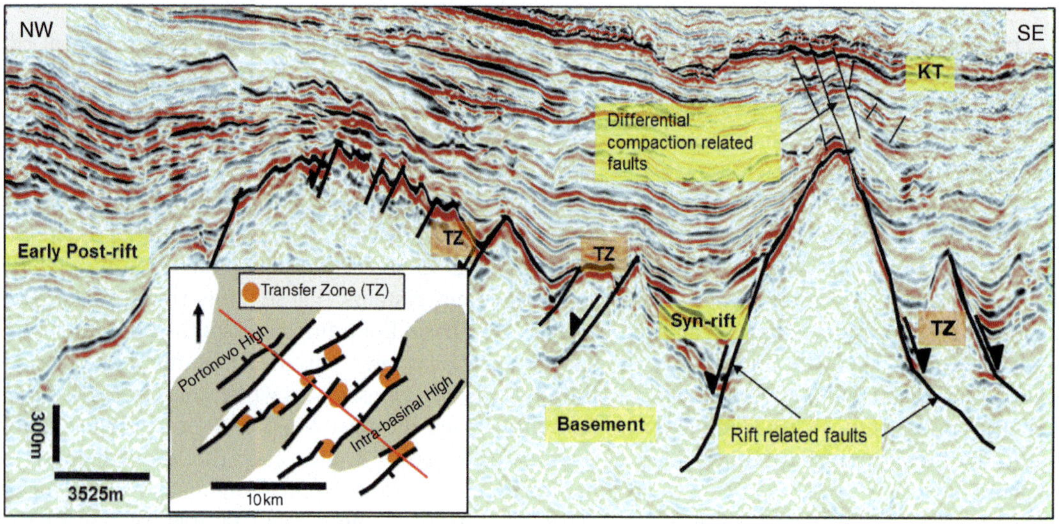

Figure 21.2 Interpreted section of Figure 21.1 depicting a set of relay faults and some differential compaction related faults over the intrabasinal high. Inset: fault map at basement level with the line location. Seismic data courtesy Reliance Industries Ltd. Reproduced with permission from the Directorate General of Hydrocarbons (DGH), India.

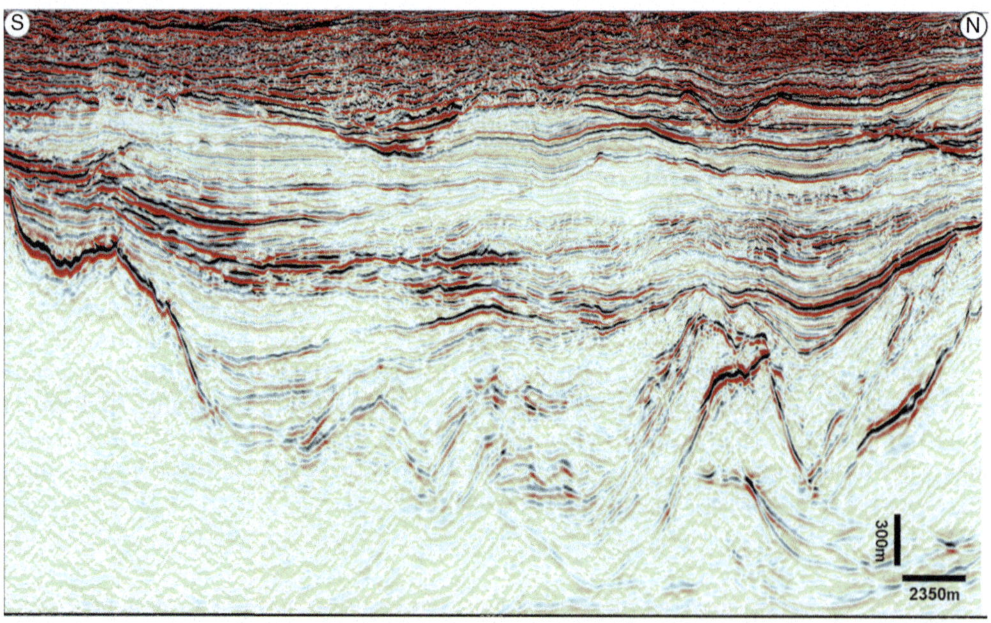

Figure 21.3 S–N seismic section adjacent to the Karaikal high in the offshore Cauvery Basin. Seismic data courtesy Reliance Industries Ltd. Reproduced with permission from the Directorate General of Hydrocarbons (DGH), India.

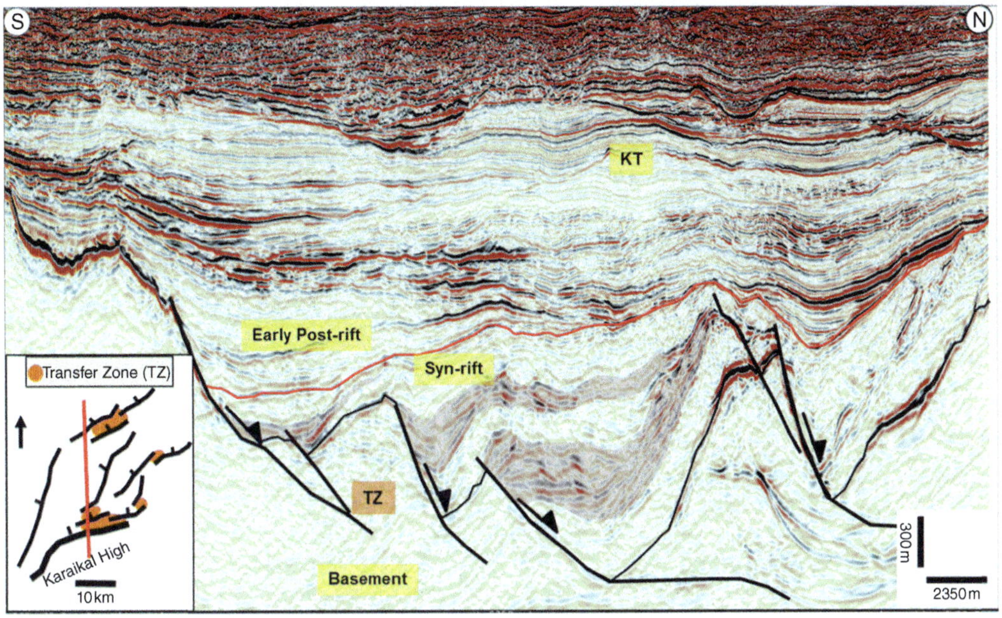

Figure 21.4 Interpreted seismic section of Figure 21.3 showing synthetic relay faults with variation of syn-rift sediment growth as indicated by shaded portions. Inset: fault map at basement level with the line location. Seismic data courtesy Reliance Industries Ltd. Reproduced with permission from the Directorate General of Hydrocarbons (DGH), India.

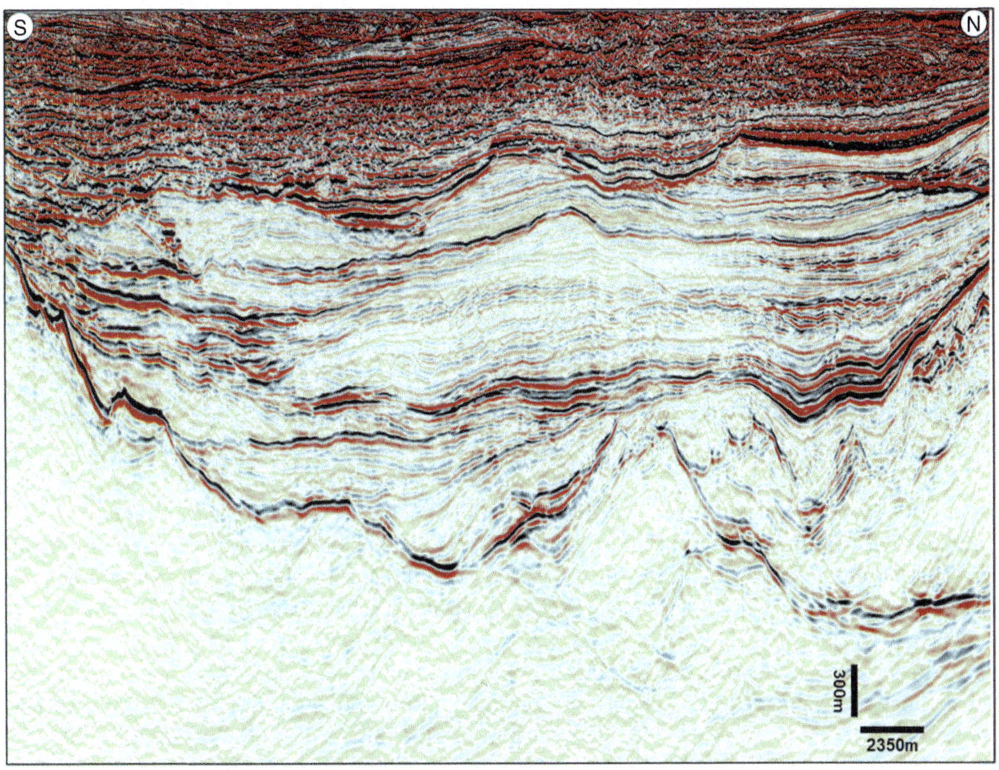

Figure 21.5 S–N seismic section adjacent to the Karaikal high, but further west, in the offshore Cauvery Basin. Seismic data courtesy Reliance Industries Ltd. Reproduced with permission from the Directorate General of Hydrocarbons (DGH), India.

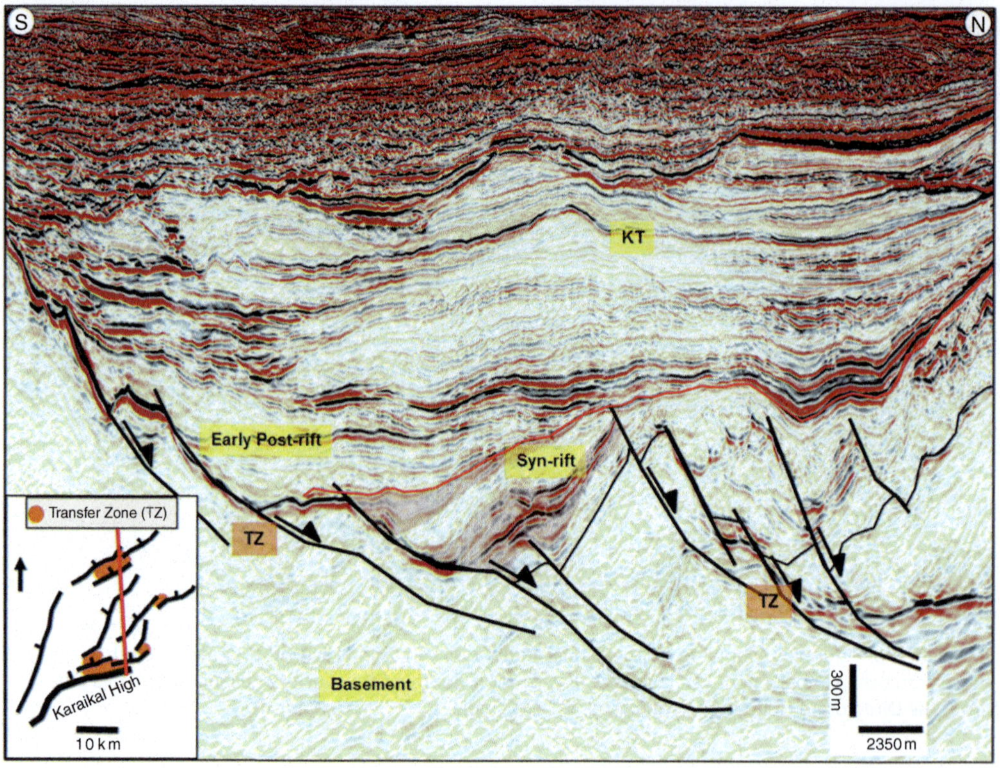

Figure 21.6 Interpreted section of Figure 21.5 depicting a set of synthetic relay faults with lower dip amount. Shaded portions within the syn-rift indicate the sedimentary growth packages. Inset: fault map at basement level with the line location. Seismic data courtesy Reliance Industries Ltd. Reproduced with permission from the Directorate General of Hydrocarbons (DGH), India.

References

Biswas, S.K., Bhasin, A.L. and Ram, J. 1993. Classification of Indian sedimentary basins in the framework of plate tectonics. In: Biswas, S.K., Pomdery, J., Dave, A. *et al.*, (eds) *Proceedings of the 2nd Seminar on Petroliferous Basins of India*, vol. 1, Indian Petroleum Publishers, Dehradun, pp. 1–46.

Chaudhuri, A., Rao, M.V., Dobriyal, J.P. *et al.* 2010. Prospectivity of Cauvery Basin in Deep Syn-rift Sequences, SE India. *Search and Discovery*, Article #10232 (2010).

Dasgupta, S. and Mukherjee, S. 2017. Brittle shear tectonics in a narrow continental rift: asymmetric non-volcanic Barmer basin (Rajasthan, India). *The Journal of Geology*, **125**, 561–591.

Faulds, J.E. and Varga, R.J. 1998. The role of accommodation zones and transfer zones in the regional segmentation of extended terranes. Special Paper 323, The Geological Society of America.

Gaina, C., Müller, R.D., Brown, B. *et al.* 2007. Breakup and early seafloor spreading between India and Antarctica, *Geophysical Journal International*, **170**, 151–169.

Gawthorpe, R.L. and Hurst, J.M. 1993. Transfer zones in extensional basins: their structural style and influence on drainage development and stratigraphy. *Journal of the Geological Society*, **150**, 1137–1152.

Gibbs, A.D. 1984. Structural evolution of extensional basin margins. *Journal of the Geological Society*, **141**, 609–620.

Lal, N.K., Siwal, A. and Kaul, A.K. 2009. Evolution of East Coast of India – A plate tectonic reconstruction. *Journal of the Geological Society of India*, **73**, 249–260.

Misra, A.A., Bhattacharya, G., Mukherjee, S. *et al.* 2014. Near N-S paleo-extension in the western Deccan region in India: Does it link strike-slip tectonics with India-Seychelles rifting? *International Journal of Earth Sciences*, **103**, 1645–1680.

Misra, A.A. and Mukherjee, S. 2015. Tectonic Inheritance in Continental Rifts and Passive Margins. Springerbriefs in Earth Sciences. ISBN 978-3-319-20576-2.

Misra, A.A., Sinha, N. and Mukherjee, S. 2015. Repeat ridge jumps and microcontinent separation: insights from NE Arabian Sea. *Marine and Petroleum Geology*, **59**, 406–428.

Morley, C.K., Nelson, R.A., Patton, T.L. and Munn, S.G. 1990. Transfer zones in the East African rift systems and their relevance to hydrocarbon exploration in rifts. *AAPG Bulletin*, **74**, 1234–1253.

Morley, C.K. 1999. Aspects of transfer zone geometry and evolution in East African rifts. In: Morley C.K. (ed) *Geoscience of Rift Systems – Evolution of East Africa*. AAPG Studies in Geology No. 44, pp. 161–171.

Morley, C.K. 1999a. How successful are analogue models in addressing the influence of pre-existing fabrics on rift structure? *Journal of Structural Geology*, **21**, 1267–1274.

Morley, C.K., Haranya, C., Phoosongsee, W., Pongwappe, S., Kornsawan, A., Wonganan, N., 2004. Activation of rift oblique and rift parallel pre-existing fabrics during extension and their effect on deformation style: examples from the rifts of Thailand. *Journal of Structural Geology*, **26**, 1803–1829.

Nemčok, M., Sinha, S.T., Stuart, C.J. *et al.* 2013. East Indian margin evolution and crustal architecture: integration of deep reflection seismic interpretation and gravity modelling. In: Mohriak, W.U., Danforth, A., Post, P.J. *et al.* (eds) *Conjugate Divergent Margins*, Special Publications vol. 369, The Geological Society, London, pp. 477–496.

Rosendahl, B.R. 1987. Architecture of continental rifts with special reference to East Africa. *Annual Review of Earth and Planetory Science Letters*, **15**, 445–504.

Sastri, V.V., Venkatachala, B.S. and Narayanan, V. 1981. The evolution of the east coast of India. *Palaeogeography Palaeoclimatology Palaeoecology*, **36**, 23–54.

Sinha, S.T., Nemčok, M., Choudhuri, M. *et al.* 2010. The crustal architecture and continental break up of East India Passive margin: an integrated study of deep reflection seismic interpretation and gravity modeling. *Search and Discovery*, Article #40611.

Veevers, J.J. and Tewari, R.C. 1995. *Gondwana Master Basin of Peninsular India Between Tethys and the Interior of the Gondwanaland Province of Pangea.* Geological Society of America Memoirs vol. 187.

Veevers, J.J. 2009. Palinspastic (pre-rift and -drift) fit of India and conjugate Antarctica and geological connections across the suture. *Gondwana Research*, **16**, 90–108.

Veevers, J.J. 2012. Reconstructions before rifting and drifting reveal the geological connections between Antarctica and its conjugates in Gondwanaland. *Earth Science Reviews*, **111**(3), 249–318.

22

Synrift Grabens and Inversion in the Cauvery Offshore Basin on the East Coast of India

Abhimanyu Maitra

Reliance Industries Ltd, Mumbai, India
abhimanyumaitra@gmail.com

Faults can be reactivated often and when the later shear sense is opposite to the earlier, for example a normal fault becomes a reverse/thrust fault, it is termed as inversion (Holdsworth *et al.*, 1997; Davis *et al.*, 2012). Inverted rift basins can be compressed or transpressed to generate uplift and extrusion of the sedimentary fill (Cooper *et al.*, 1989). Figures 22.1 and 22.2 show N–S trending seismic strike sections from the offshore Cauvery Basin on the east coast of India. The Cauvery Basin evolved as India separated from Antarctica (Katz, 1978; Gaina *et al.*, 2007; Veevers, 2009). The Cauvery Basin, at the basement level, exhibits a highly segmented horst and graben/half-graben structure characteristic of synrift extension (Chari *et al.*, 1995; also see Mazumdar *et al.*, 2018).

In the seismic line (Figure 22.1), two major basin bounding faults B1 and B2 are observed. Several sub-basin bounding faults (e.g. SBF1, SBF2 and SBF3) are also observed, which has created the internal architecture of the basin. SBF faults are synthetic to the B2 fault and give rise to smaller sub-basinal grabens. Major faults are recognized by termination of basement reflection and offset in stratigraphy markers. Basement is identified on the basis of chaotic/transparent reflection and in some places by high impedance contrast with overlying sediments. Fault plane reflections are prominent in the case of SBF1 and SBF2 faults. Significant displacement of the basement section (b1–b1*) can be seen along SBF1 whereas SBF2 shows a moderate displacement of the basement (b2–b2*). Growth of the synrift section (s1–s1*) along the low-angle fault. BF2 is well observed. The gentle BF2 is thought to have formed by the rotation of initially high-angle normal fault. Rotation of fault planes has been well described in structural geology (Mukherjee 2014). SBF2, BF1 and BF2 become gentler in the lower part and connected to a detachment surface, as shown in the interpreted section Figure 22.1b.

Synrift sedimentary packs are identified by the growth in the reflections along the fault planes in the subgrabens. Synrift related sediment thickness is greater towards subgraben 4 and less towards subgraben 1. A significant growth is seen along the gentle fault in subgraben 3 (Figure 22.2). A substantial amount of uplift/inversion of synrift sediments is observed in subgraben 4 (Figure 22.3). The upper part of the folded synrift sediment is absent due to erosional truncation of the sediment fill above the erosional base level. The hinge (H) of the fold is observed in the middle to lower part of the synrift section with folding related extensional faults. The seismic section is an example of compressional activity in a rifted extensional tectonic set up. Similarly, finding extensional tectonics from compressional regimes (Mukherjee and Koyi, 2010a,b; Mukherjee *et al.*, 2012; Mukherjee, 2013; Mukherjee *et al.*, 2013, 2015) have also intrigued geoscientists!

Atlas of Structural Geological Interpretation from Seismic Images, First Edition. Edited by Achyuta Ayan Misra and Soumyajit Mukherjee.

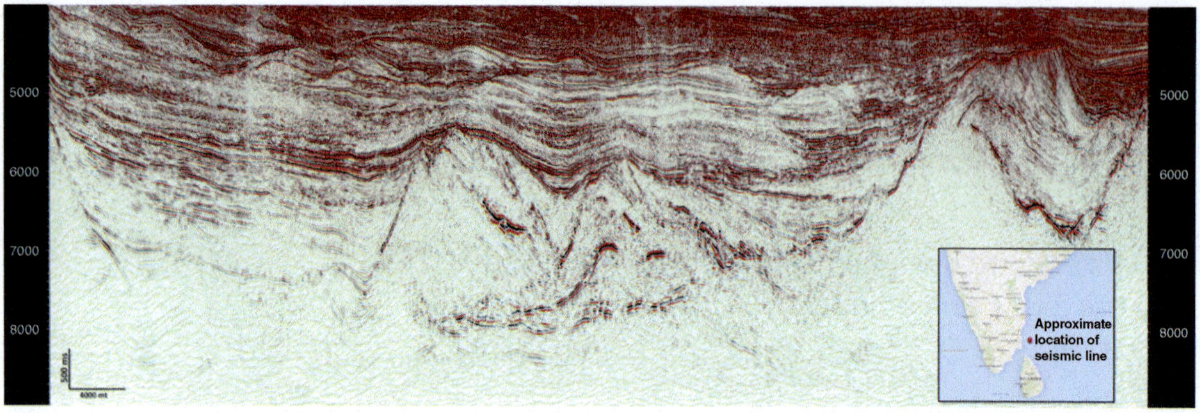

Figure 22.1a N–S seismic section from offshore Cauvery Basin, India. Seismic data courtesy Reliance Industries Ltd. Reproduced with permission from the Directorate General of Hydrocarbons (DGH), India.

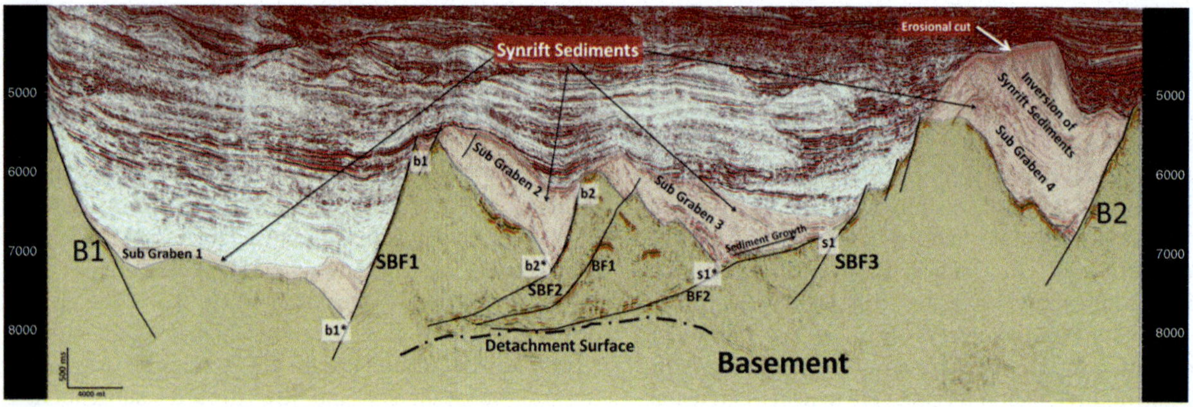

Figure 22.1b Interpretation of seismic section in Figure 22.1a. Seismic data courtesy Reliance Industries Ltd. Reproduced with permission from the Directorate General of Hydrocarbons (DGH), India.

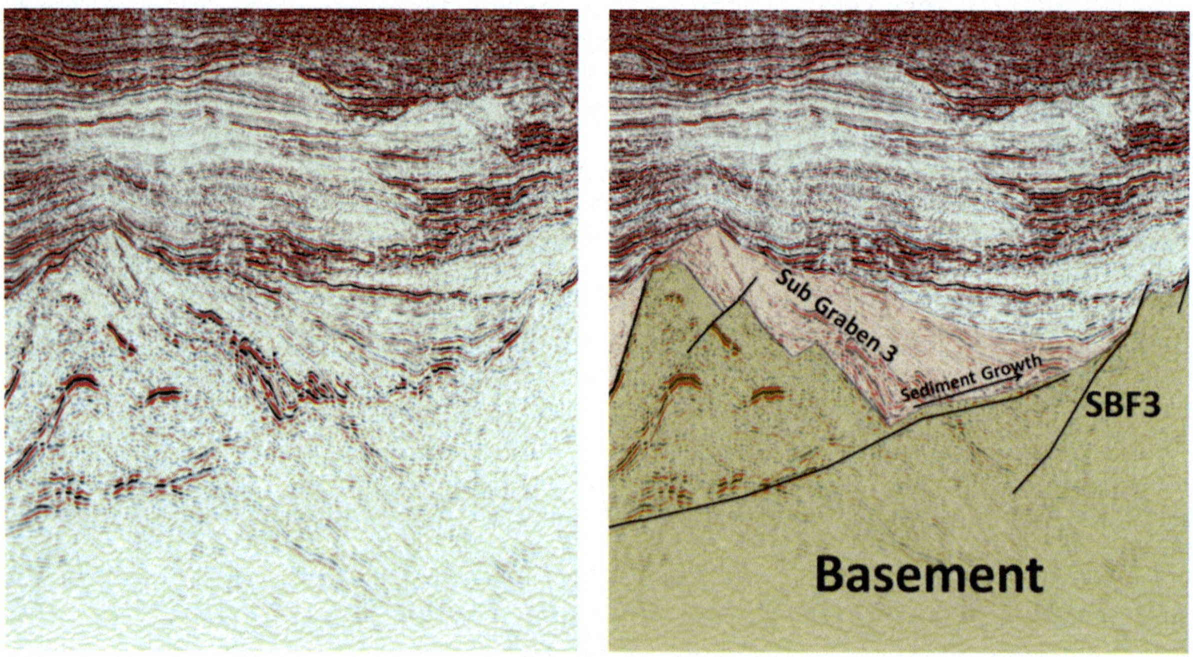

Figure 22.2 Left: enlarged section of subgraben 3; right: interpretation of the enlarged section. Seismic data courtesy Reliance Industries Ltd.Reproduced with permission from the Directorate General of Hydrocarbons (DGH), India.

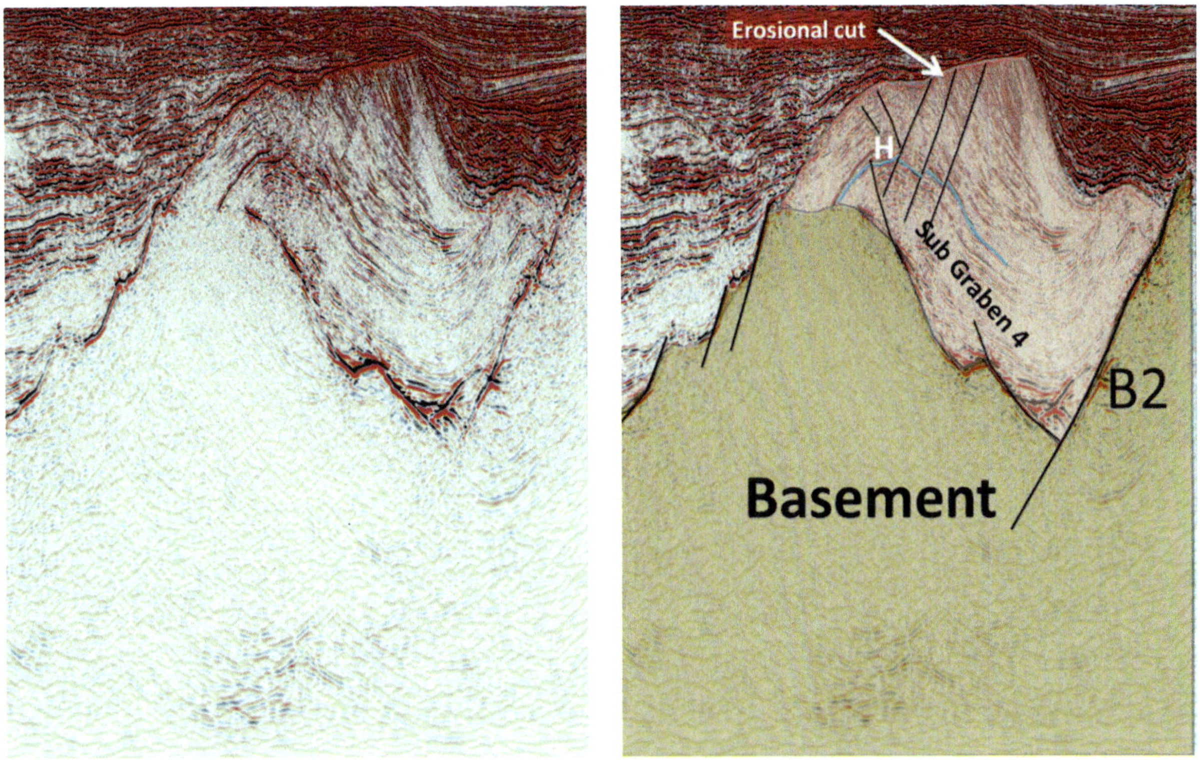

Figure 22.3 Left: enlarged section of subgraben 4; right: interpretation of the enlarged section. Seismic data courtesy Reliance Industries Ltd. Reproduced with permission from the Directorate General of Hydrocarbons (DGH), India.

References

Chari, M.N., Sahu, J.N., Banerjee, B. *et al.* 1995. Evolution of the Cauvery basin, India from subsidence modelling. *Marine and Petroleum Geology*, **12**, 667–675.

Cooper, M.A., Williams, G.D., De Graciansky, P.C. *et al.* 1989. Inversion tectonics – a discussion. In: Cooper, M.A. and Williams, G.D. (eds) *Inversion Tectonics*, Special Publications vol. 44,The Geological Society, London, pp.335–347.

Davis, G.H., Reynolds, S.J. and Kluth, C. 2012. *Structural Geology of Rocks and Regions*, 3rd edn. John Wiley & Sons Ltd, Hoboken, NJ.

Gaina, C., Müller, R.D., Brown, B. *et al.* 2007. Breakup and early seafloor spreading between India and Antarctica. *Geophysical Journal International*, **170**, 151–169.

Holdsworth, R.E., Butler, C.A. and Roberts, A.M. 1997. The recognition of reactivation during continental deformation. *Journal of the Geological Society*, **154**, 73–78.

Katz, M.B. 1978. Sri Lanka in Gondwanaland and the evolution of the Indian Ocean. *Geological Magazine*, **115**, 237–244.

Mukherjee, S. 2014. Review of flanking structures in meso- and micro-scales. *Geological Magazine*, **151**, 957–974.

Mukherjee, S., Koyi, H.A., 2010a. Higher Himalayan Shear Zone, Zanskar Section- Microstructural Studies & Extrusion Mechanism by a Combination of Simple Shear & Channel Flow. *International Journal of Earth Sciences*, **99**, 1083–1110.

Mukherjee, S., Koyi, H.A. 2010a. Higher Himalayan Shear Zone, Sutlej Section- Structural Geology & Extrusion Mechanism by Various Combinations of Simple Shear, Pure Shear & Channel Flow in Shifting Modes. *International Journal of Earth Sciences*, **99**, 1267–1303.

Mukherjee, S., Koyi, H.A. and Talbot, C. J. 2012. Implications of channel flow analogue models in extrusion of the Higher Himalayan Shear Zone with special reference to the out-of-sequence thrusting. *International Journal of Earth Sciences*, **101**, 253–272.

Mukherjee, S. 2013. Higher Himalaya in the Bhagirathi section (NW Himalaya, India): its structures, backthrusts and extrusion mechanism by both channel flow and critical taper mechanisms. *International Journal of Earth Sciences*, **102**, 1851–1870.

Mazumdar, S., Tep, B., Pangtey, K.K.S. *et al.* 2018. Basement tectonics and shear zones in Cauvery basin: Implications in hydrocarbon exploration. In: Mukherjee, S. (Ed) *Tectonics and Structural Geology: Indian Context.* Springer International Publishing AG, Cham. Springer.

Mukherjee, S., Mukherjee, B. and Thiede, R. 2013. Geosciences of the Himalaya-Karakoram-Tibet Orogen. *International Journal of Earth Sciences*, **102**, 1757–1758.

Mukherjee, S., Carosi, R., van der Beek, P.A. *et al.* 2015. Tectonics of the Himalaya: an introduction. In: Mukherjee, S., Carosi, R., van der Beek, P., Mukherjee, B.K., Robinson, D (Eds) Geological Society, *London, Special Publications*, **412**, 1–3.

Veevers, J.J. 2009. Palinspastic (pre-rift and-drift) fit of India and conjugate Antarctica and geological connections across the suture. *Gondwana Research*, **16**, 90–108.

23

Pull-Apart Basin in the Offshore Cauvery–Palar Basin, India

Swagato Dasgupta

Reliance Industries Ltd, Mumbai, India
Swagato.Dasgupta@ril.com, swagato.dg@gmail.com

Pull-apart basins are sigmoid to rhombic shaped extensional basins formed in areas where a strike-slip fault bends or two adjacent strike slip faults step over. Such releasing and restraining bends or step-over of adjacent strike-slip faults ideally generates extensional or compressional stress, thereby forming transtensional or transpressional basins, respectively. Pull-apart basins develop in releasing fault bends on oblique segments (Mann *et al.*, 1983; Frisch *et al.*, 2000). These basins are short-lived with fast subsidence rate comprising of syn-sedimentary deformation structures (Waldron, 2004; Strauss *et al.*, 2001). In general they have a low length-to-width ratio (Mann *et al.*, 1983), representing a typical rhombic shape in a horizontal section. Pull-apart basins form along strike-slip margins in different settings: (i) within a continental crust; (ii) at transform margins between two adjacent continental or continental-oceanic lithospheric plates; (iii) in 'escaping' continental blocks in the vicinity of the convergent boundary (Mann *et al.*, 1983, Scharf *et al.*, 2013).

A typical pull-apart basin is identified adjacent to the Coromandal transform margin in the offshore deepwater Cauvery–Palar Basin, in the eastern side of Tamil Nadu state, India (Figure 23.1). The pull-apart basin is formed due to dextral separation of India and Antarctica plate along the Coromandal strike-slip margin (Nemčok *et al.*, 2013) during the Early Cretaceous Period. It emplaced above the thinned continental crust along this strike-slip margin (Figure 23.2). The continental and oceanic crustal boundary is towards the east of the pull-apart basin. The maximum width and length of the basin are 12 and 35 km, respectively, depicting elongate lozenge-shaped geometry (Corti and Dooley, 2015; Ustaszewski *et al.*, 2004). Structural inversion with fault reactivation is observed in the syn-rift, associated with a distinct down-lapping stratal package of the early post-rift section above the syn-rift unit (Figure 23.3).

Atlas of Structural Geological Interpretation from Seismic Images, First Edition. Edited by Achyuta Ayan Misra and Soumyajit Mukherjee.

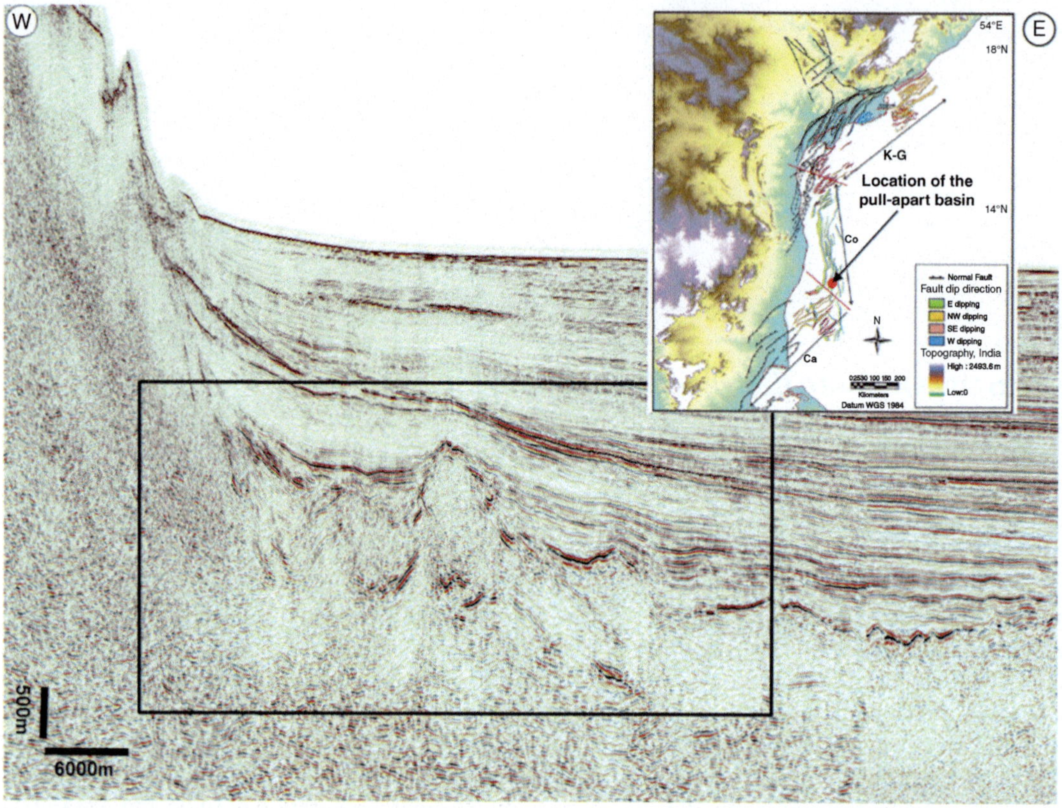

Figure 23.1 W–E seismic section across a pull-apart structure in the Cauvery-Palar offshore deepwater basin. Inset: location of the pull-apart basin on the tectonic fault map of east coast of India (adapted from Nemčok *et al*., 2013). The black rectangle indicates the position of the pull-apart basin as shown in Figure 23.3. Seismic data courtesy Reliance Industries Ltd. Reproduced with permission from the Directorate General of Hydrocarbons (DGH), India.

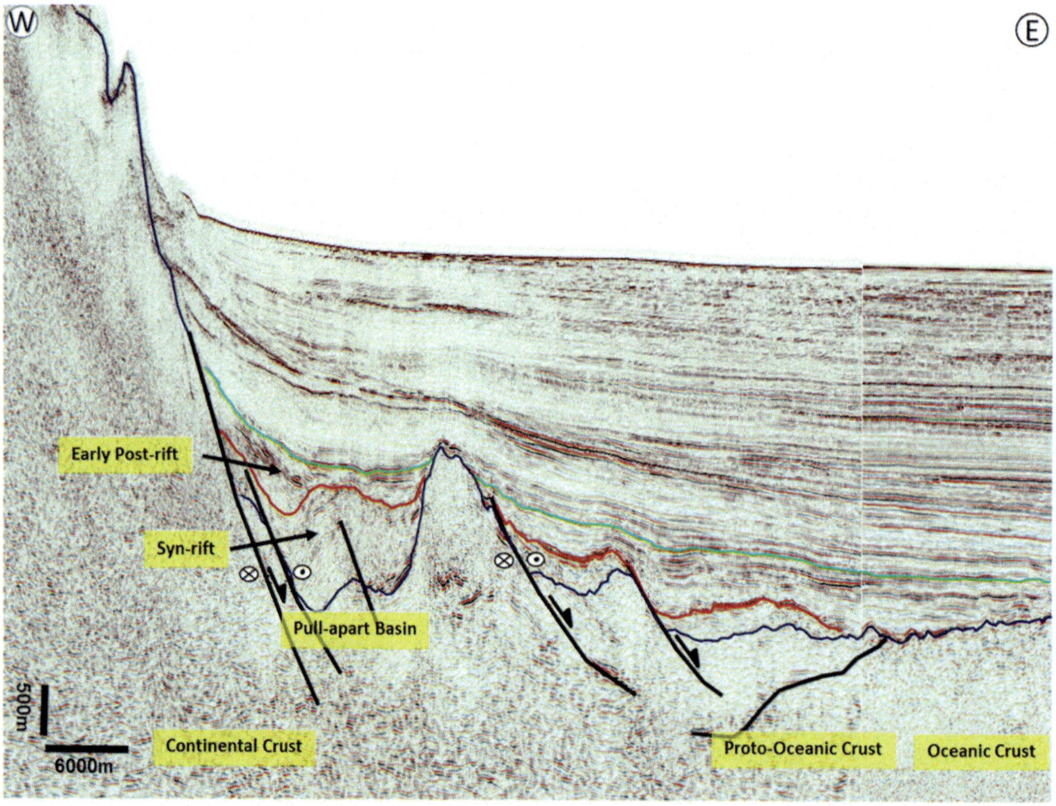

Figure 23.2 Interpreted seismic section of Figure 23.1 indicating the pull-apart basin adjacent to the dextral transform margin and the continental and oceanic crustal boundary. Seismic data courtesy Reliance Industries Ltd. Reproduced with permission from the Directorate General of Hydrocarbons (DGH), India.

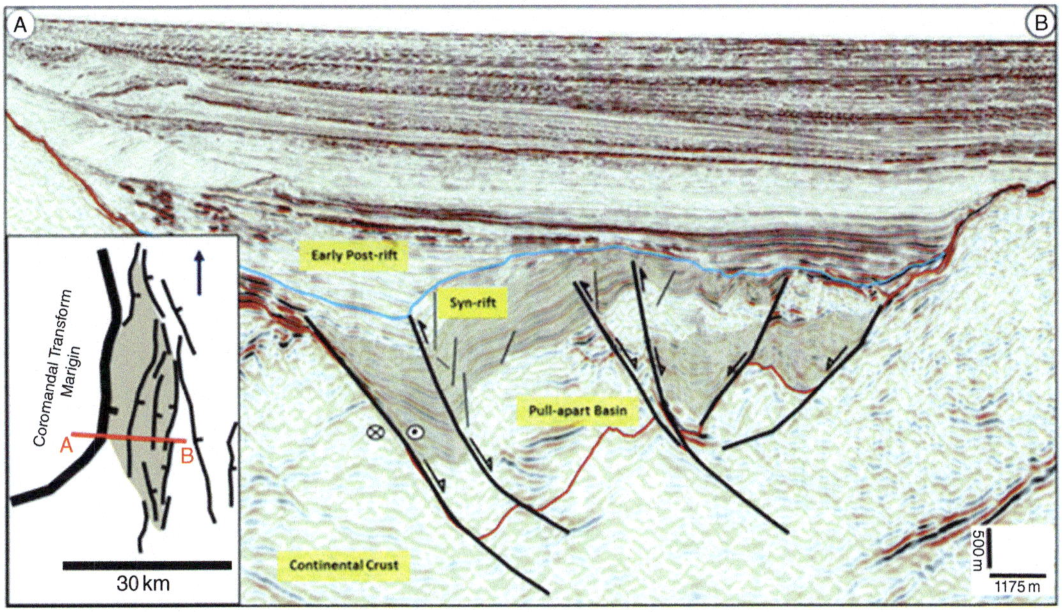

Figure 23.3 Enlarged section of the pull-apart basin. Syn-sedimentray growth package is identified within the syn-rift section. Note the inverted faults. Seismic data courtesy Reliance Industries Ltd. Reproduced with permission from the Directorate General of Hydrocarbons (DGH), India. Inset: Fault map of the pull-apart structure, in grey shade. Arrow indicates north.

References

Corti G. and Dooley T.P. 2015. Lithospheric-scale centrifuge models of pull-apart basins. *Tectonophysics*, **664**, 154–163.

Frisch, W., Dunkl, I. and Kuhlemann, J. 2000. Post-collisional orogeny parallel large-scale extension in the Eastern Alps. *Tectonophysics*, **327**, 239–265.

Mann, P., Hempton, M., Bradley, D. and Burke, K. 1983. Development of pull-apart basins. *Journal of Geology*, **91**, 529–554.

Nemčok, M., Sinha, S.T., Stuart, C.J. *et al.* 2013. East Indian margin evolution and crustal architecture: integration of deep reflection seismic interpretation and gravity modelling. In: Mohriak, W.U., Danforth, A., Post, P.J. *et al.* (eds) *Conjugate Divergent Margins*, Special Publications vol. 369, The Geological Society, London, pp. 477–496.

Scharf, A., Handy, M.R., Favaro, S. *et al.* 2013. Modes of orogen-parallel stretching and extensional exhumation in response to microplate indentation and roll-back subduction(Tauern Window, Eastern Alps). *International Journal of Earth Sciences*, **102** (6), 1627–1654. doi: 10.1007/s00531-013-0894-4.

Strauss, P., Wagreich, M., Decker, K. and Sachsenhofer, R.F. 2001. Tectonicsnand sedimentation in the Fohnsdorf-Seckau Basin (Miocene, Austria): from a pull-apart basin to a half-graben, *International Journal of Earth Sciences*, **90**, 549–559. doi: 10.1007/s005310000180.

Ustaszewski K., Schumacher M.E., Schmid S.M. and Nieuwland D. 2004. Fault reactivation in brittle–viscous wrench systems–dynamically scaled analogue models and application to the Rhine–Bresse transfer zone. *Quaternary Science Reviews*, **24**, 365–382. doi:10.1016/j.quascirev.2004.03.015.

Waldron, J.W.F. 2004. Anatomy and evolution of a pull-apart basin, Stellarton, Nova Scotia. *GSA Bulletin*, **116**, 109–127. doi: 10.1130/B25312.1.

24

Sedimentary Deformation Features Produced by Differential Compaction and Buoyancy in the Offshore Palar Basin, East Coast of India

Swagato Dasgupta

Reliance Industries Ltd, Mumbai, India
Swagato.Dasgupta@ril.com, swagato.dg@gmail.com

Fluid migration pathways like mud diapirs/volcanoes and gas chimneys can be closely related to particular tectonic settings. The sedimentation rate and pressure–temperature regimes are important parameters that guide such migration. Mud volcanoes, in general, occur in the convergent plate boundaries, suggesting that compressional stress is the primary driving mechanism for such extrusive features (Dimitrov, 2002; Yusifov and Rabinowitz, 2004; Basu *et al.*, 2012). However, fast sedimentation associated with typical high formation pressures in passive margins also leads to the formation of such mud volcanoes (Yin *et al.*, 2003; Milkov, 2000). Vide Mukherjee *et al.* (2010), Mukherjee (2011) and Mukherjee and Mulchrone (2012) for few diapiric mechanisms.

The seismic line is from the offshore Palar Basin in the southern part of the east coast of India (Figures 24.1 and 24.2). The Palar Basin is a pericratonic basin, juxtaposed between the Cauvery Basin in the south and the Krishna–Godavari Basin towards the north. The basin started as an intracratonic rift basin during the Early Jurassic. At a later stage, it developed into a pericratonic rift basin as a result of the continental breakup of Gondwanaland, which rifted and drifted the Indian landmass from the Antarctica Plate during the Early Cretaceous. Like other east coast basins, the Palar Basin trends NE and straddles the coast. All the east coast basins have a similar evolution and geological history (Veevers and Tewari, 1995; Gaina *et al.*, 2007; Veevers, 2009; Sinha *et al.*, 2010; Nemčok *et al.*, 2013).

This W–E trending line is near orthogonal to the Mesozoic rift trend (Figure 24.2). The seismic line depicts three synthetic rift-related faults that have affected the basement topography. Prominent growth is seen along fault A. Fault B is adjacent to the pinnacle-shaped basement ridge. The vertical slip between faults B and C is guided by a set of synthetic relay ramps (Figures 24.2 and 24.3).

A number of sediment deformation features are observed in this seismic section (Figure 24.2). These are: (i) sagging of early post-rift strata overlying the basement; (ii) a set of normal faults within Cretaceous sediments over the pinnacle-shaped basement high formed due to differential compaction of sediments (Figures 24.2 and 24.3); (iii) a vertical narrow zone of seismic wipe-out with velocity pull-up features at its edges is distinctly observed in Tertiary sequence, leading to formation of a mud volcano at the sea floor.

The late rift stage faulting and subsequent thermal subsidence during early post-rift sagged the post-rift strata, thereby generating a synclinal pattern towards the western side of the pinnacle basement high. Syn-sedimentary normal faulting is observed on either side of this synclinal feature. These are formed due to differential compaction of the Cretaceous post-rift sediments. Such faults are also visible over the pinnacle basement high (Figure 24.3). These diverging-up normal faults are formed due to variation of sediment thickness, and thus compaction, on either side on the basement high (Figure 24.3). These faults have buckled the uppermost strata.

The erosional Cretaceous–Tertiary (KT) boundary is formed as a result of easterly tilting of the basin due to Reunion plume-related basinal uplift towards the west-

Atlas of Structural Geological Interpretation from Seismic Images, First Edition. Edited by Achyuta Ayan Misra and Soumyajit Mukherjee.

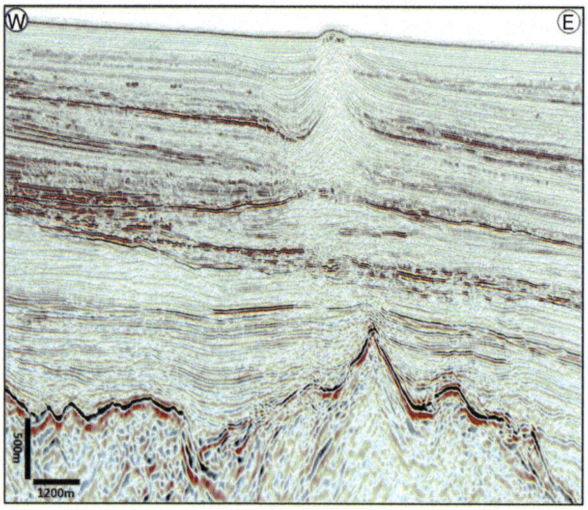

Figure 24.1 Seismic section across the offshore Palar Basin. Seismic data courtesy Reliance Industries Ltd. Reproduced with permission from the Directorate General of Hydrocarbons (DGH), India.

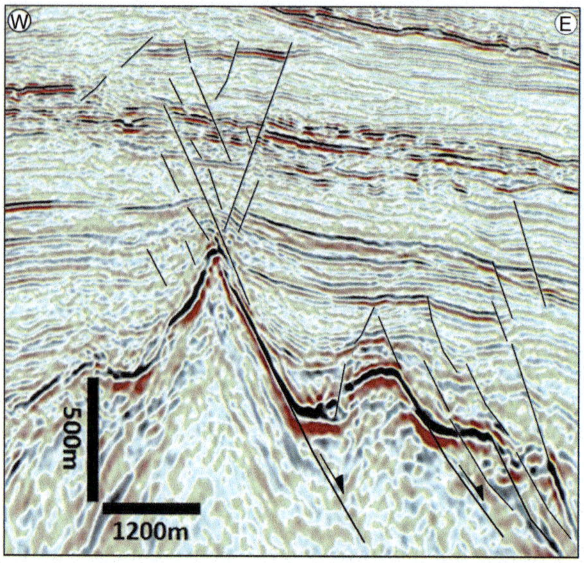

Figure 24.3 Enlargeded section of the pinnacle basement high associated with rift related faults as well as the overlying differential compaction related faulting. Seismic data courtesy Reliance Industries Ltd. Reproduced with permission from the Directorate General of Hydrocarbons (DGH), India.

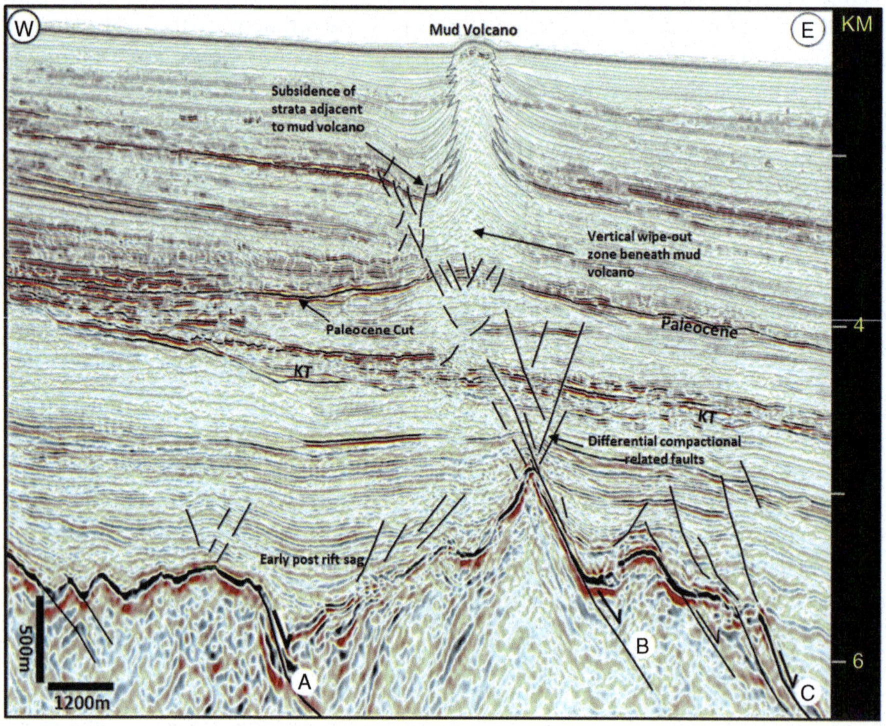

Figure 24.2 Interpreted seismic section across the Palar Basin depicting rift related faults, differential compaction related sedimentary deformations and buoyancy driven structures interpreted as mud volcano. Seismic data courtesy Reliance Industries Ltd. Reproduced with permission from the Directorate General of Hydrocarbons (DGH), India.

ern peninsular India. This is followed by rapid sedimentation during the Early Tertiary. Prominent Late Paleocene channel cuts are observed in the seismic section. A vertical narrow zone of seismic wipe-out associated with velocity pull-up features at its edges is distinctly observed in the Tertiary sequence overlying the Paleocene shale (Figure 24.2) having a typical Christmas-tree structure. Sedimentary deformation features associated with normal faulting is seen at the base of this wipe out zone (Figure 24.2).

It protrudes at the sea floor surface with a bulging conical shape comprising a central crater with a height of 50–60 m from the sea floor and diameter of about 1 km. Expression of this wipe-out zone is very well visible in the

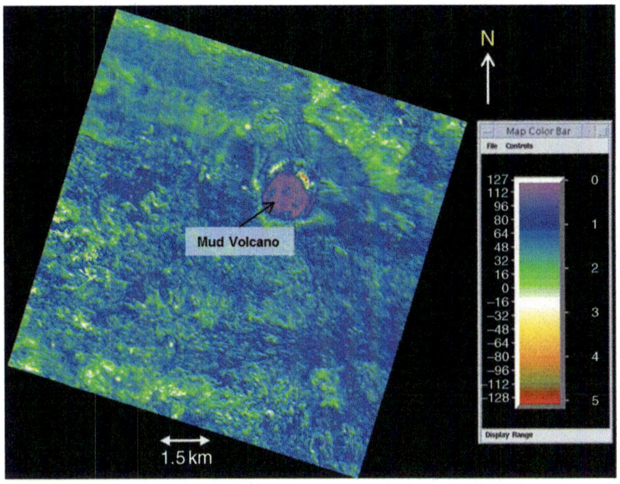

Figure 24.4a Wipe-out zone related to the mud volcano feature is indicated in an RMS amplitude map at shallow Tertiary level. Seismic data courtesy Reliance Industries Ltd. Reproduced with permission from the Directorate General of Hydrocarbons (DGH), India.

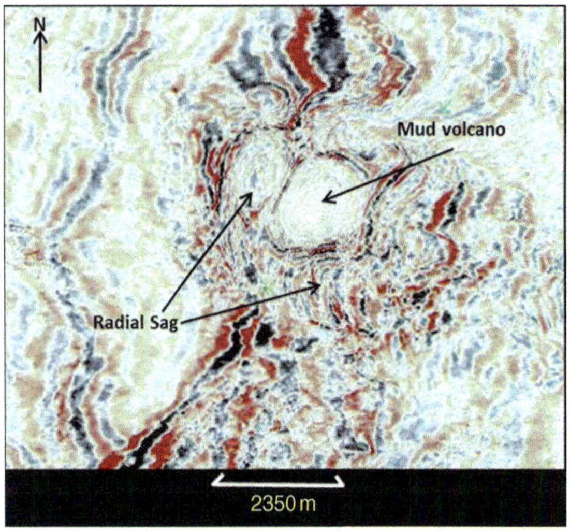

Figure 24.4b Seismic wipe-out zone due to mud-volcano associated with radial sagging is observed in a time slice at shallow Tertiary level. Seismic data courtesy Reliance Industries Ltd. Reproduced with permission from the Directorate General of Hydrocarbons (DGH), India.

seismic section and time slice map (Figure 24.2 and 24.4b). An RMS amplitude map of a shallow level also bears the signature of this feature (Figure 24.4a). All these observations point out that this vertical seismic wipe-out zone is due to mud volcanism. The undercompacted, overpressurized shale of the Late Paleocene is the key guiding factor for the formation of this mud volcano. Fluid saturated shale and siltstone bodies have low shear strength, due to an increase in pore pressure, as a result of which this buoyancy-driven structure is formed in the passive margin. Subsidence of Late Tertiary sedimentary layers adjacent to this mud volcano is observed at a 500–1200 m depth below the sea bottom (Figure 24.2). This radial sagging of strata around the mud volcano (Figure 24.4a and 24.4b) and has probably formed due to local fluid accumulation.

References

Basu, P., Verma, R., Paul, R., Viswanath, K. 2012. Mud volcanoes in deep water of Andaman Forearc Basin. *Geohorizons*, 26–31.

Dimitrov, L.I. 2002. Mud volcanoes—the most important pathway for degassing deeply buried sediments. *Earth Science Review*, **59**, 49–76.

Gaina, C., Müller, R.D., Brown, B. *et al.* 2007. Breakup and early seafloor spreading between India and Antarctica. *Geophysical Journal International*, **170**, 151–169.

Milkov, A.V. 2000. Worldwide distribution of submarine mud volcanoes and associated gas hydrates. *Marine Geology*, **167**, pp.29–42.

Mukherjee, S. 2011. Estimating the Viscosity of Rock Bodies- a Comparison Between the Hormuz- and the Namakdan Salt Domes in the Persian Gulf, and the Tso Morari Gneiss Dome in the Himalaya. *The Indian Journal of Geophysical Union*, **15**, 161–170.

Mukherjee, S. and Mulchrone, K. 2012. Estimating the viscosity and Prandtl number of the Tso Morari Gneiss Dome, western Indian Himalaya. *International Journal of Earth Sciences*, **101**, 1929–1947.

Mukherjee, S., Talbot, C. J. and Koyi, H. A. 2010. Viscosity estimates of salt in the Hormuz and Namakdan salt diapirs, Persian Gulf. *Geological Magazine*, **147**, 497–507.

Nemčok, M., Sinha, S.T., Stuart, C.J. *et al.* 2013. East Indian margin evolution and crustal architecture: integration of deep reflection seismic interpretation and gravity modelling. In: Mohriak, W.U., Danforth, A., Post, P.J. *et al.* (eds) *Conjugate Divergent Margins*, Special Publications vol. 369, Geological Society London, pp. 477–496.

Sinha, S.T., Nemčok, M., Choudhuri, M. *et al.* 2010. The crustal architecture and continental break Up of East India Passive Margin: an integrated study of deep reflection seismic interpretation and gravity modeling; *Search and Discovery*, Article 40611.

Veevers, J J. 2009. Palinspastic (pre-rift and -drift) fit of India and conjugate Antarctica and geological connections across the suture. *Gondwana Research*, **16**, 90–108.

Veevers, J.J. and Tewari, R.C. 1995. *Gondwana Master Basin of Peninsular India Between Tethys and the Interior of the Gondwanaland Province of Pangea*. Geological Society of America Memoir vol. 187, Boulder, CO, pp. 72–187.

Yin, P., Berne, S., Vagner, P. *et al*. 2003. Mud volcanoes at the shelf margin of the East China Sea. *Marine Geology*, **194**, 135–149.

Yusifov, M., Rabinowitz, P.D. 2004. Classification of mud volcanoes in the South Caspian Basin, offshore Azerbaijan. *Marine and Petroleum Geology*, **21**, 965–975.

Part II

Folds

Folds are structures in which the initial curvature of a layer/foliation has been increased by deformation. In this section, examples are provided of:

- Compressional folds: antiforms and synforms (Chapters 25–30)
- Sediment interactions with folds (Chapter 27)

25

Compression Folds and a Strike-Slip Fault in Flowing Evaporites, Red Sea

Neil C. Mitchell[1]* *and Marco Ligi*[2]

[1] *University of Manchester, United Kingdom*
[2] *ISMAR-CNR, Bologna, Italy*
* *neil.mitchell@manchester.ac.uk*

In the Red Sea, evaporites deposited from the Mid Miocene up to the end of the Miocene (5.3 Ma) (Hughes and Beydoun, 1992). They vary in thickness up to about 4 km (Egloff *et al.*, 1991). Away from the coasts, the evaporites are covered by only 200–300 m of hemipelagic sediment of density similar to that of the evaporites, so that diapirism is not as strongly promoted by sedimentary loading as in typical continental marginal settings. Consequently, 25-m-resolution multibeam echo-sounder data (Figure 25.1) reveal giant flows of evaporites, where buttressing of the evaporites has been reduced by active rifting in the axial spreading centre running down the centre of the Red Sea, allowing the evaporites to flow into the axial valley (Ligi *et al.*, 2012; Augustin *et al.*, 2014; Mitchell *et al.*, 2010, 2015).

In seismic reflection images (Figure 25.2), the top of the evaporites in the Red Sea is marked by a prominent reflection called 'S' (Ross and Schlee, 1973). A 2D seismic line shown here from the central Red Sea (line TL-25M of Mitchell *et al.* (2010)) is interpreted with the aid of folds and faults visible in multibeam sonar data. This image has been migrated, so that 'smiles' in lower parts of the section represent noise and apparently crossing reflections (such as around shot point (SP) 280 in the upper section) are likely due to 3D structure being inadequately accounted for by the 2D migration method. In the image, a reflective Plio-Pleistocene (PP) sequence overlies a prominent reflector S. The base of the evaporites was not imaged by the data, although away from the spreading centre it is expected to lie at generally >0.5 s two-way seismic time below the seabed, based on results of a seismic refraction experiment here (Tramontini and Davies, 1969).

The top of the evaporites, the S reflection, is abruptly offset by approximately 200 ms at SP 495 and the seabed is marked by an angular ridge. In the multibeam sonar data, this feature corresponds to a boundary within the evaporites, where they have apparently been moving towards the spreading centre at differing speeds. Differing speeds are most likely caused by different parts of the evaporites being variedly impeded by basement relief (Mitchell *et al.*, 2010).

Left of SP 495, the PP layer thins out towards SP 700. The ridge at SP 740 has a morphology in multibeam sonar data of a neovolcanic ridge associated with the spreading centre. As dissolution of carbonate pelagic sediment is limited in the Red Sea, the PP thinning has instead been interpreted as being caused by extension, where the underlying evaporites have run out into the volcanic valley of the spreading centre (Mitchell *et al.*, 2010).

Folds in the PP sediments occur between SP 200 and 300. The thickness of the PP interval appears to almost double from SP 200 towards SP 280, in part because layers within the PP thicken. This, therefore, is a tectonic growth structure (developed progressively over the PP). At the edge of this folded zone at SP 280 a possible reverse fault occurs, suggesting the folding has occurred because of local compression of the mobile evaporites, in practice, perhaps due to a reverse component on a strike-slip fault.

Atlas of Structural Geological Interpretation from Seismic Images, First Edition. Edited by Achyuta Ayan Misra and Soumyajit Mukherjee.

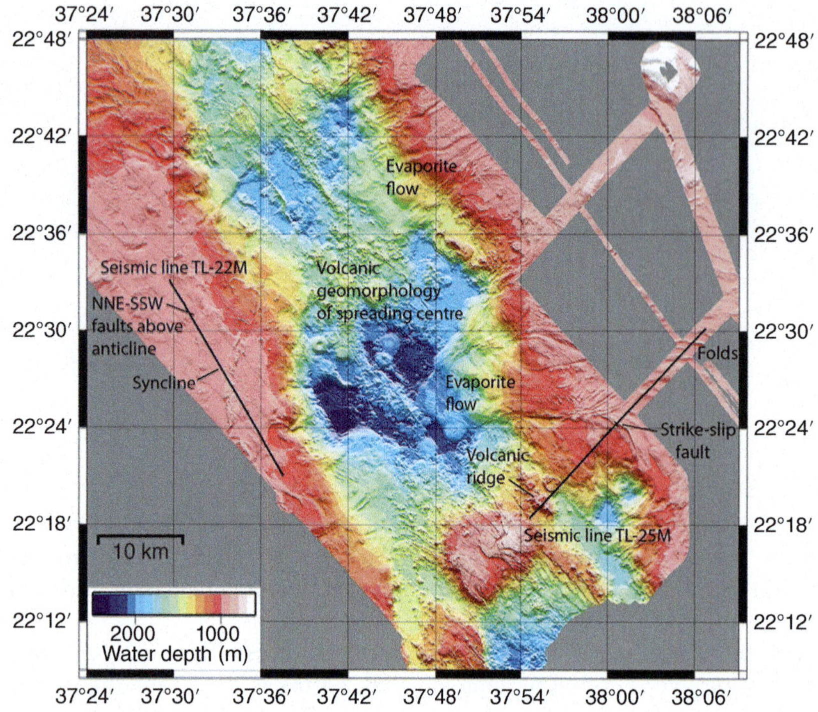

Figure 25.1 Map of the study area showing interpreted multibeam sonar bathymetry with seismic reflection lines marked.

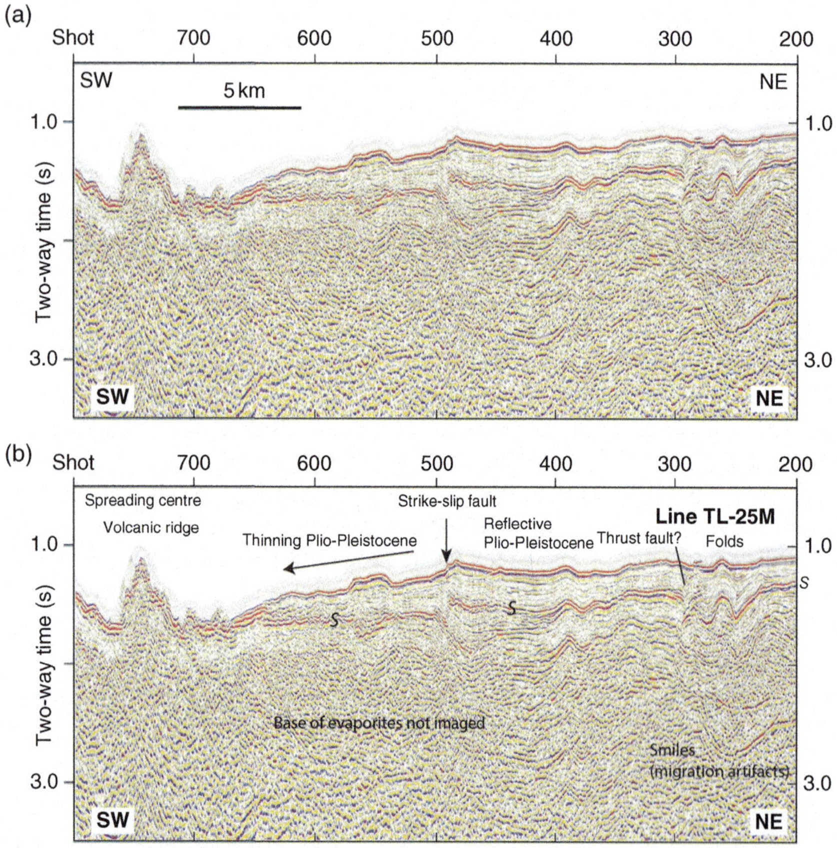

Figure 25.2 (a) Uninterpreted and (b) interpreted reflection seismic section.

References

Augustin, N., Devey, C.W., van der Zwan, F.M. *et al.* 2014. The transition from rifting to spreading in the Red Sea. Earth Planet. *Science Letters*, **395**, 217–230.

Egloff, F., Rihm, R., Makris, J. *et al.* 1991, Contrasting structural styles of the eastern and western margins of the southern Red Sea: the 1988 SONNE experiment. *Tectonophysics*, **198**, 329–353.

Hughes, G.W. and Beydoun, Z.R. 1992. The Red Sea–Gulf of Aden: biostratigraphy, lithostratigraphy and palaeoenvironments. *Journal of Petroleum Geology*, **15**, 135–156.

Ligi, M., Bonatti, E., Bortoluzzi, G. *et al.* 2012. Birth of an ocean in the Red Sea: Initial pangs. *Geochemistry, Geophysics, Geosystems*, **13**, Paper Q08009.

Mitchell, N.C., Ligi, M., Farrante, V. *et al.* 2010. Submarine salt flows in the central Red Sea. *Geological Society of America Bulletin*, **122**, 701–713.

Mitchell, N.C., Ligi, M., Feldens, P. and Hübscher, C. 2015. Deformation of a young salt giant: regional topography of the Red Sea Miocene evaporites. *Basin Research*, **29** (S1), 352–369. doi: 10.1111/bre.12153.

Ross, D.A. and Schlee, J. 1973. Shallow structure and geologic development of the southern Red Sea. *Geological Society of America Bulletin*, **84**, 3827–3848.

Tramontini, C. and Davies, D. 1969. A seismic refraction survey in the Red Sea. *Geophysical Journal of the Royal Astronomical Society*, **17**, 225–241.

26

Didigori Syncline and Anticline: Kura Foreland Fold and Thrust Belt, Georgia

V. Alania[1], A. Chabukiani[1,2], O. Enukidze[1], K. Gogrichiani[2], Z. Mikeladze[2] and G. Tatishvili[2]*

[1] *I. Javakhishvili State University, Institute of Geophysics, Tbilisi, Georgia*
[2] *State Agency of Oil and Gas, Ministry of Energetics, Tbilisi, Georgia*
* *v_alania2002@yahoo.com*

The Didigori syncline and Didigori anticline is part of the Kura foreland fold and thrust belt. This belt is located between Lesser and Greater Caucasus. An interpretation of the structure, is presented based on surface data, a seismic profile, and data from the T-14 well (Figure 26.1). Seismic reflection data reveal the presence of shallow south-vergent thrusts and duplex above a structural wedge (or duplex). Fault-related folding theories (Suppe, 1983; Shaw *et al.*, 2005) were used for structural interpretation of the S–N trending seismic profile (Figure 26.1). Upper thrusts (F_1, F_2 and F_3) and duplex are represented by Upper Miocene and Pliocene strata. The Lower north-vergent duplex sequence consists of Paleogene strata that formed by passive-roof duplexing. Upper Miocene and Pliocene are represented by thick continental and shallow marine sediments. In the Kura foreland fold and thrust belt, the geometry of Upper Miocene–Pliocene growth strata is associated with footwall syncline and piggyback basins (Adamia *et al.*, 2010).

Atlas of Structural Geological Interpretation from Seismic Images, First Edition. Edited by Achyuta Ayan Misra and Soumyajit Mukherjee.

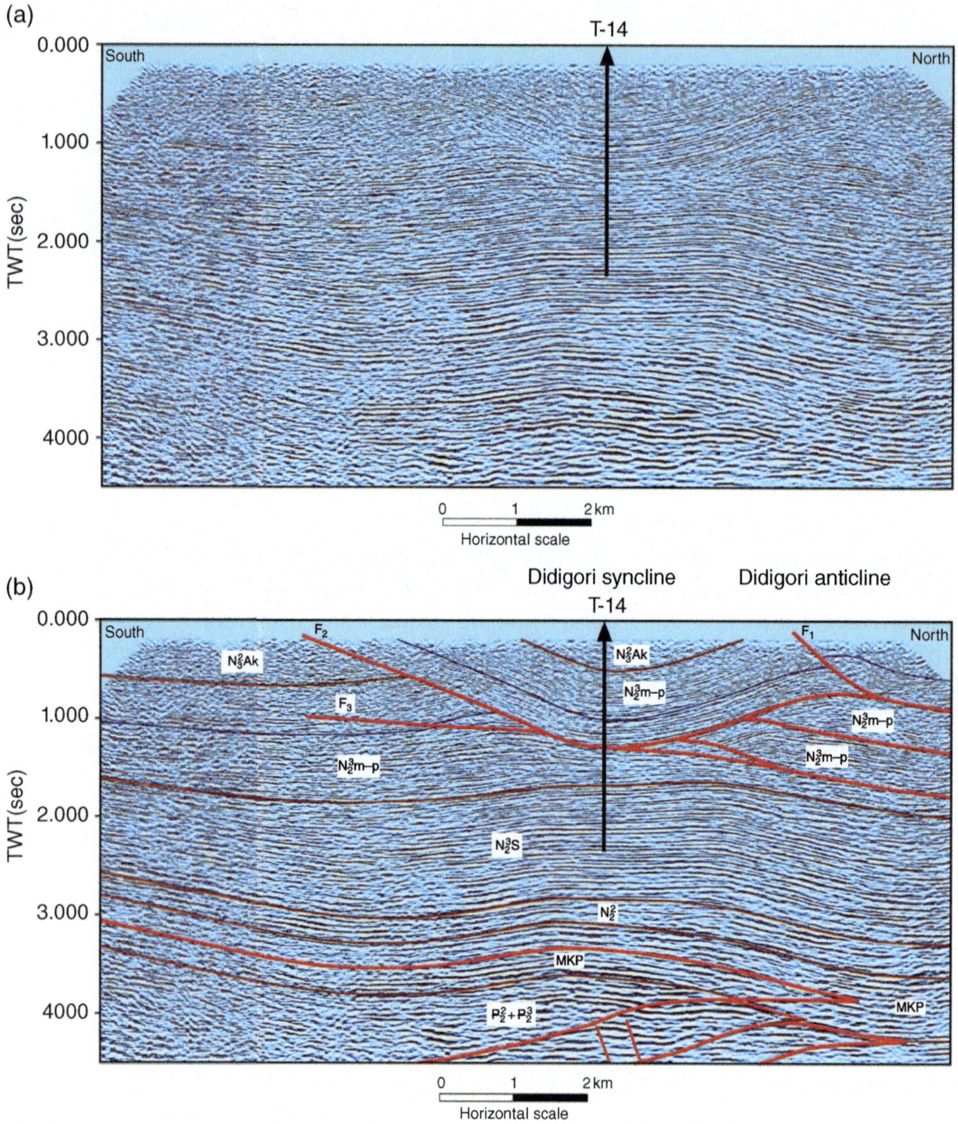

Figure 26.1 (a) Uninterpreted and (b) interpreted 2D time-migrated seismic profile through the Didigori syncline and Didigori anticline: Kura foreland fold and thrust belt (Georgia). Abbreviations: F1, F2 and F3 is south-vergent thrust; P22+ P23 – Middle Eocene–Upper Eocene; MKP – Oligocene–Lower Miocene (Maikopian); N22 – Middle Miocene; N23S – Upper Miocene (Sarmatian); N23 m-p – Upper Miocene (Meotis–Pontian); N23 Ak – Pliocene (Akhagilian).

References

Adamia Sh., Alania, V., Chabukiani, A. *et al.* 2010. Evolution of the Late Cenozoic basins of Georgia (SW Caucasus): a review. In: Sosson, M., Kaymakçı, N., Stephenson, R. and Bergerat, F. (eds) *Sedimentary basin tectonics from the Black Sea and Caucasus to the Arabian Platform*, Special Publications vol. 340, The Geological Society, London, pp. 239–259.

Shaw, J.H., Connors, C.D. and Suppe, J. 2005. *Seismic Interpretation of Contractional Fault-Related Folds*. AAPG Studies in Geology No. 53, American Association of Petroleum Geologists, Tulsa, OK.

Suppe, J. 1983. Geometry and kinematics of fault-bend folding. *American Journal of Science*, **283**, 684–721.

27

Compressional Folds in the Active North Ecuador–South Colombia Margin

Achyuta Ayan Misra

Reliance Industries Ltd, Mumbai, India
Achyuta.Misra@ril.com; achyutaayan@gmail.com

The active offshore North Ecuador–South Columbia (NESC) margin, also called the Pacific Frontal Basin (Bueno Salazar, 1989), extends between the trench and the coastline from the equator to 5.5°N (Figure 27.1). It formed as the Nazca Plate subducted underneath the South American Plate since 23 Ma. The most important tectonic elements in the NESC margin are the Garrapatas fault system, the Buenaventura fault system and, the Gorgona terrane, which are exposed subaerially as the Gorgona Island, and the Dagua Piñón terrane (Figure 27.1) (Cediel *et al.*, 2003). The approximately 90 Ma old Dagua Piñón- and Gorgona terranes (Kerr and Tarney, 2005) are traces of an oceanic hotspot (Kerr *et al.*, 2002) that were carried on the Farallon Plate to be docked against the South American landmass between 45 and 37 Ma. The Buenaventura fault system is a suture zone between those two terranes. It developed when the Gorgona terrane docked against the Dagua Pinon Terrane during 37–40 Ma. The Garrapatas fault system is a boundary between the western limit of the Gorgona terrane and the oceanic sediments and later terrigenous sediments after the Gorgona terrane docked (Figure 27.1).

Two seismic lines are presented here (Figures 27.2 and 27.3), which image the Gorgona Terrane, fore arc basin and a part of the Buenaventura and Garrapatas fault systems besides the compressional folds that characterise the region. The Buenaventura fault system is a suture between the Gorgona terrane and the Dagua Piñón terrane and closely follows the ~ NE-SW trending coast line (Figure 27.1). It is characterized by large folds and thrusts, a part of which is imaged in the SE side of Figure 27.3, which is the western-most part of the suture. The deformation is expected to strengthen towards the east as one moves towards the main part of the Buenaventura fault system. The anticlines are open to close and the corresponding thrusts are not convincingly imaged in the seismic data (Ramsay and Huber, 1987) (Figures 27.3b and 27.3d). Figures 27.3a and 27.3b show a train of three anticlines. The hinge of the westernmost one is normal faulted (Figure 27.3b). Such normal faults may occur during tectonic wedge relaxation due to episodic subduction. They may also occur due to change in the local stress field. The actual reason needs to be investigated. There are smaller thrusts in the foreland of the large folds.

The onlapping of the sediments is very conspicuous in all the seismic sections (Figures 27.3a–27.3d). The first onlap is observed on the Oligocene reflection (Unit D and yellow line in Figures 27.3b and 27.3d). Well data indicate conglomerates at the base of Unit D. All reflections older than Oligocene (Units E and F) are subparallel. From the above evidence, it can be confirmed that these folds started developing around that time (see also Marcaillou and Collot, 2008). The onlaps on the fold limb followed by conformable sequences on the hinge indicated that the uplift gradually slowed down compared to the sedimentation (Nemčok *et al.*, 2009: Chapter 9). The top of crust in this seismic section is that of the Gorgona terrane, which may also be thrust/folded. However, due to the nonoptimal data quality, it cannot be confirmed. Even, the top of the crust cannot be mapped with confidence. There are a few NW dipping normal faults towards the NW portion of the seismic section. These faults are related to the Garrapatas fault system.

The fore arc basin, Garrapatas fault system and the Gorgona terrane are imaged (Figures 27.4a and 27.4b). In this section the sedimentary units D, E and F can be separated based on their seismic facies. The sediments in the fore arc basin downlap on the Oligocene reflection, which is the base of Unit D (Figures 27.4b and 27.4d). The Garrapatas fault system is a NW dipping right-lateral fault separating the oceanic crust and accretionary complex sediments on the NW (Figure 27.4b; see also Figure 8 of Cediel *et al.*, 2003). Within Unit E, one high amplitude reflection shows

Atlas of Structural Geological Interpretation from Seismic Images, First Edition. Edited by Achyuta Ayan Misra and Soumyajit Mukherjee.

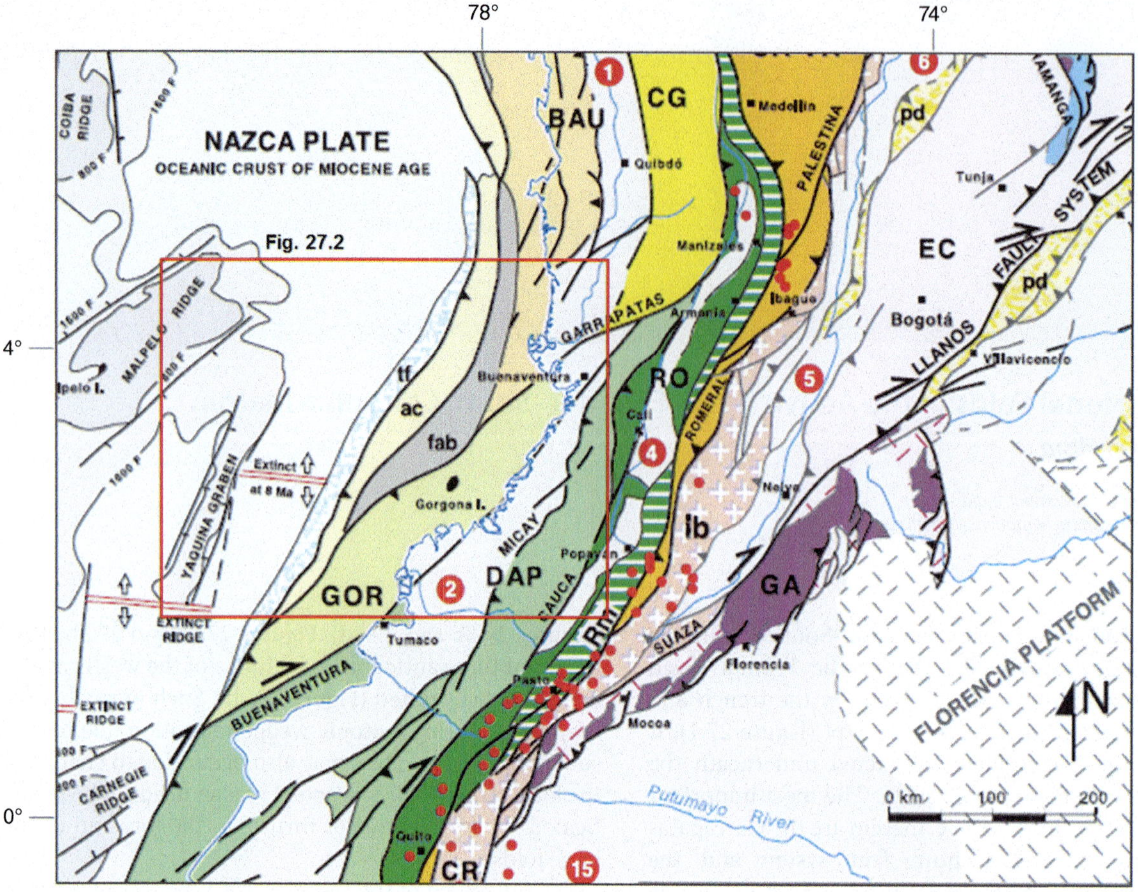

Figure 27.1 Geological map of the NESC margin. BAU: Baudo Terrane; CG: Cañas Gordas terrane; EC: Eastern Cordillera; GA: Garzón massif; Ib: Ibagué block; RO: Romeral terrane; Rm: Romeral mélange; CR: Cordillera Real; DAP: Dagua–Piñón terrane; GOR: Gorgona terrane; fab: fore arc basin; ac: accretionary prism; tf: trench fill; pd: piedmonte; 1: Atrato (Chocó) Basin; 2: Tumaco Basin; 4: Cauca–Patía Basin; 5: Upper Magdalena Basin; 6: Middle Magdalena Basin; 15: Putumayo–Napo Basin. Detail in Cediel *et al.* (2003). Red rectangle: Extent of area in Figure 27.2. Cediel *et al.* (2003). AAPG © 2003, reprinted by permission of the AAPG whose permission is required for further use.

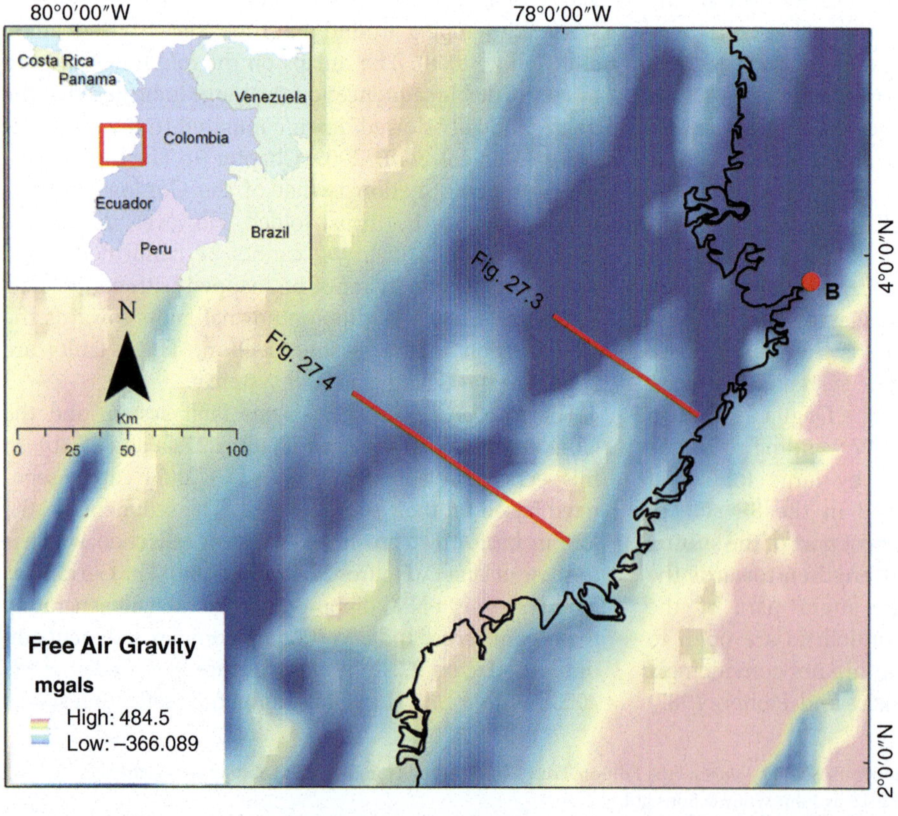

Figure 27.2 Location of the seismic profiles in the NESC margin overlaid on a hill-shaded free-air gravity anomaly map. B: Buenaventura city.

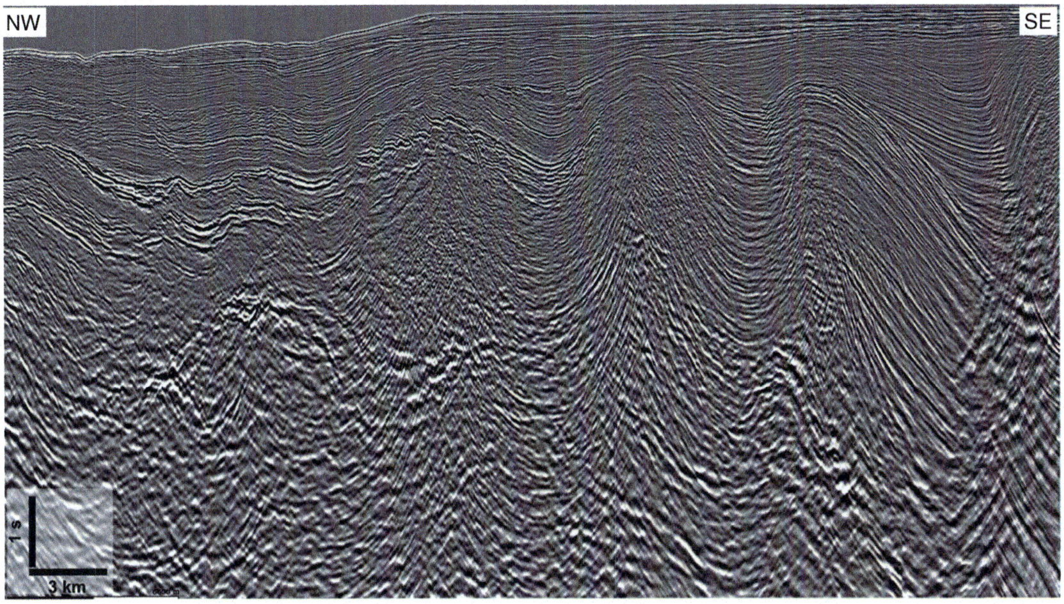

Figure 27.3a Uninterpreted section in the NESC margin. Note the noise content of the data in the deeper parts. See Figure 27.2 for location of the section. Seismic data courtesy Reliance Industries Ltd.

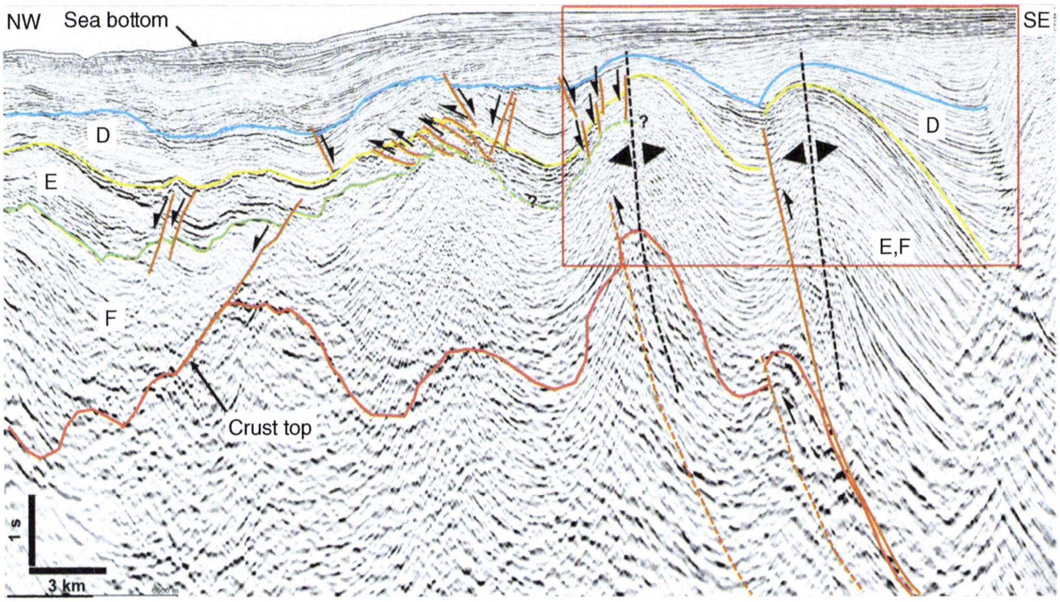

Figure 27.3b Interpretation of the seismic line as in Figure 27.3a. The boundary between Units E and F cannot be identified in the SE. Note the numerous normal faults on the hinge of the fold on the NW. Red rectangle: extent of the seismic section in Figures 27.3c and 27.3d. Seismic data courtesy Reliance Industries Ltd.

numerous small-scale thrusts (Figure 27.4d). There are a few folds and thrusts in the older section of Unit E also. Unit F also shows folding and thrusting, but their mapping or proper identification is limited by the data quality.

The seismic section in Figures 27.4e and 27.4f show a few normal faults. The hinge of the anticline collapsed because of the difference in compaction on the southeast of the basement high (Figures 27.4e and 27.4f). However, there may be other reasons for the genesis of these normal faults.

A very prominent feature in the second seismic section (Figures 27.4c–27.4f) is a seismic reflection that cross-cuts other reflections and is near the sea bottom. This reflection is known as the bottom simulating reflector (BSR). BSRs indicate the presence of free gas (usually methane) beneath the sea bed. The reflection is most

Figure 27.3c Uninterpreted enlarged seismic section of the area marked in Figure 27.3b. Seismic data courtesy Reliance Industries Ltd.

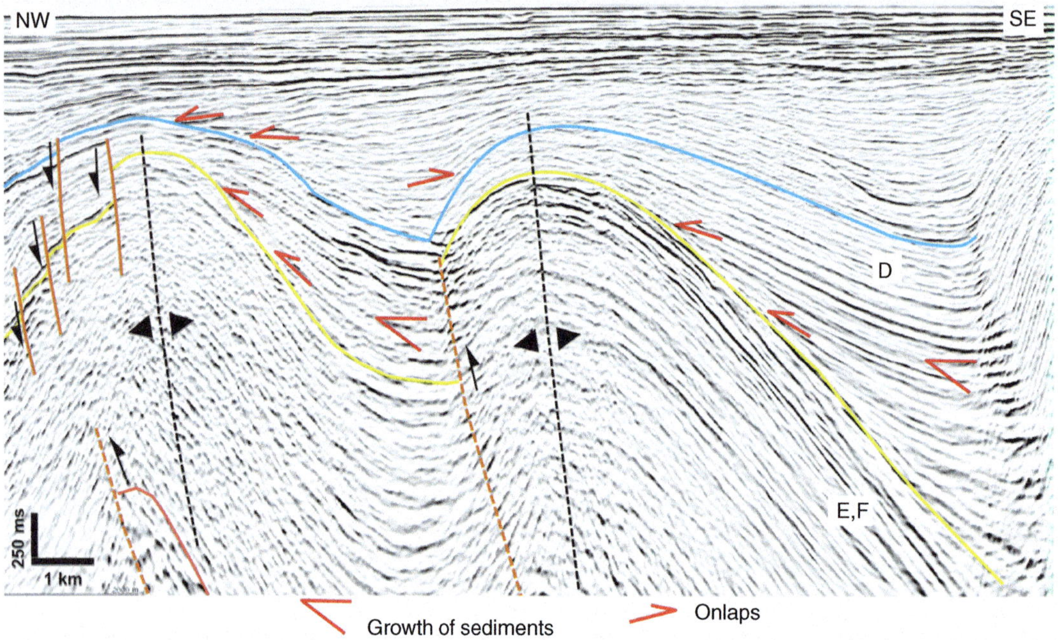

Figure 27.3d Interpretation of Figure 27.3c. Note the growth of sediments and the onlaps, which indicate the start of deformation. The onlaps gradually forming conformable beds on the hinge of the folds indicate that the uplift was initially more than the sedimentation and later uplift slowed down to become lesser then the sedimentation (see Nemčok *et al.* (2009) for details). Seismic data courtesy Reliance Industries Ltd.

possibly caused by the impedance contrast between the overlying gas hydrate and underlying gas saturated sediments. BSRs generally subparallel the sea bottom. However, in one instance here, they are not (Figures 27.4e and 27.4f). This is because, since the margin is active, the sea bottom configuration changes very fast and the BSR needs time to restabilize. The depth of the BSRs from the sea is a very important tool to study the heat flow or geothermal gradient in the area (Marcaillou *et al.*, 2006, 2008).

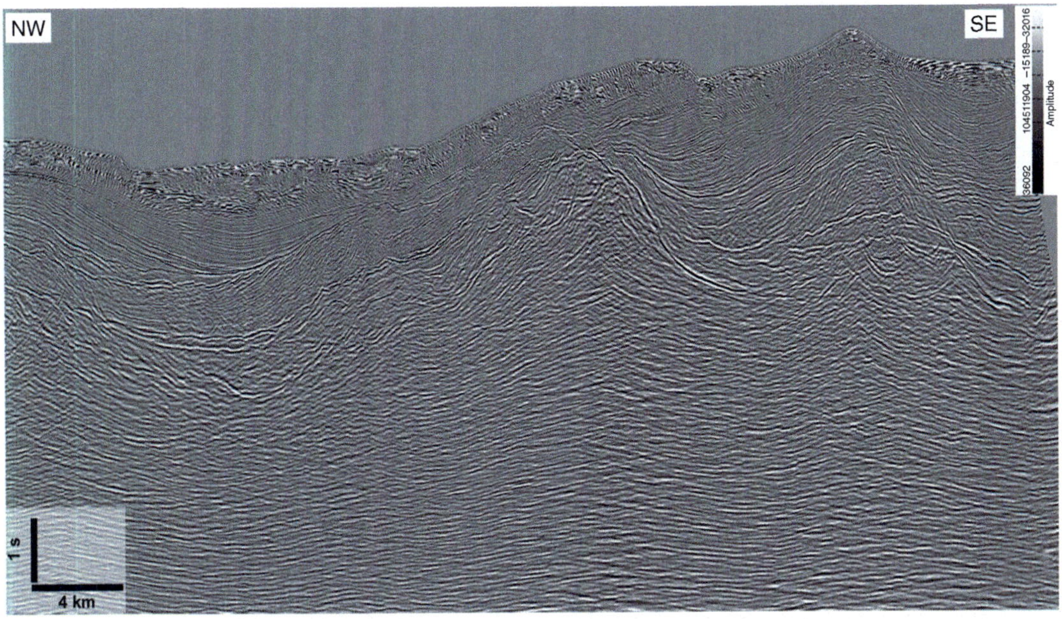

Figure 27.4a Uninterpreted seismic section in NESC margin. See Figure 27.2 for location of the section. Seismic data courtesy Reliance Industries Ltd.

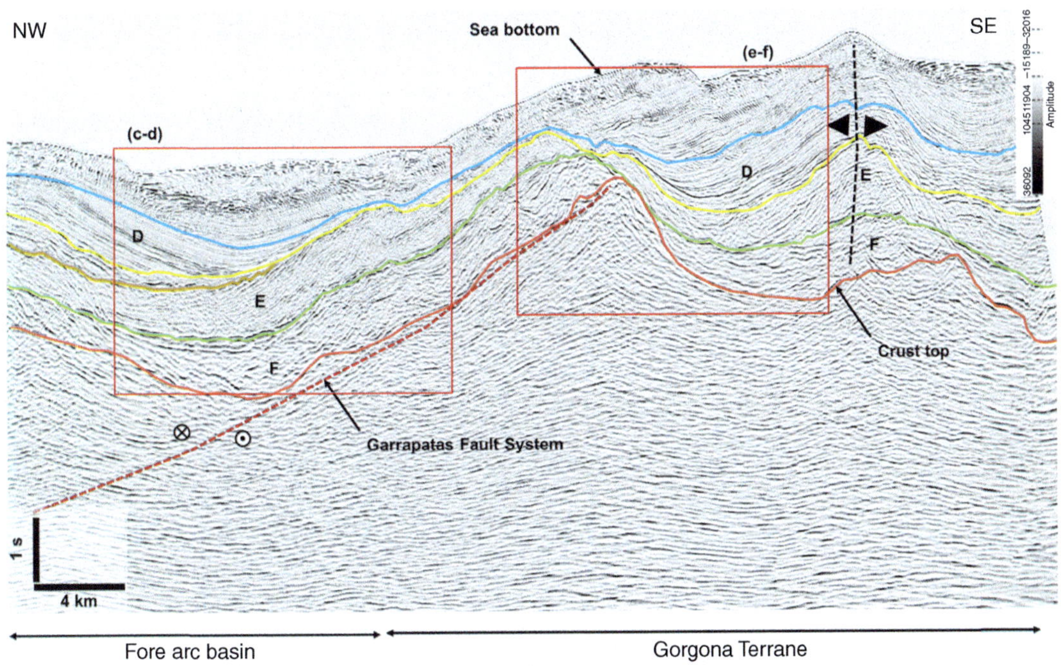

Figure 27.4b Interpretation of the seismic section in Figure 27.4a. Note the Garrapatas fault system, Gorgona terrane and the fore arc basin. The fold in the SE of the section is observed till the sea bottom, which indicates that the deformation is active at present. Seismic data courtesy Reliance Industries Ltd.

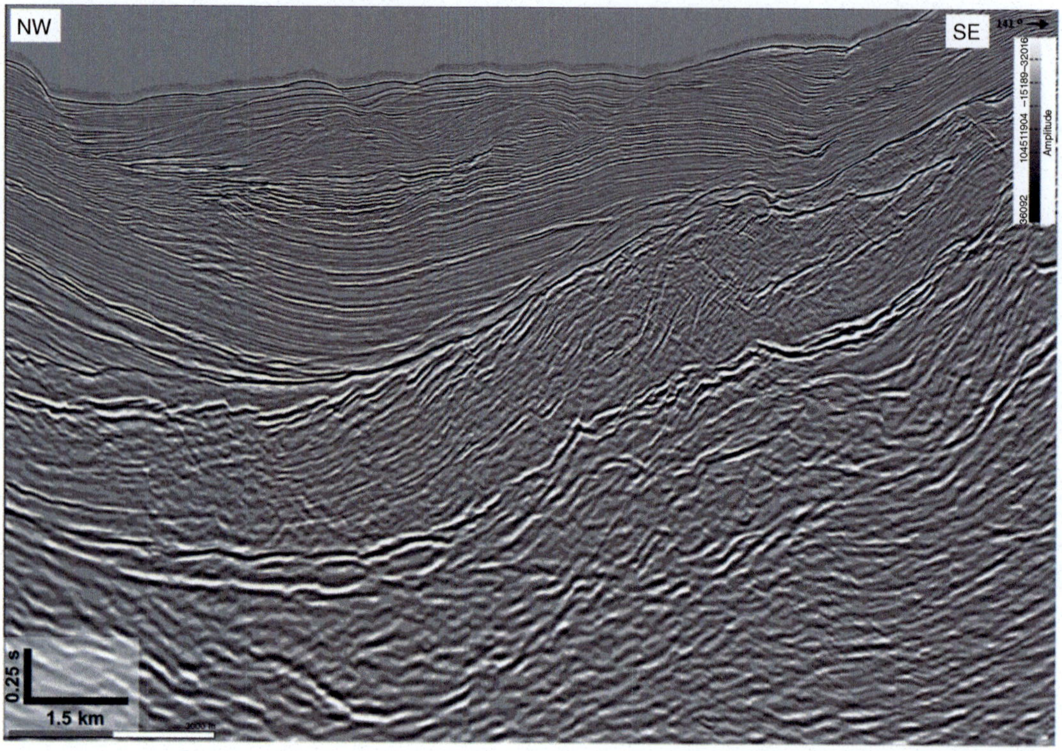

Figure 27.4c Uninterpreted section showing the fore arc basin. Location of the section shown in figure 27.4b. Seismic data courtesy Reliance Industries Ltd.

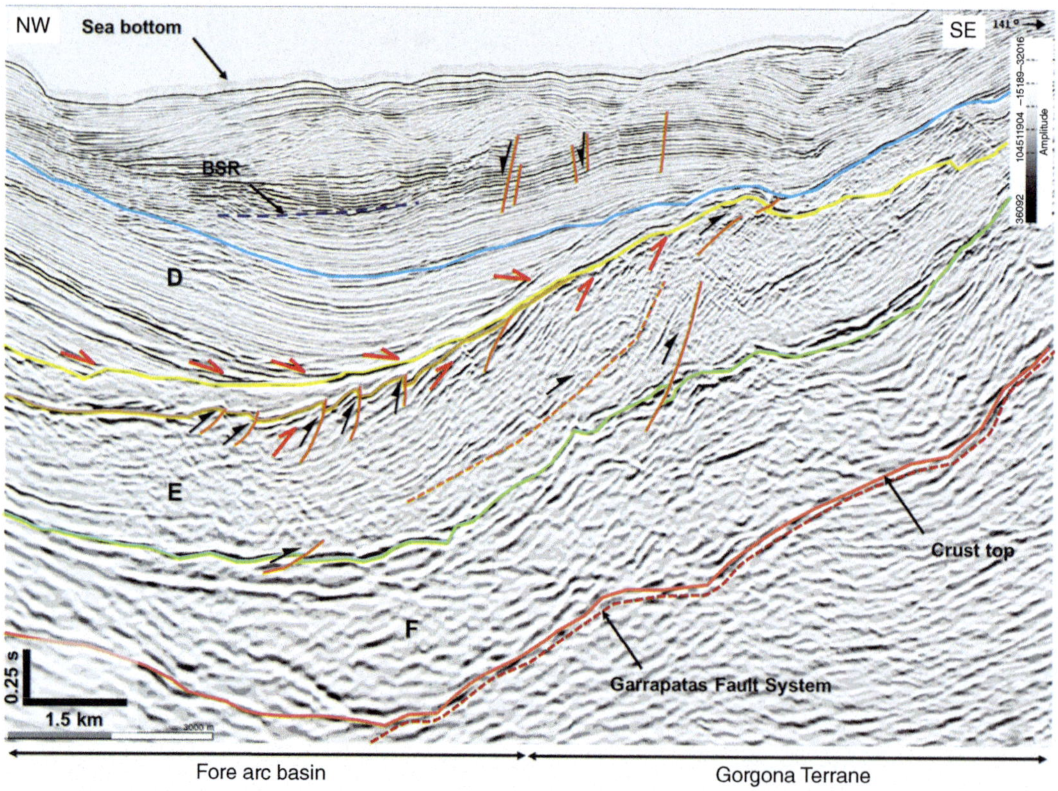

Figure 27.4d Interpreted section of the seismic section in Figure 27.4c showing the fore arc basin with the underlying pre-tectonic strata. Note small-scale thrusts in E. F are also folded but the data is suboptimal to identify them. The BSR is marked in broken blue line. Seismic data courtesy Reliance Industries Ltd.

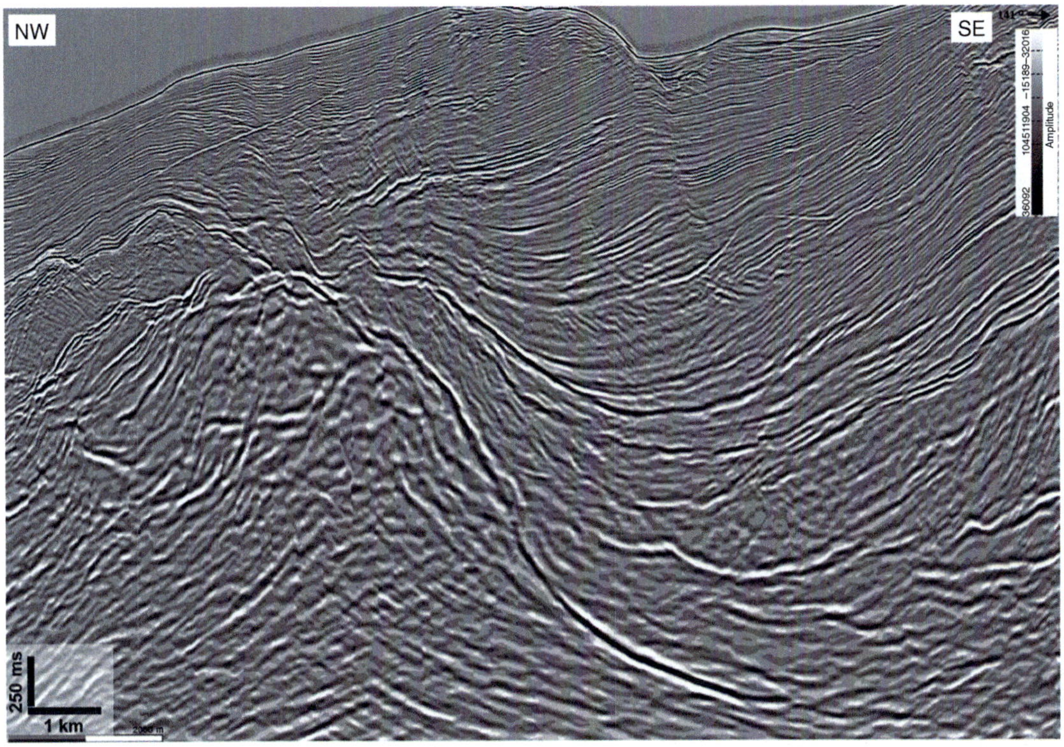

Figure 27.4e Uninterpreted seismic section showing part of the sedimentary sequence atop the Gorgona terrane. Location of the section shown in Figure 27.4b. Seismic data courtesy Reliance Industries Ltd.

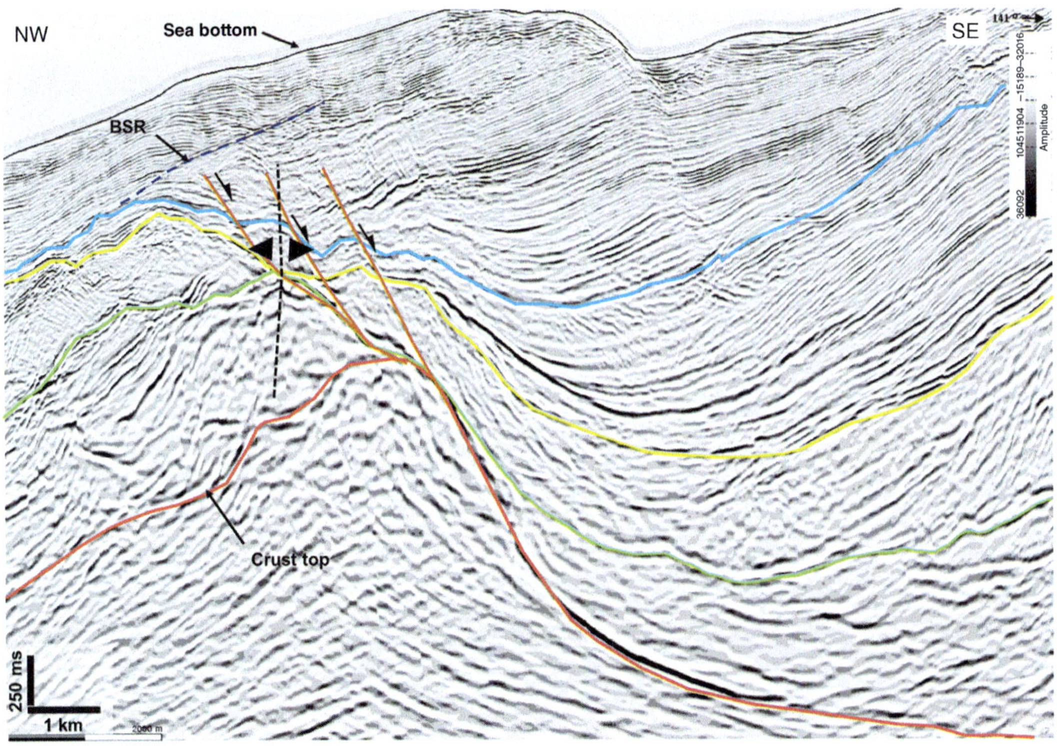

Figure 27.4f Interpretation of the seismic section shown in Figure 27.4e. Note the collapsed hinge of the fold, possibly due to larger compaction on the SE side of the high. Seismic data courtesy Reliance Industries Ltd.

References

Bueno Salazar, R. 1989. Hydrocarbon exploration and potential of the Pacific coastal basin of Colombia. In: Ericksen, G.E., Canas Pinochet, M.T. and Reinemund, J.A. (eds) *Geology of the Andes and its Relation to Hydrocarbon and Mineral Resources*. Earth Sciences Series, Circum-Pacific Council for Energy and Mineral Resources, Houston, TX, pp. 335–343.

Cediel, F., Shaw, R.P. and Ca'ceres, C. 2003. Tectonic assembly of the Northern Andean Block. In: Bartolini, C., Buffler, R.T. and Blickwede, J. (eds) *The Circum-Gulf of Mexico and the Caribbean: Hydrocarbon habitats, Basin Formation, and Plate Tectonics*. AAPG Memoir 79, American Association of Petroleum Geologists, Tulsa, OK, pp. 815–848.

Kerr, A.C. and Tarney, J. 2005. Tectonic evolution of the Caribbean and northwestern South America: The case for accretion of two Late Cretaceous oceanic plateaus. *Geology*, **33**, 269–272.

Kerr, A.C., Aspden, J.A., Tarney, J. and Pilatasig, L.F. 2002. The nature and provenance of accreted oceanic terranes in western Ecuador: geochemical and tectonic constraints. *Journal of the Geological Society*, **159**, 577–594.

Marcaillou, B. and Collot, J.Y. 2008. Chronostratigraphy and tectonic deformation of the North Ecuadorian–South Colombian offshore Manglares fore arc basin. *Marine Geology*, **255**, 30–44.

Marcaillou, B., Spence, G., Collot, J.Y. and Wang, K. 2006. Thermal regime from bottom simulating reflectors along the north Ecuador–south Colombia margin: Relation to margin segmentation and great subduction earthquakes. *Journal of Geophysical Research: Solid Earth*, **111**, B12407.

Marcaillou, B., Spence, G., Wang, K. *et al*. 2008. Thermal segmentation along the N. Ecuador–S. Colombia margin (1–4N): Prominent influence of sedimentation rate in the trench. *Earth and Planetary Science Letters*, **272**, 296–308.

Nemcok, M., Schamel, S. and Gayer, R. 2009. *Thrustbelts: Structural Architecture, Thermal Regimes and Petroleum Systems*. Cambridge University Press, Cambridge, pp. 541.

Ramsay, J.G. and Huber, M.I. 1987. *The Techniques of Modern Structural Geology*, vol. **2**, Academic Press, Oxford, pp. 700.

28

Fault Propagation Folds in the Active North Ecuador–South Colombia Margin

Achyuta Ayan Misra

Reliance Industries Ltd, Mumbai, India
Achyuta.Misra@ril.com; achyutaayan@gmail.com

Fault propagation folds form on the tips of the fault planes and commonly remain blind. They are commonly imaged by seismic reflection data. The most common characteristics that are exhibited by fault propagation folds are (Davis *et al.*, 2013):

- the folds are commonly asymmetric, with forelimbs steeper and narrower than their backlimbs;
- the fold tightens with depth;
- progressively smaller slip accommodate upward on the fault plane and become zero within the fold;
- synclines are present at the fault tips.

The North Ecuador–South Colombia (NESC) margin also has fault propagation folds developed in its Tertiary sequence due to subduction-related compression of the Farallon- and the Nazca plates beneath the South American plate. (See Chapter 27 and Cediel *et al.* (2003) for tectonics.)

The seismic section presented here trends NE (Figure 28.1) and shows a fault propagation fold, which depicts the above criterion (Figure 28.2a). The fault in this fault propagation fold too is conspicuous. On the NE side of the section there is some noise. This may arise from reflections from the adjacent syncline. Five horizons could be interpreted. The imaging is suboptimal for one horizon (dotted purple line in Figure 28.2b). The topmost horizons of the fold are eroded and channel cuts can be interpreted on them (Figure 28.2b). The interpreted horizons show that the thickness of the individual sedimentary packs thickens towards the southwest. This indicates that there was possibly a depocentre towards the southwest, which later folded. Onlaps in the sedimentary sequence progress into conformable beds upwards. This indicates that uplift due to folding was faster than the sedimentation in the earlier stages and subsequently slowed down.

Atlas of Structural Geological Interpretation from Seismic Images, First Edition. Edited by Achyuta Ayan Misra and Soumyajit Mukherjee.

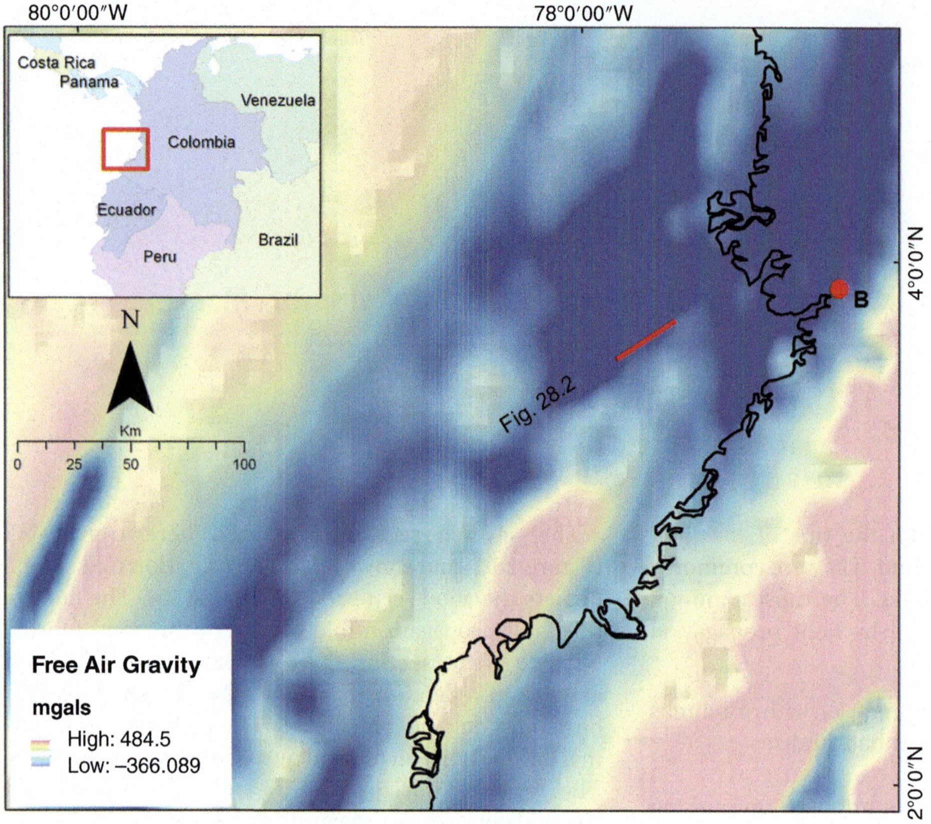

Figure 28.1 Location of the seismic profile in the NESC margin overlaid on a hill-shaded free-air gravity anomaly map. B: Buenaventura city.

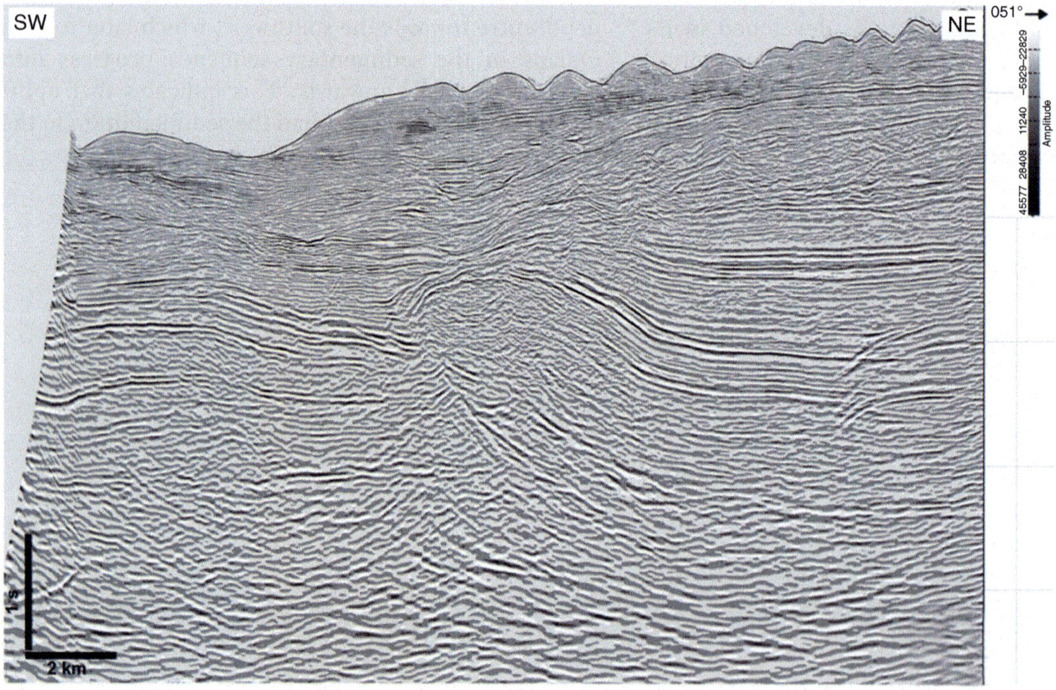

Figure 28.2a Uninterpreted seismic section showing a fault propagation fold in the NESC margin. For location of the line see Figure 28.1. Seismic data courtesy Reliance Industries Ltd.

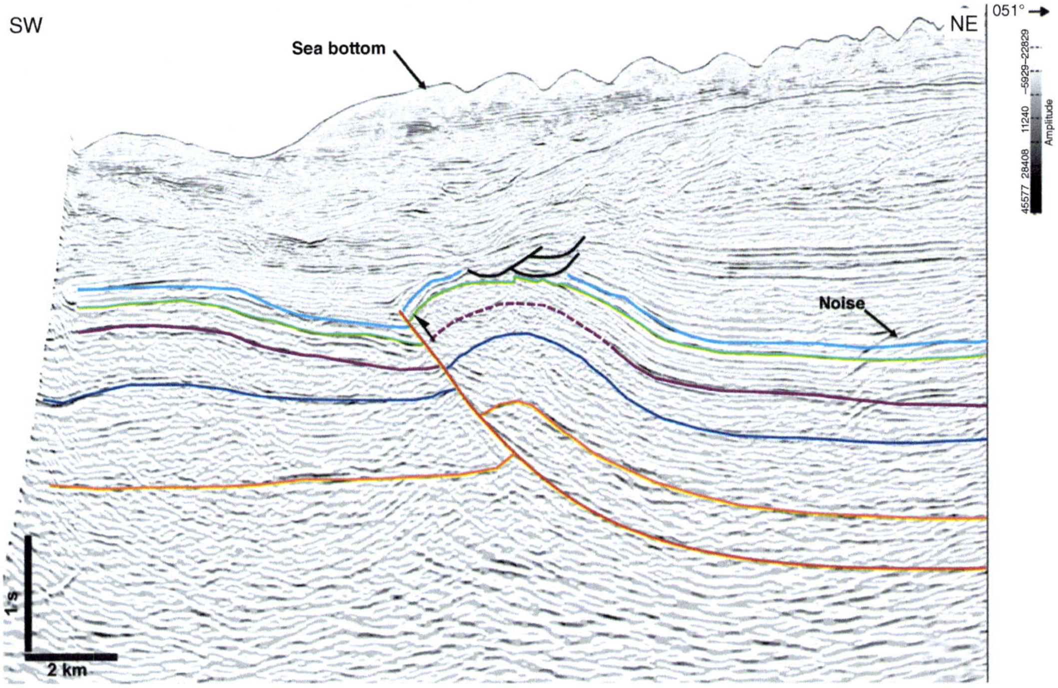

Figure 28.2b Interpreted seismic profile shown in Figure 28.2a. Note the noise in the NE and the possible channel cuts on the crest of the fold. The hanging wall block shows stronger drag folds (Mukherjee 2014) than the footwall block. Seismic data courtesy Reliance Industries Ltd.

References

Cediel, F., Shaw, R.P. and Ca'ceres, C. 2003. Tectonic assembly of the Northern Andean Block. In: Bartolini, C., Buffler, R.T. and Blickwede, J. (eds) The Circum-Gulf of Mexico and the Caribbean: Hydrocarbon habitats, Basin Formation, and Plate Tectonics. AAPG Memoir 79, American Association of Petroleum Geologists, Tulsa, OK, pp. 815–848.

Davis, G.H., Reynolds, S.J. and Kluth, C. 2013. *Structural Geology of Rocks and Regions*, 3rd edn, John Wiley & Sons, Inc, Hoboken, NJ.

Mukherjee, S. 2014. Review of flanking structures in meso- and micro-scales. *Geological Magazine*, **151**, 957–974.

29

Antiformal Stacks and Hinterland Dipping Duplexes in the North Ecuador–South Colombia Margin

Achyuta Ayan Misra

Reliance Industries Ltd, Mumbai, India
Achyuta.Misra@ril.com; achyutaayan@gmail.com

The active offshore North Ecuador–South Colombia margin (detail in Chapter 27 and Cediel *et al.*, 2003) provides excellent opportunity to study various structures formed during compression. Presented here are two sets of seismic sections (Figure 29.1) that show (i) an antiformal stack (Figure 29.2) and (ii) a hinterland dipping duplex (Figure 29.3).

29.1 Antiformal Stack (Figure 29.2)

An antiformal stack forms when a duplex develops with an antiform on the hangingwall side and associated with thrusts splaying from the main floor thrust (Buttler, 1982; McClay, 1992). Their trailing branch lines may coincide. A series of antiforms may stack one above the other (Mukherjee and Koyi, 2010; Mukherjee 2012a, 2012b, 2013 as microscale examples). Individual horses move near to or immediately beyond the start of the next horse to produce a stack (Ramsay and Huber, 1987). The lowermost imbricate thrusts are youngest and become progressively older upward. Figure 29.2 shows an antiformal stack, which is about 1 km high at the highest point. It forms within the fore arc basin atop a basement high on the Gorgona Terrane (Figures 29.2a and 29.2b). The Garrapatas fault system acts as the ramp of the floor thrust and the antiformal stack develops on the flat. The base thrusts and the lowermost imbricate thrusts, that is the youngest ones, are low frequency, high amplitude reflections (Figures 29.2c–29.2e). The younger imbricates are interpreted from the gentler limb of the antiforms. Consecutive growth packages with onlaps on the gentler limb indicate episodic uplift of the antiformal stack. The beds are conformable in the shallow stratigraphies, which indicate that the tectonic uplift was more than sedimentation initially. The uplift rate was lower than the sedimentation rate in the late stages. There are possibly two hangingwall drop faults (Butler, 1982) in the southeast of the antiformal stack. These faults dip ~40°. There is also a bottom simulating reflector (BSR) (see Chapter 27) with very high amplitude and low frequency (Figures 29.2c–29.2e).

29.2 Hinterland Dipping Duplex (Figure 29.3)

Here, the thrusts and the marker beds both dip towards the hinterland (McClay, 1992). The elevation of the roof thrust in this example also varies smoothly. So, it can also be designated as a smooth roof duplex. The dips of the thrusts decrease towards the foreland.

Atlas of Structural Geological Interpretation from Seismic Images, First Edition. Edited by Achyuta Ayan Misra and Soumyajit Mukherjee.

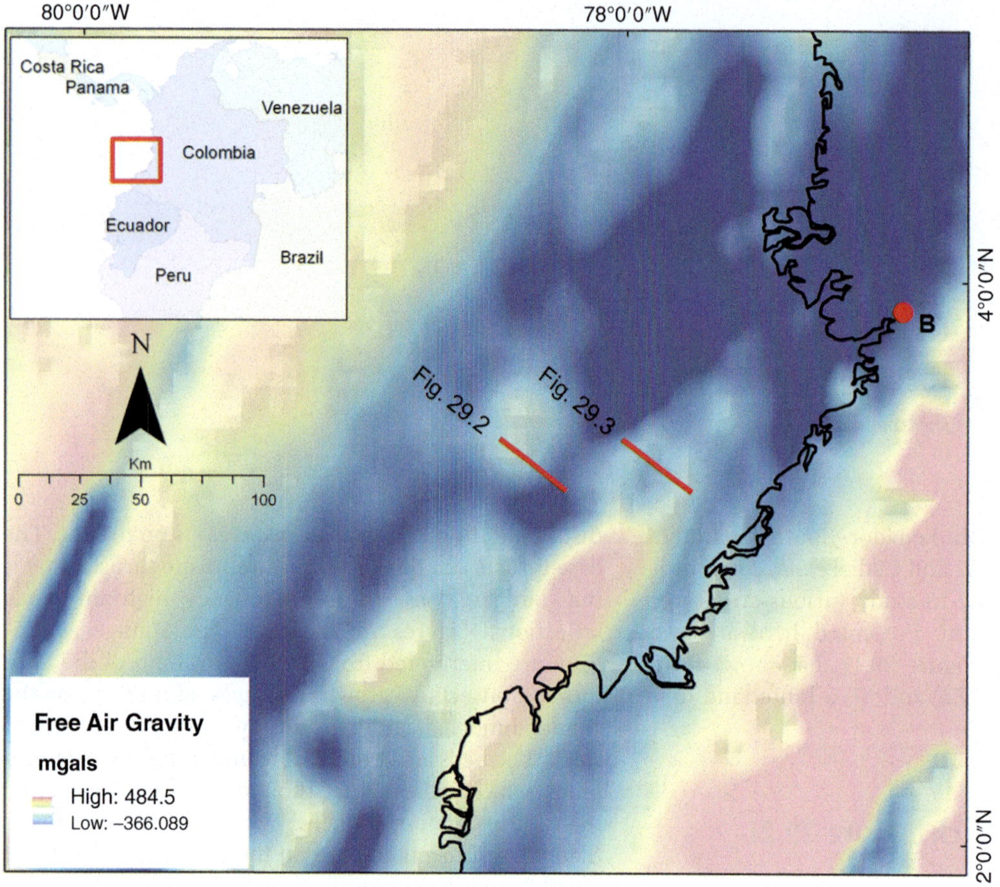

Figure 29.1 Location of the seismic profiles in the NESC margin overlaid on a hill-shaded free-air gravity anomaly map. B: Buenaventura city.

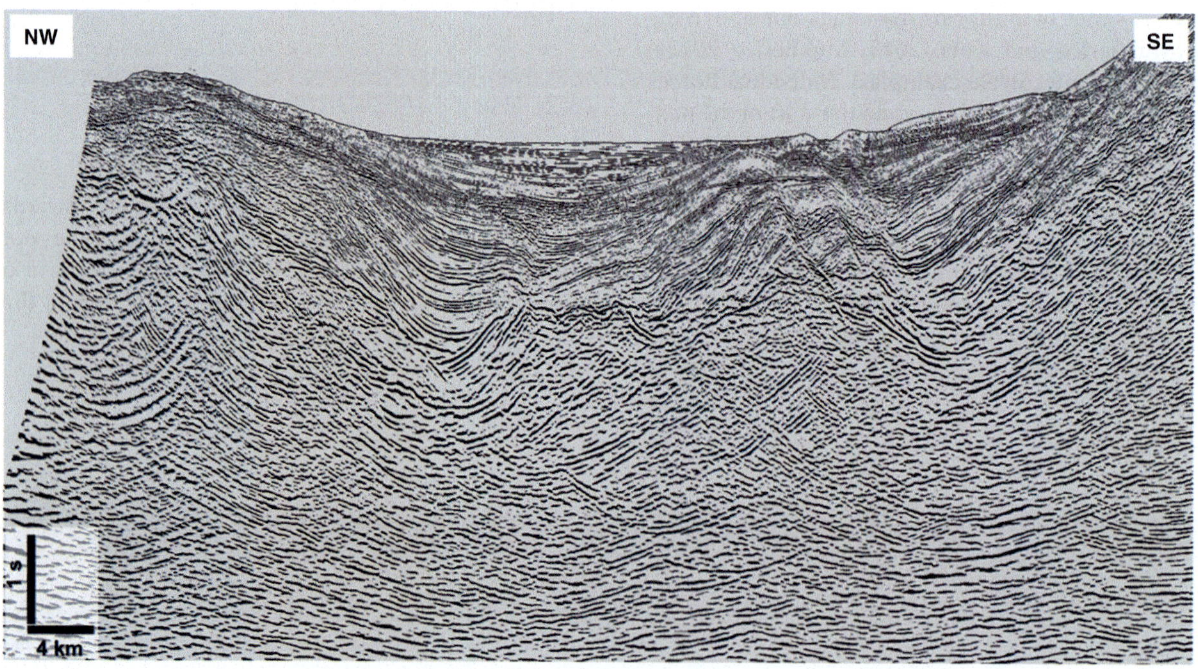

Figure 29.2a Uninterpreted seismic section in the fore arc basin. Seismic data courtesy Reliance Industries Ltd.

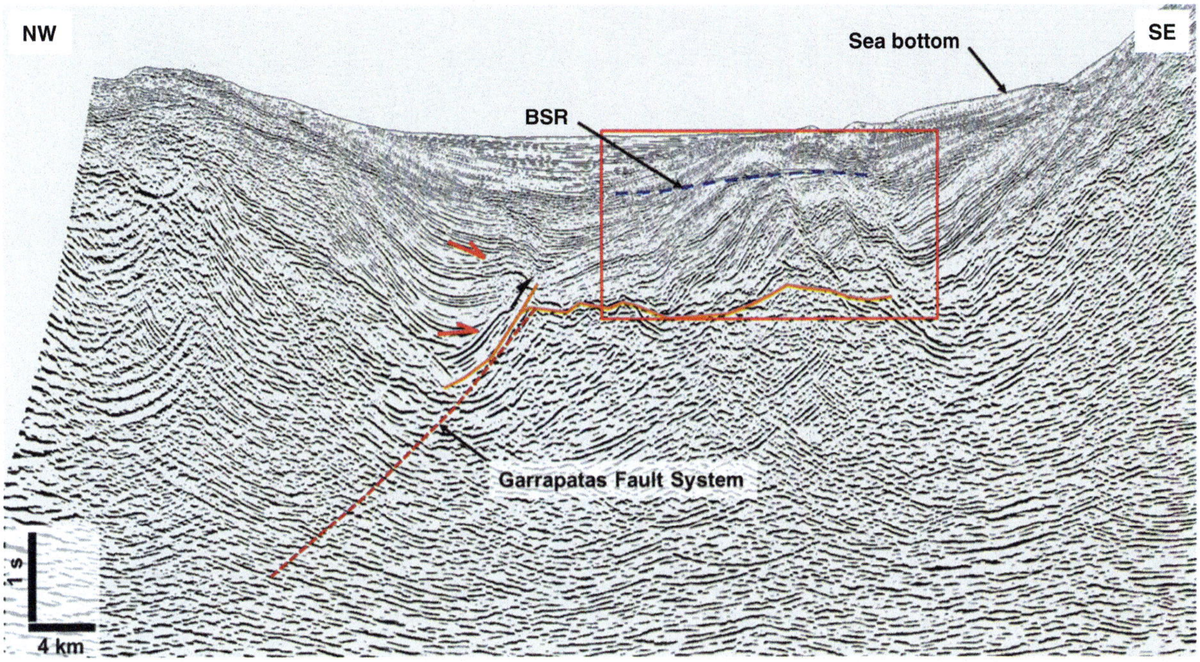

Figure 29.2b Interpreted seismic section of Figure 29.2a showing the antiformal stack and the Garrapatas fault system. Red barbed arrows: onlaps and downlaps. Red rectangle: extent of sections shown in Figures 29.2c– 29.2e. Seismic data courtesy Reliance Industries Ltd.

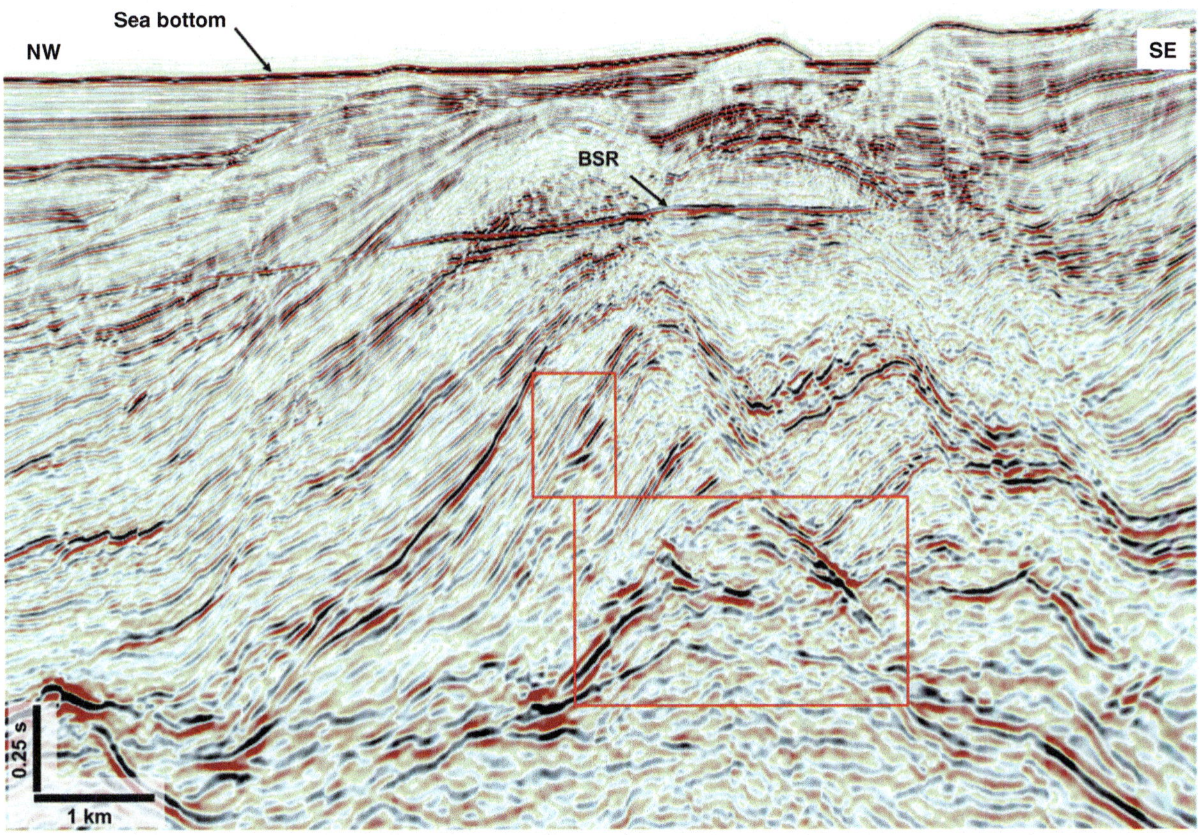

Figure 29.2c Red-white-black display of the antiformal stack. Red rectangles: seismic facies that help to identify the thrusts. Larger rectangle shows the high amplitude thrusts and smaller rectangle shows an asymmetric antiform, whose gentler limb indicates the fault plane. Seismic data courtesy Reliance Industries Ltd.

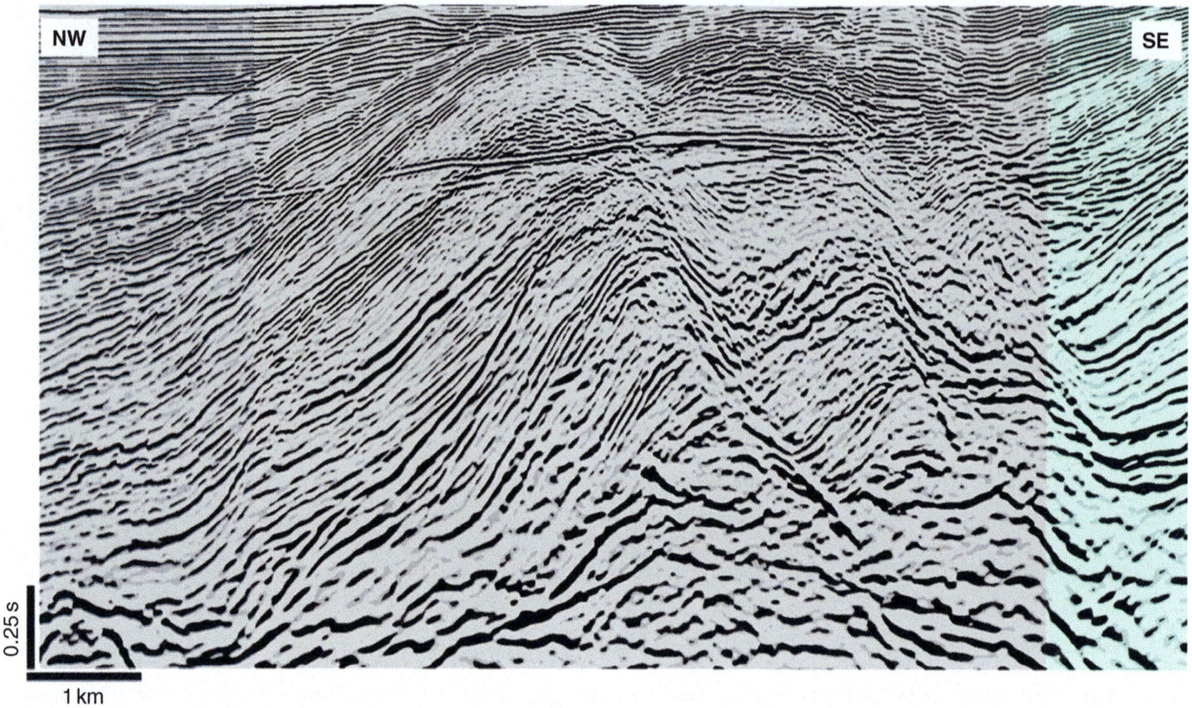

Figure 29.2d Uninterpreted grey-scale display for the antiformal stack. Seismic data courtesy Reliance Industries Ltd.

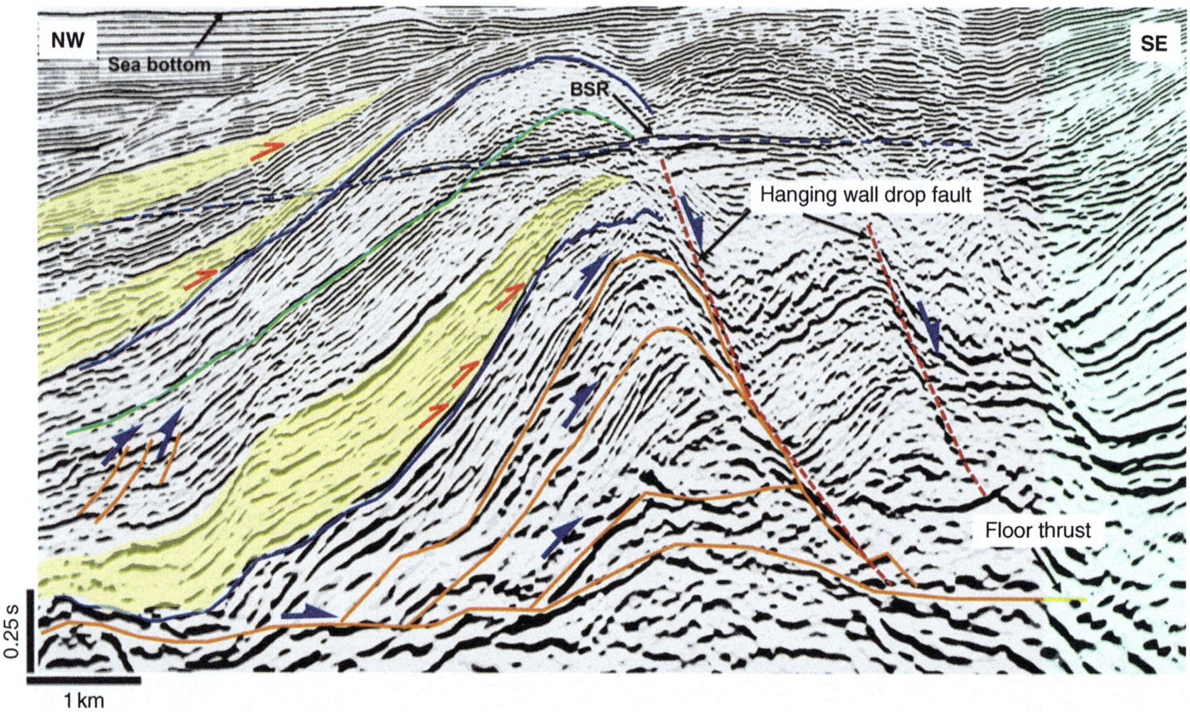

Figure 29.2e Interpreted seismic section of figure 29.2d. Yellow polygons: growth packages; barbed red arrows: onlaps; blue lines: onlap surfaces, green line: possible unconformity. Note an earlier BSR just below the BSR. Compare with Figure 29.2c. Seismic data courtesy Reliance Industries Ltd.

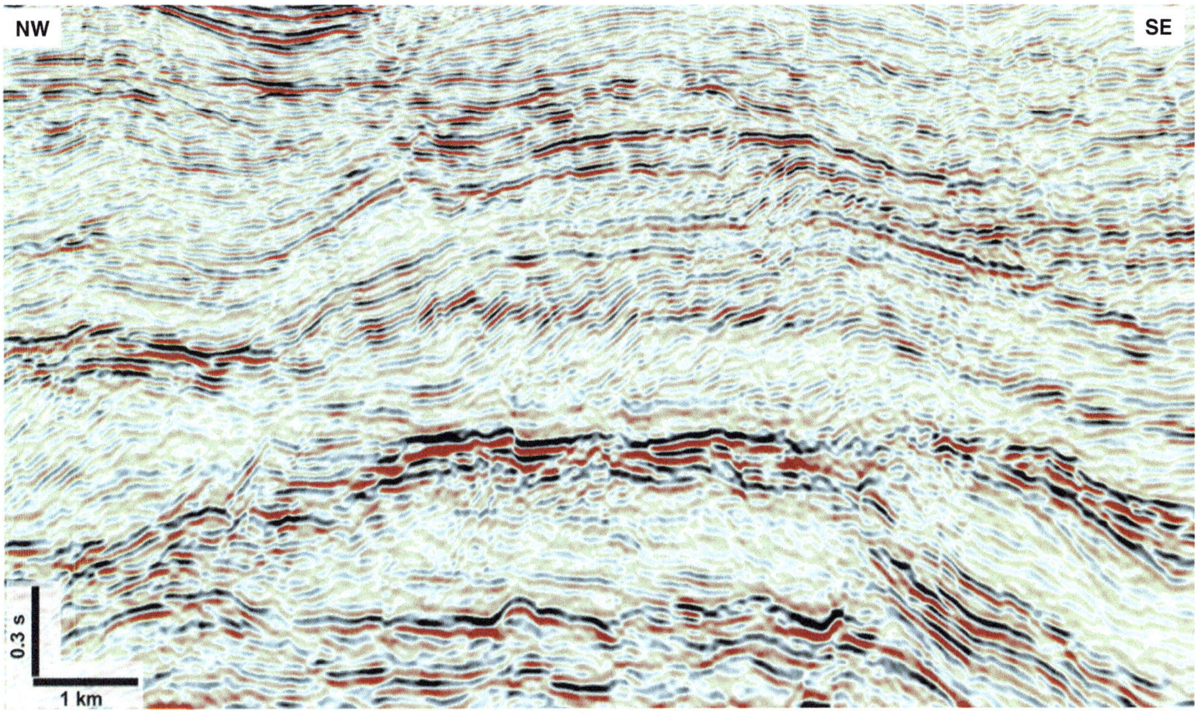

Figure 29.3a Uninterpreted seismic section showing the hinterland dipping duplex. Seismic data courtesy Reliance Industries Ltd.

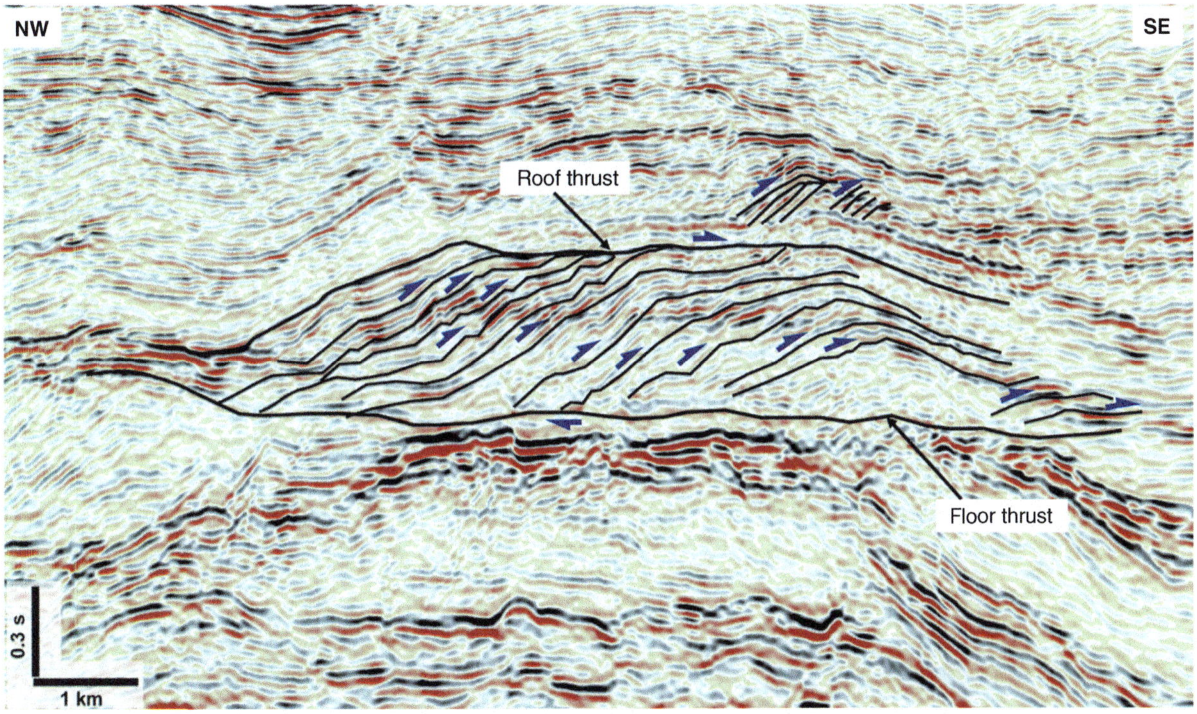

Figure 29.3b Interpreted seismic section of Figure 29.3a. The high amplitude reflection within the roof and floor thrust is the marker horizon here. Note the small thrusts and marker dip towards the hinterland. There is another duplex developing above the well-developed one. Seismic data courtesy Reliance Industries Ltd.

References

Butler, R.W. 1982. The terminology of structures in thrust belts. *Journal of Structural Geology*, **4**, 239–245.

Cediel, F., Shaw, R.P. and Ca'ceres, C. 2003. Tectonic assembly of the Northern Andean Block. In: Bartolini, C., Buffler, R.T. and Blickwede, J. (eds) The Circum-Gulf of Mexico and the Caribbean: Hydrocarbon habitats, Basin Formation, and Plate Tectonics. AAPG Memoir 79, American Association of Petroleum Geologists, Tulsa, OK, pp. 815–848.

McClay, K.R. 1992. Glossary of thrust tectonics terms. In: McClay, K.R. (ed) *Thrust Tectonics*, Chapman and Hall, London, pp. 419–433.

Mukherjee S. 2012a. Tectonic implications and morphology of trapezoidal mica grains from the Sutlej section of the Higher Himalayan Shear Zone, Indian Himalaya. *The Journal of Geology*, **120**, 575–590.

Mukherjee S. 2012b. A micro-duplex. *International Journal of Earth Sciences*, **101**, 503.

Mukherjee S. 2013. *Deformation Microstructures in Rocks*. Springer Geochemistry/Mineralogy. Berlin. pp. 1–111.

Mukherjee, S. and Koyi, H.A. 2010. Higher Himalayan shear zone, Zanskar Indian Himalaya: microstructural studies and extrusion mechanism by a combination of simple shear and channel flow. *International Journal of Earth Sciences*, **99**, 1083–1110.

Ramsay, J.G. and Huber, M.I. 1987. *The Techniques of Modern Structural Geology*, vol. 2. Academic Press, Oxford, pp. 700.

Part III

Mobile Substrates

Shale and salt may be mobile and induce a passive flow of overlying layers. In this section, examples are provided of:

- Salt structures (Chapters 31, 32, 38)
- Shale Growth Fault Thrusts Belts (GFTBs) (Chapters 35–37)
- Complications in salt and subsalt interpretations (Chapters 33, 34)

30

Compressional Salt-Related Structures in the Western Quele Area, Kuqa Depression, Tarim Basin, China

Zhenyun Wu[1], Hongwei Yin[2], and Hemin A. Koyi[3]*

[1] *Earthquake Administration of Jiangsu Province, Nanjing, China*
[2] *Nanjing University, China*
[3] *Department of Earth Sciences, Uppsala University, Uppsala Villavägen 16, Sweden*
* *zyw5918@163.com.* *This research was supported by China National Science fund (No.41602208, No.41572187 and No.41272227)*

The Quele salt nappe and the Misikantake anticline are located in the Qiulitage structural belt of the Kuqa Depression (Figure 30.1).The overburden deformation is concentrated in the Quele salt nappe and the Misikantake anticline, while the subsalt structures have accumulated mainly in the hinterland in the northern part of the section (Wu *et al.*, 2014, 2015). The Quele salt nappe is derived from the pre-existing Quele diapir with a long-distance allochthonous salt sheet. The Misikantake anticline is an asymmetric fold with a large Kumugeliemu salt core. The salt underlying the anticline was thinning to its two flanks, given that salt can have rather low viscosity (Mukherjee *et al.* 2010; Mukherjee 2011), and even salt weld was formed. Moreover, the overburden thickness of the anticline is roughly isopachous until the Upper Kuqa Formation (N_2k). A fan-shaped growth stratum formed gradually since the late Pliocene, which denotes the initial formation time of the Misikantake anticline. In this section, we hold the view that the Misikantake anticline was not only a simple decollement fold that occurred in the overburden by the compression stress but also controlled by the subsalt structure (a subsalt ramp interpreted in this section) (Figure 30.2). A good vertical spatial relationship can be indicated clearly between the core of the Misikantake salt anticline and the subsalt basement ramp (Figure 30.2) (Tang *et al.*, 2015).

Unlike the section in the eastern Quele area (Figure 30.3), the rock salt was flowing laterally in the western Quele area and converged from east to west (Tang *et al.*, 2012; Wu *et al.*, 2014), which swelled the rock salt. The salt layer is >5 km thick in this area and the Awate sag (a narrow and thin synclinal salt-withdrawl minibasin) (Wu *et al.*, 2014) uplifted wholly by the forward-vergent imbricates in the subsalt basement strata (here the F_1 fault is younger than the F_2 fault) (Wang *et al.*, 2011; Li *et al.*, 2012). In the subsalt basement strata, a northward uplifted imbricate thrust structure, consisting of several N-dipping reverse faults, formed in the northern hinterland. Furthermore, several folds (Awate folds) could be interpreted along the Awate fault in the front of the Awate sag (southward) according to the fieldwork data, seismic data and the analogue experiments (Wu *et al.*, 2015).

From the observations several conclusions are made.

- The interpreted seismic profiles show that the structural styles and features, presented from east to west of the Qiulitage structural belt in the Quele area, are obviously different. Moreover, due to the existance of the huge rock salt layer, the obvious differential structural deformation occurred in the overburden and the basement strata in the Quele area.
- The Misikantake anticline is a typical compressional salt-related asymmetrical fold with a large Kumugeliemu salt core. The interpretation of the seismic profile indicates that a good vertical spatial relationship can be identified clearly between the core of this salt anticline and the subsalt basement ramp.
- The Awate sag is a thin-thickness synclinal salt-withdrawl minibasin uplifted by huge rock salt layer. The structural deformation (rim folds) occurred mainly along the Awate fault in the front of the Awate sag.

Atlas of Structural Geological Interpretation from Seismic Images, First Edition. Edited by Achyuta Ayan Misra and Soumyajit Mukherjee.

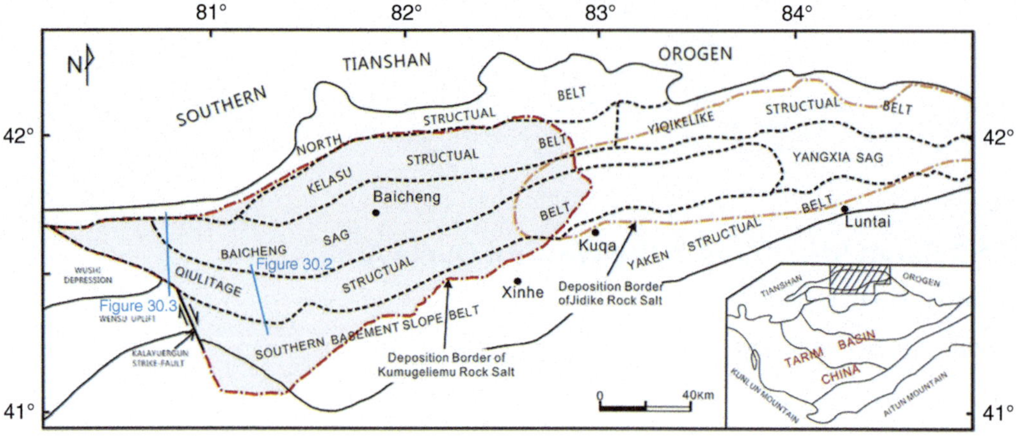

Figure 30.1 Location of the seismic profiles in the Kuqa Depression.

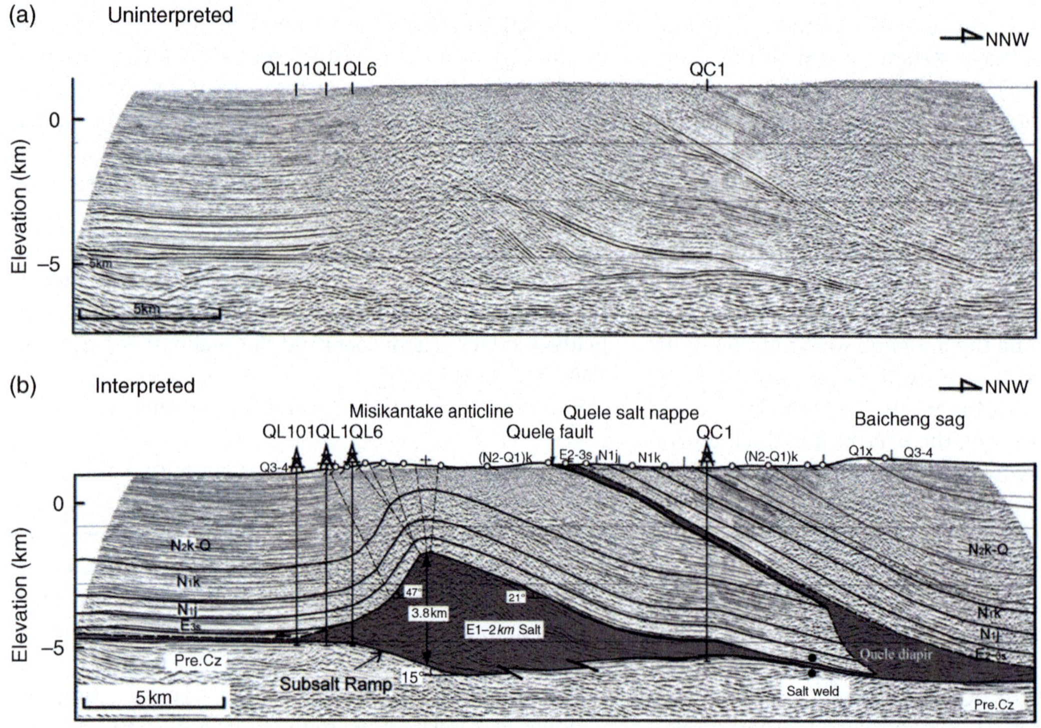

Figure 30.2 (a) Uninterpreted and (b) interpreted seismic section showing salt structures (Quele salt nappe and Misikantake anticline) of the eastern Quele area in the western Kuqa Depression (see the location in Figure 30.1) (Li *et al*., 2012; Wu *et al*., 2014). The strata of Suweiyi Formation (E2–3s) to Kangcun Formation (N1k) are pre-kinematic strata and the strata of Kuqa Formation (N2k) to Xiyu Formation (Q1x) are syn-kinematic strata. Pre-Cz is the abbreviation of Pre-Cenozoic strata in the Kuqa Depression. (Wu *et al*., 2014).

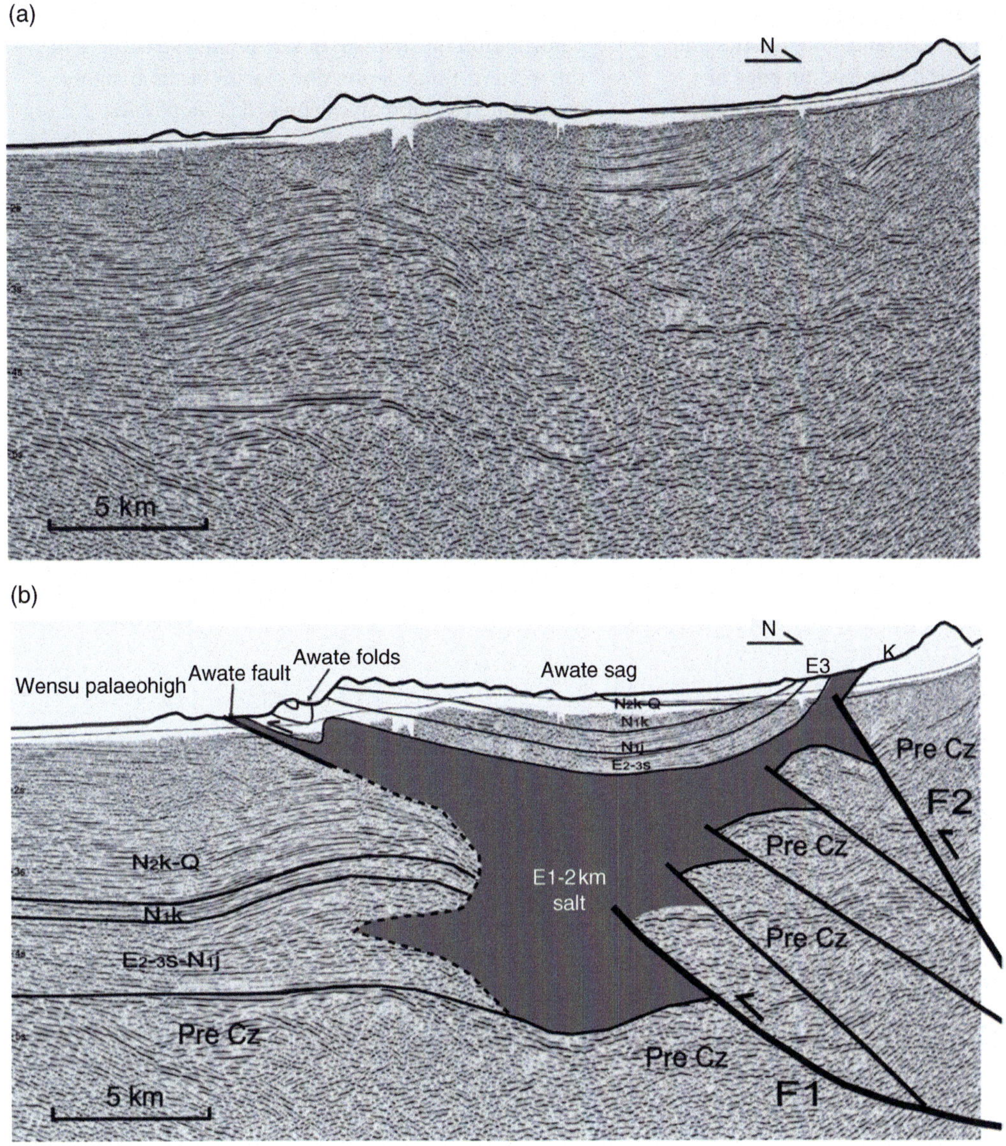

Figure 30.3 (a) Uninterpreted and (b) interpreted seismic section showing salt-related structures (Awate sag and its surrounding structures) of the western Quele area (the boundary area of the Kumugeliemu rock salt) in the western Kuqa Depression (see the location in Figure 30.1). The strata of the Suweiyi Formation (E2–3s) to the Kangcun Formation (N1k) are pre-kinematic strata and the strata of Kuqa Formation (N2k) to Xiyu Formation (Q1x) are syn-kinematic strata. Pre-Cz is the abbreviation of Pre-Cenozoic strata in the Kuqa Depression. (Wu *et al.*, 2014)

References

Li, S., Wang, X. and Suppe, J. 2012. Compressional salt tectonics and synkinematic strata of the western Kuqa foreland basin, southern Tian Shan, China. *Basin Research*, **23**, 1–23.

Mukherjee, S. 2011. Estimating the viscosity of rock bodies – a comparison between the Hormuz- and the Namakdan Salt domes in the Persian Gulf, and the Tso Morari Gneiss dome in the Himalaya. *The Indian Journal of Geophysical Union*, **15**, 161–170.

Mukherjee, S., Talbot, C.J. and Koyi, H.A. 2010. Viscosity estimates of salt in the Hormuz and Namakdan salt diapirs, Persian Gulf. *Geological Magazine*, **147**, 497–507.

Tang, P., Li, C., Lei, G. and Huang, S. 2012. Characteristics and formation of salt structures in Baicheng sag, Kuqa fold-and-thrust belt. *Earth Science - Journal of China University of Geosciences*, **37**, 69–76 (in Chinese with English abstract).

Tang, P., Rao, G., Li S. and Wang, R. 2015. The impact of salt layer thickness on the structural characteristics and evolution of detachment folds in the leading edge of Kuqa fold and thrust belt. *Earth Science Frontiers*, **22**, 312–327 (in Chinese with English abstract).

Wang, X., Suppe, J., Guan, S. *et al.* 2011, Cenozoic structure and tectonic evolution of the Kuqa fold belt, southern Tianshan, China. In: McClay, K., Shaw, J.H. and Suppe, J. (eds) *Thrust Fault-Related Folding*, AAPG Memoir 94, American Association of Petroleum Geologists, Tulsa, OK, pp. 215–243.

Wu, Z., Yin, H., Wang, X. *et al.* 2014. Characteristics and deformation mechanism of salt-related structures in the western Kuqa depression, Tarim basin: Insights from scaled sandbox modeling. *Tectonophysics*, **612**, 81–96.

Wu, Z., Yin, H., Wang, X. and Xu, S. 2015. The structural features and formation mechanism of exposed salt diapirs in the front of fold-thrust belt, Western Kuqa depression. *Journal of Nanjing University (Natural Science)*, **51**, 612–625 (in Chinese with English abstract).

31

Reactive Salt Diapir, Southern Precaspian Basin, Middle Asia

Zhenyun Wu[1], Hongwei Yin[2], and Junzhang Zheng[3]*

[1] *Earthquake Administration of Jiangsu Province, Nanjing, China*
[2] *Nanjing University, China*
[3] *PetroChina Research Institute of Petroleum and Development, Beijing, China*
** zyw5918@163.com.* *This research was supported by China National Science fund (No.41602208, No.41572187 and No.41272227)*

Shown in this example from the Southern Precaspian Basin (Figure 31.1) are salt structures in seismic section and their interpretation (Figures 31.2 and 31.3). A buried triangle salt diapir (left) and an indented salt diapir (right) can be interpreted clearly. In addition, a progradational front can be inferred at the bottom of the indented salt diapir. In the overburden strata, a minor graben bound by V-shaped conjugate normal faults on the top of the buried triangle salt diapir and a crestal sag (salt-withdrawal minibasin) occurred on the top of the indented salt diapir

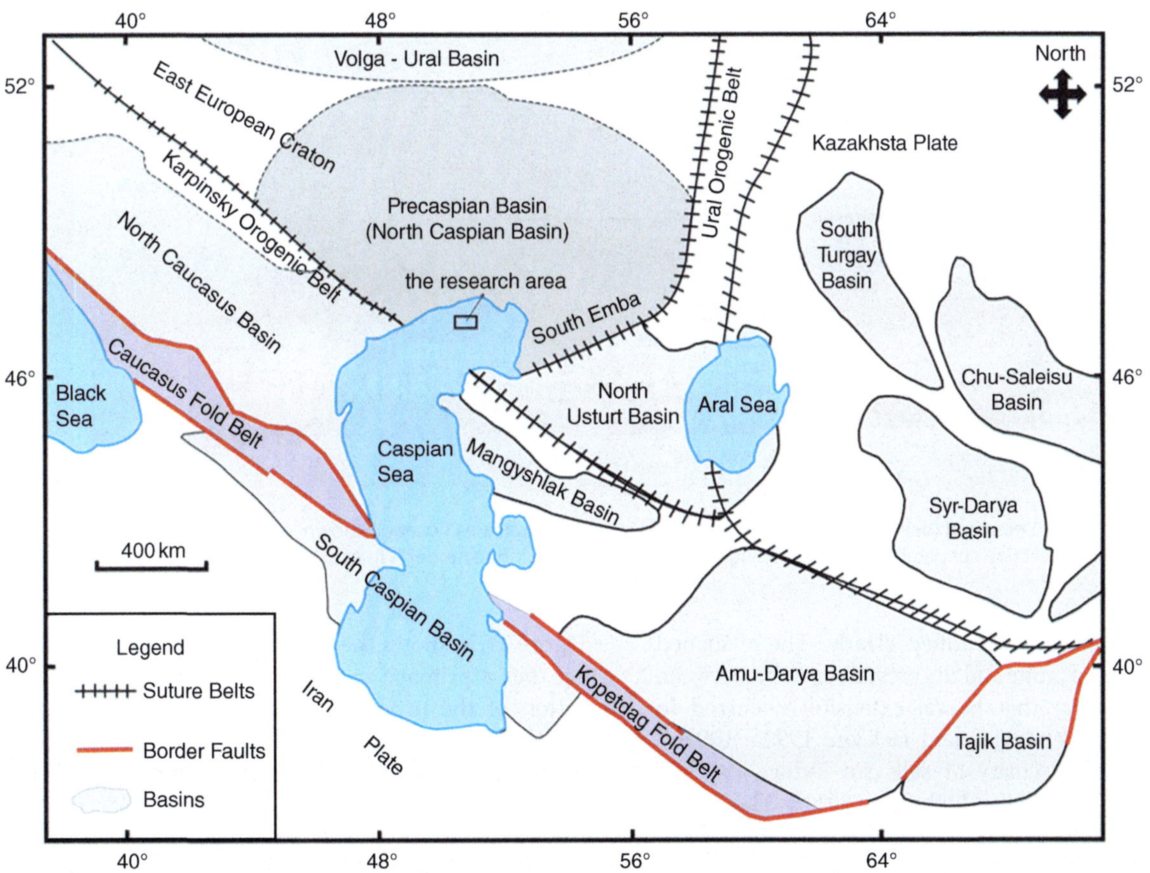

Figure 31.1 Location of the seismic profiles in the PreCaspian Basin (North Caspian).

Atlas of Structural Geological Interpretation from Seismic Images, First Edition. Edited by Achyuta Ayan Misra and Soumyajit Mukherjee.

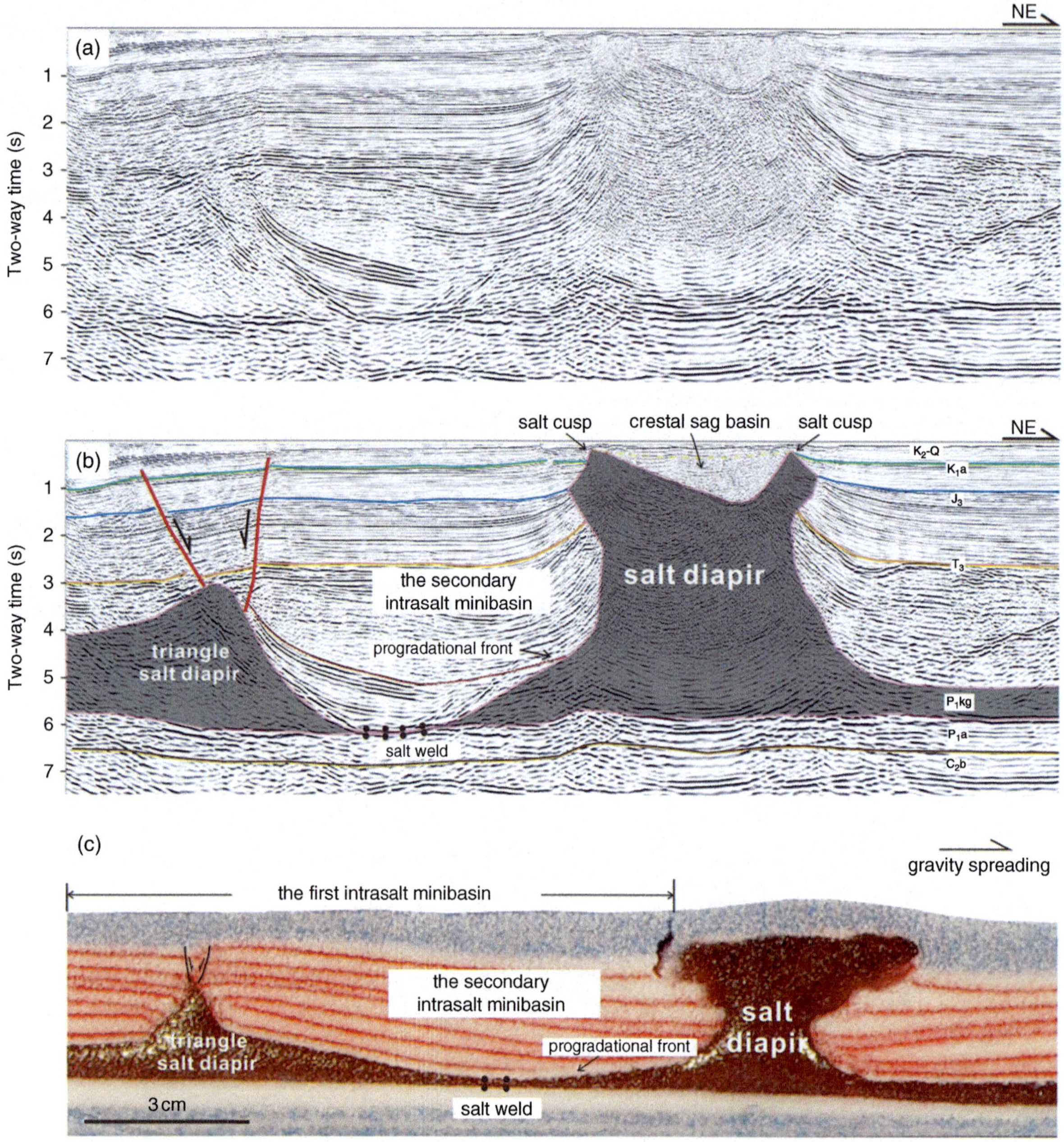

Figure 31.2 (a) Uninterpreted and (b) interpreted seismic section 01 in the research area (location shown in Figure 31.1) and (c) the photograph of a cross-section cut parallel in the spreading direction from an analogue model (Wu et al., 2015).

(Figure 31.2b) can be identified clearly. The V-shaped conjugate normal faults and the crestal sag basin between salt cusps suggest that lateral extension occurred in the study area (Vendeville and Jackson, 1992a, 1992b). A rather lower viscosity of salt can induce its flow (Mukherjee *et al.* 2010; Mukherjee, 2011). The lateral extension induced spreading and downbuilding of the main salt diapir along with the development of the crestal sag basin between salt cusps, and triggered a minor triangle reactive salt diapir (left in Figure 31.2).

Figure 31.2c shows the result from the analogue modeling that a primary salt diapir (right) occurred at the location of the progradational front and a secondary triangle salt diapir (left) occurred in the first intrasalt minibasin. Finally, a large salt diapir (indented diapir, right) developed at the progradational front.

Figures 31.3a and 31.3b show the interpretation of salt structures on the uplift basement in the Southern Precaspian Basin. Some triangle salt structures occur on the subsalt uplift basement. An obvious relationship

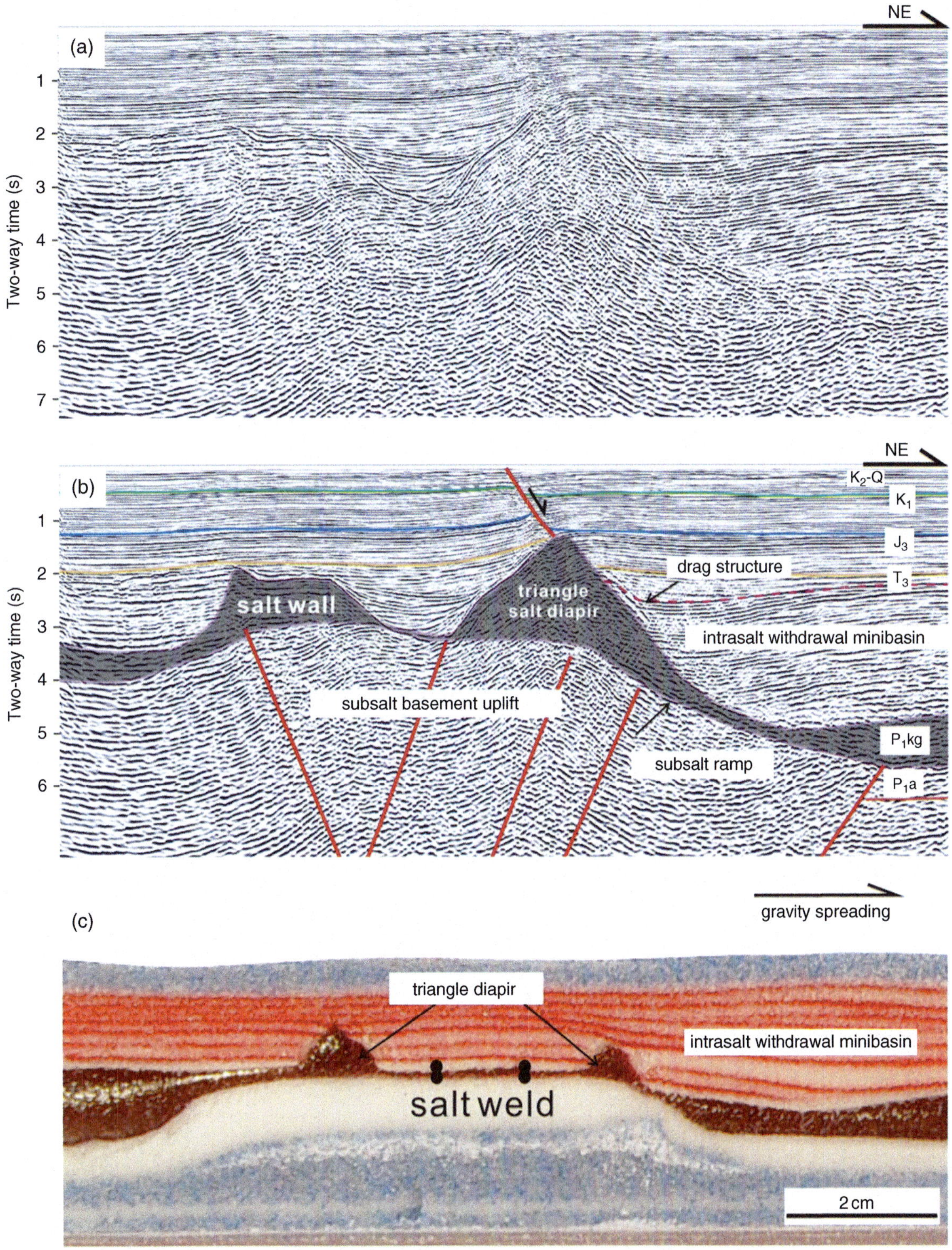

Figure 31.3 (a) Uninterpreted and (b) interpreted seismic section 02 in the research area (location shown in Figure 31.1) and (c) the photograph of a cross-section cut parallel in the spreading direction from an analogue model (Wu et al., 2015).

between the salt diapir (right) and the uplift can be observed. The seismic profile shows that the margin of the subsalt basement uplift (the hinge between the uplift and the slope) induced a salt diapir. Notice that salt welds invariably occur in the flanks of these salt diapirs. Moreover, the thickness of the overburden in the slope flank is far thicker than that in the uplift flank. Based on the relationship of the salt diapir and the subsalt uplift presented above, we think that the pattern of the subsalt basement structures is probably a main influence factor on the formation and evolution of the salt structures in the study area. It is proved by analogue modelling that a triangle salt structure develops on the subsalt uplift basement (Figure 31.3c) (Wu *et al.*, 2015).

From the observations we conclude:

- in the Southern Precaspian Basin, the typical styles of salt structures interpreted from the seismic profiles mainly include salt diapir (wall) with different sizes and shapes;
- according to the interpretation of seismic profiles and the analogue modelling, we suggest that the formation and distribution of these salt (salt-related) structures are related with differential loading and gravity spreading triggered by progradation and aggradation, and also the patterns of the subsalt basement structures.

References

Mukherjee, S. 2011. Estimating the viscosity of rock bodies – a comparison between the Hormuz- and the Namakdan Salt domes in the Persian Gulf, and the Tso Morari Gneiss dome in the Himalaya. *The Indian Journal of Geophysical Union*, **15**, 161–170.

Mukherjee, S., Talbot, C.J. and Koyi, H.A. 2010. Viscosity estimates of salt in the Hormuz and Namakdan salt diapirs, Persian Gulf. *Geological Magazine*, **147**, 497–507.

Vendeville B.C. and Jackson M.P.A. 1992a. The rise of diapirs during thin-skinned extension. *Marine and Petroleum Geology*, **9**(4), 331–354.

Vendeville B.C. and Jackson M.P.A. 1992b. The fall of diapirs during thin-skinned extension. *Marine and Petroleum Geology*, **9**(4), 354–371.

Wu Z., Yin H., Wang X. *et al.* 2015. The structural styles and formation mechanism of salt structures in the Southern Precaspian Basin: insights from seismic data and analog modeling. *Marine and Petroleum Geology*, **62**, 58–76.

32

Detection of Potential Slide Planes in Onshore Urban Areas Using Shear Wave Reflection Seismics

Charlotte M. Krawczyk[1]* *and U. Polom*[2]

[1] *GFZ German Research Centre for Geosciences, Potsdam, Germany; Technische Universität Berlin, Germany*
[2] *Leibniz Institute for Applied Geophysics, Hannover, Germany*
* *lotte@gfz-potsdam.de*

The safe development of urban areas requires a good understanding of soil conditions, particularly in areas prone to land sliding or other critical ground movements. A geophysical technique that may provide both structural image and shear wave velocity of the subsurface, an elastic parameter related to soil stiffness, is active shear wave reflection seismics (Polom *et al.*, 2008; Krawczyk *et al.*, 2013; Polom and Krawczyk, 2014).

In the bay of Trondheim, on the western coast of Norway, successions of fjord-deltaic deposits on top of marine and glaciomarine clays cover the bedrock. Glaciomarine deposits exposed locally above the sea level following deglaciation and the river outlet in the fjord moved continuously northwards in phase with Holocene delta progradation (see the summary in Hansen *et al.*, 2013). The harbour area of Trondheim suffered offshore slumping in historical times, while strong onshore subsidence is also deciphered. To detect potential slide planes in the onshore urban area, a shear wave reflection seismic survey was undertaken (Polom *et al.*, 2010; L'Heureux *et al.*, 2013).

The shear wave reflection seismic sections are presented in Figure 32.1 as, respectively, the stacked time section down to 1800 ms TWT (two-way travel time) (top), the time-migrated stacked section overlain by shear wave interval velocities (colour coded; middle), and the interpreted depth section cut at about 1500 ms, 250 m depth (bottom). Polom *et al.* (2010) detail acquisition and processing. These are interpreted in Hansen *et al.* (2013), where the main lithologies and seismic facies are also inferred from offshore sediment stratigraphy, onshore drillings and CPTs.

The seismic image of the sediments, overlying the bedrock at about 800–1200 ms TWT or 150–200 m depth, exhibits distinct sequences that can be tied seismo-stratigraphically to the well-calibrated marine sequences (Hansen *et al.*, 2013). Strong amplitude, continuous reflectors can be traced across the entire section, separating well-layered (400–800, 1000–1200 ms) and rather transparent sequences (50–400, 800–1000 ms). 10–25 m deep incoherent delta deposits overlie layered marine and prodeltaic deposits (25–90 m depth) with an intercalated turbidite sequence (50–60 m depth) that exposes fine internal roll-over structures (100–400 m distance, 400–600 ms, 50–60 m depth). Glaciomarine sediments level out the basement topography, formed by reverse and normal faulting.

Shear-wave velocities of about 100 $m\,s^{-1}$ occurring in the upper 50 m of the fjord-deltaic sediment succession suggest low sediment stiffness directly below 10–15 m thick and stiffer man-made fill. In addition, a possible future slip plane may be interpreted below this layer.

Compared to well-known P-wave data, the strong difference between TWT and depth values in the S-wave sections results from the shear wave velocities. For P-wave sections in such settings it is common that TWT is nearly the same as depth assuming a P-wave velocity of nearly ±2000 $m\,s^{-1}$ (if the wave propagation is below the ground water level in the saturated zone). This does not hold for S-waves, because they are always slower (Vs = max. 0.59*Vp in case of an ideal solid rock). Furthermore, S-waves propagate only in the matrix and not in the pore fluid like P-waves. This effect is visible in the upper part of the section, where 200 ms translate to only 20 m depth in the Holocene.

The low S-wave velocities in the sediments, therefore, cause a strong compression effect comparing time and depth sections. However, the much smaller wavelength improves the resolution up to a factor of 6–10. On the other hand, in the bedrock section, the Vp/Vs relationship is

Atlas of Structural Geological Interpretation from Seismic Images, First Edition. Edited by Achyuta Ayan Misra and Soumyajit Mukherjee.

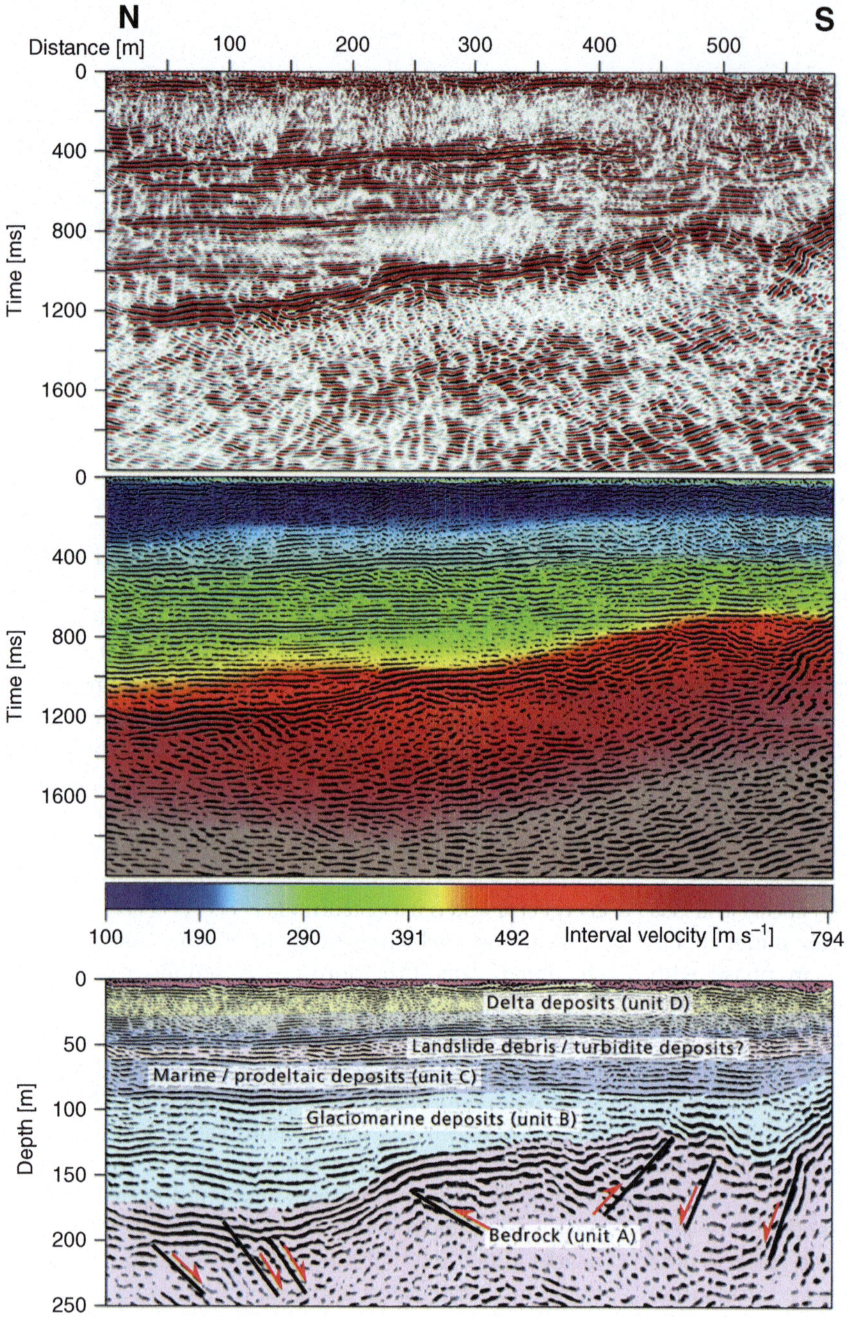

Figure 32.1 (top) Uninterpreted shear wave time section, (middle) shear velocity, (bottom) interpreted shear wave depth section showing normal and reverse faults. Polom *et al*., 2010. Reproduced with permission of Society of Exploration Geophysicists.

smaller, thus yielding less compression, which looks like a change in frequency on a first view. In reality, there is no change in frequency, as frequency is only valid in time sections, while we can only notice wavelengths in depth sections.

The strong benefit of S-waves over P-waves is the high resolution in the sedimentary part of the section: a vertical resolution of nearly 0.5 m is reached using S-waves, while for P-waves the best possible resolution would be nearly 5 m only.

References

Hansen, L., L'Heureux, J.S., Sauvin, G. *et al.* 2013. Effects of mass-wasting on the stratigraphic architecture of a fjord-valley fill: correlation of onshore, shear-wave seismic and marine seismic data at Trondheim, Norway. *Sedimentary Geology*, **289**, 1–18.

Krawczyk, C.M., Polom, U., Malehmir, A. and Bastani, M. 2013. Quick-clay landslides in Sweden – insights from shear wave reflection seismics and geotechnical integration. *Near Surface Geoscience 2013*, 19th EAGE European Meeting of Environmental and Engineering Geophysics.

L'Heureux, J.S., Long, M., Vanneste, M. *et al.* 2013. On the prediction of settlement from high-resolution shear-wave reflection seismic data: The Trondheim harbour case study, mid Norway. *Engineering Geology*, **167**, 72–83.

Polom, U., Krawczyk, C.M. and Malehmir, A. 2014. Shear wave reflection seismics image internal structure of quick-clay landslides in Sweden. American Geophysical Union Fall Meeting, 15–19 December, San Francisco, CA, abstract #NS41B-3842.

Polom, U., Arsyad, I. and Kümpel, H.-J., 2008. Shallow shear-wave reflection seismics in the tsunami struck Krueng Aceh river basin, Sumatra. *Advances in Geosciences*, **14**, 135–140.

Polom, U., Hansen, L., Sauvin, G. *et al.* 2010. High-resolution SH-wave seismic reflection for characterization of onshore ground conditions in the Trondheim harbor, central Norway. In: Miller, R.D., Bradford, J.D. and Holliger, K. (eds) *Advances in Near-Surface Seismology and Ground-Penetrating Radar*, The Society of Exploration Geophysicists, Tulsa, OK, pp. 297–312.

33

Detection of Mobile Quick-Clay Layers Using Shear Wave Reflection Seismics

Charlotte M. Krawczyk[1] *and U. Polom*[2]

[1] *GFZ German Research Centre for Geosciences, Potsdam, Germany; Technische Universität Berlin, Germany*
[2] *Leibniz Institute for Applied Geophysics, Hannover, Germany*
* *lotte@gfz-potsdam.de*

Covering many different sizes of scale, landslides are widespread and pose a severe hazard in many areas, as soon as humans or infrastructure are affected. In order to provide geophysical tools and techniques to better characterize sites prone to sliding, a geophysical assessment was made in Sweden (Malehmir *et al.*, 2013, 2016), supplemented by borehole sediment sampling and downhole physical property measurements (Salas-Romero *et al.*, 2015).

The study site, Lilla Edet, lies close to Gothenburg, Sweden, in an area where quick-clay landslides are known. Quick-clays are originally marine deposited clays with a pronounced salt content that, if leached by freshwater, lose their strength and may lead to land sliding. At the site of the investigation, a mixture of silty clay to clayey silt and gravel is reported; this was uplifted after deglaciation and allowed freshwater intrusion (Lundström *et al.*, 2009).

Aiming at high-resolution imaging of Quaternary sediments and quick-clays at shallow depth, the presented section was measured using an active reflection seismic survey set up using shear waves (Krawczyk *et al.*, 2013a), allowing for a vertical resolution of 0.5–1 m. The presented profile is given as a migrated depth section (Figure 33.1) and interpreted depth section including borehole control (Figure 33.2). Details can be found elsewhere (Krawczyk *et al.*, 2013b; Polom *et al.*, 2013).

In the upper 40 m of the SW–NE profile, sub-/horizontally layered sediments are visible above the top basement. The top basement, which crops out at both profile ends, is imaged between approximately 10 and 40 m depth with a bright amplitude pattern and bowl-shape morphology centring almost in the middle of the profile. Several distinct sediment sequences are separated by high amplitude unconformities that were later calibrated by borehole data (Salas-Romero *et al.*, 2015). The structural seismic image shows an unprecedented continuity of reflectors and sequence patterns, matching well to the borehole-evidenced lithology.

Apparently, different sequences show different degrees of seismic transparency; however, they do not indicate a certain lithology. With respect to the quick-clays, their sequences seem to be characterized by a certain degree of transparency but different level of reflector continuity. The uppermost sand–silt–gravel layer below the first quick-clay layer can be clearly traced along the whole profile marked by a reflection pattern of higher amplitudes and a transgressive cut of the structures below. This layer is a potential path for freshwater intrusion between the two quick-clay layers and may support salt leaching and quick-clay destabilization.

In summary, the profile exhibits:

- clear internal structuring of the up to 40 m thick marine sediments, with quick-clay intercalations at about 10, 15 and 25 m depths, as well as strong undulations of top basement underneath;
- a considerably improved structural resolution, up to a factor of 10 higher than P-wave data usually allow.

Atlas of Structural Geological Interpretation from Seismic Images, First Edition. Edited by Achyuta Ayan Misra and Soumyajit Mukherjee.

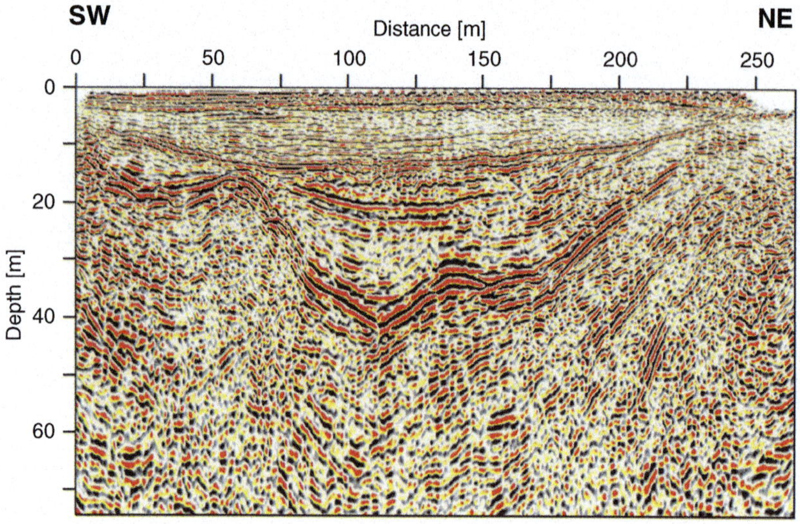

Figure 33.1 Migrated depth section. Note depth of the section and the resolution. Modified from Krawczyk *et al.*, 2013b.

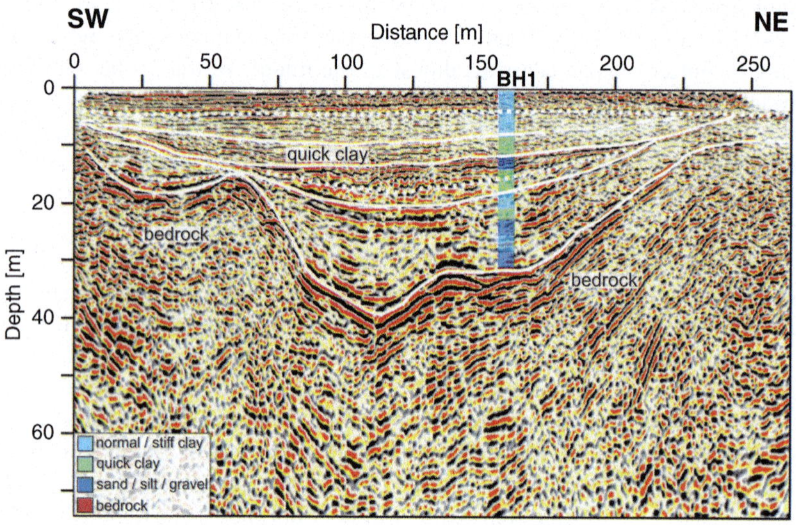

Figure 33.2 Interpreted depth section including borehole control. Modified from Krawczyk *et al.*, 2013b.

References

Krawczyk, C.M., Polom, U. and Beilecke, T. 2013a. Shear wave reflection seismics as valuable tool for near-surface urban applications. *The Leading Edge*, **32**, 256–263.

Krawczyk, C.M., Polom, U., Malehmir, A. and Bastani, M. 2013. Quick-clay landslides in Sweden – insights from shear wave reflection seismics and geotechnical integration. *Near Surface Geoscience 2013*, 19th EAGE European Meeting of Environmental and Engineering Geophysics.

Lundström, K., Larsson, R. and Dahlin, T. 2009. Mapping of quick-clay formations using geotechnical and geophysical methods. *Landslides*, **6**, 1–15.

Malehmir, A., Bastani, M., Krawczyk, C.M. *et al.* 2013. Geophysical assessment and geotechnical investigation of quick-clay landslides – a Swedish case study. *Near Surface Geophysics*, **11**, 341–350.

Malehmir, A., Socco, L.V., Bastani, M. *et al.* 2016. Near-surface geophysical characterization of areas prone to natural hazards: a review of the current and perspective on the future. In: Nielsen, L. (ed) *Advances in Geophysics*, vol. 57, Academic Press, pp. 51–146.

Polom, U., Bagge, M., Wadas, S. *et al.* 2013. Surveying near-surface depocentres by means of shear wave seismics. *First Break*, **31**, 63–75.

Salas-Romero, S., Malehmir, A., Snowball, I. *et al.* 2015. Identifying landslide preconditions in Swedish quick clays – insights from integration of surface geophysical, core sample- and downhole property measurements. *Landslides*, **13**, 905–923.

34

Growth Fault-Thrust Belts with Single Detachment in the Krishna–Godavari Basin, India

Achyuta Ayan Misra

Reliance Industries Ltd, Mumbai, India
Achyuta.Misra@ril.com; achyutaayan@gmail.com

Normal faults and linked thrust belts are present in many deep water deltaic systems (Cohen and McClay, 1996; Morley and Guerin, 1996; Gupta, 2006; Mourgues *et al.*, 2009; Choudhuri *et al.*, 2010). The normal faults detach over a mobile shale layer and this extension compensate by shortening in the form of folds, termed 'shale bulges', or thrusts, commonly called 'toe thrusts'. Normally, a layer of overpressurized shales is involved in such linked normal fault-thrust systems. Vide Dasgupta and Mukherjee (submitted) for detail of overpressure issues. Presented here is a seismic section from the Krishna–Godavari Basin showing a linked growth fault-thrust belt (GFTB) (Figure 34.1). See Chatterjee *et al.* (2015) and Roy *et al.* (2015) for details of the tectonics of the basin. The growth faults here occurred during the Oligocene (Choudhuri *et al.*, 2010). The Oligocene depocentre was present in this area and it shifted further south. Shale movement during this time is present in the Krishna delta area and south of it. The geometries that can be observed are large-scale normal growth faults with large throws and steep-to-gently dipping toe thrusts (Figures 34.2a–34.2h). There are antithetic faults accompanying the growth faults (Figures 34.2c and 34.2d). The faults are steeper in the proximal area than that at the distal region (Figures 34.2e and 34.2f). The faults can be identified by nearly all the seven criteria of fault identification on seismic data as detailed in Chapter 2.3.

Atlas of Structural Geological Interpretation from Seismic Images, First Edition. Edited by Achyuta Ayan Misra and Soumyajit Mukherjee.

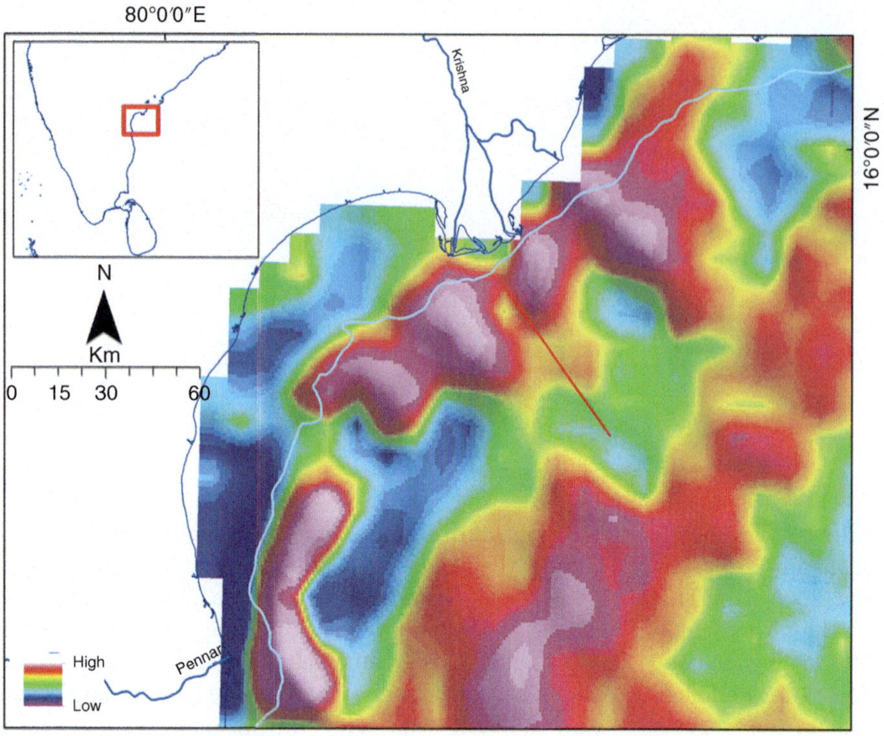

Figure 34.1 Map of the Krishna–Godavari Basin along the ECMI overlaid on hill-shaded, upward-continued (10 km) Bouguer gravity anomaly data (see the colour scale). Blue line: shelf-slope break. Red line: position of seismic section.

Figure 34.2a Uninterpreted seismic section in the Krishna–Godavari Basin showing a growth fault-thrust belt. See Figure 34.1 for location. Seismic data courtesy Reliance Industries Ltd. Reproduced with permission from the Directorate General of Hydrocarbons (DGH), India.

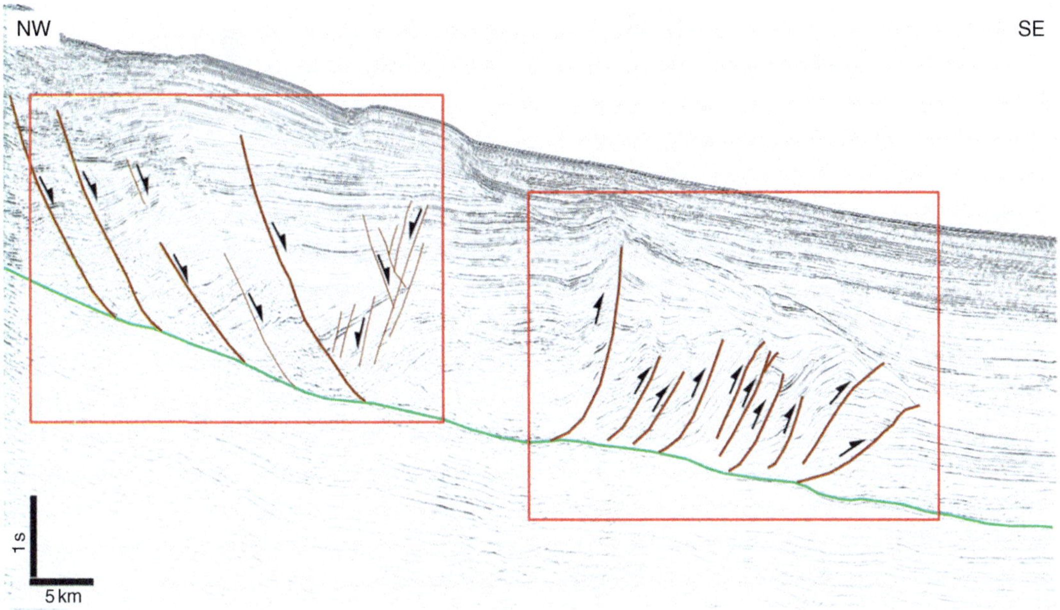

Figure 34.2b Intrepretation of seismic section shown in Figure 34.2a. Note the fault identification criteria from the normal faults and the thrusts. The area of little to no deformation between the normal and thrust faults is called the translational zone. Green horizon: detachment. Red rectangles: locations of sections in Figure 34.2c and 34.2f. Seismic data courtesy Reliance Industries Ltd. Reproduced with permission from the Directorate General of Hydrocarbons (DGH), India.

Figure 34.2c Uninterpreted seismic section showing the details of the normal faults in the GFTB. See Figure 34.2b for location of the profile. Seismic data courtesy Reliance Industries Ltd. Reproduced with permission from the Directorate General of Hydrocarbons (DGH), India.

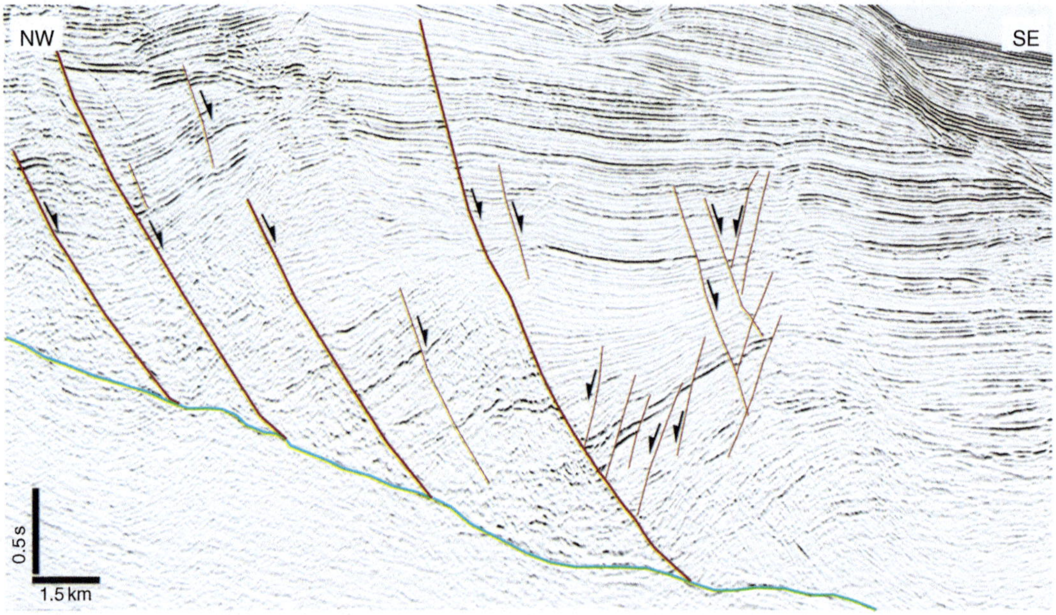

Figure 34.2d Interpretation of seismic section shown in Figure 34.2c. Note the criteria for identification of faults (Chapter 2.3) and the growth of the sediments towards the fault planes. Green horizon: detachment. Seismic data courtesy Reliance Industries Ltd. Reproduced with permission from the Directorate General of Hydrocarbons (DGH), India.

Figure 34.2e Uninterpreted seismic section showing the details of the toe thrusts in the GFTB. See Figure 34.2b for location of the profile. Seismic data courtesy Reliance Industries Ltd. Reproduced with permission from the Directorate General of Hydrocarbons (DGH), India.

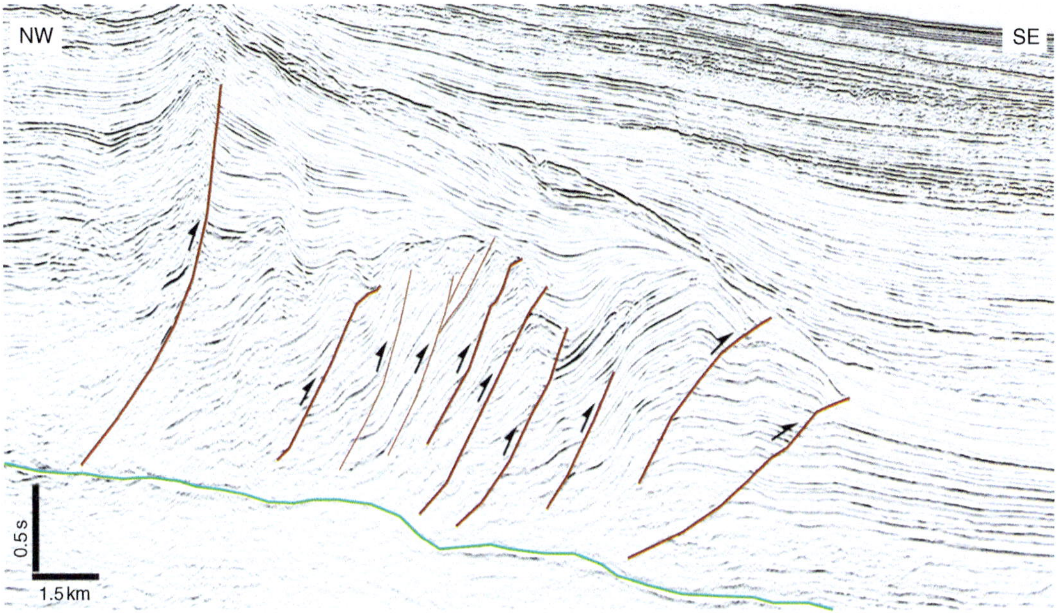

Figure 34.2f Interpretation of seismic section shown in Figure 34.2f. Note the decreasing dips of the thrusts towards the distal end of the thrust belt. Green horizon: detachment. Seismic data courtesy Reliance Industries Ltd. Reproduced with permission from the Directorate General of Hydrocarbons (DGH), India.

Figure 34.2g Uninterpreted seismic section showing the details of the toe thrusts in the GFTB without vertical exaggeration. See Figure 34.2b for location of the profile. Seismic data courtesy Reliance Industries Ltd. Reproduced with permission from the Directorate General of Hydrocarbons (DGH), India.

Figure 34.2h Interpretation of seismic section shown in Figure 34.2g. Note the visual change in geometry in the seismic section of different vertical exaggerations. Green horizon: detachment. Seismic data courtesy Reliance Industries Ltd. Reproduced with permission from the Directorate General of Hydrocarbons (DGH), India.

References

Chatterjee, R., Paul, S., Singha, D.P. and Mukhopadhyay, M. 2015. Overpressure zones in relation to in situ stress for the Krishna-Godavari Basin, Eastern Continental Margin of India: implications for hydrocarbon prospectivity. In: Mukherjee, S. (ed) *Petroleum Geosciences: Indian Contexts*, Springer Geology, pp. 127–142.

Choudhuri, M., Guha, D., Dutta, A. *et al.* 2010. Spatiotemporal variations and kinematics of shale mobility in the Krishna-Godavari basin, India. In: Wood, L. (ed) *Shale Tectonics*, AAPG Memoir 93, American Association of Petroleum Geologists, pp. 91–109.

Cohen, H.A. and McClay, K. 1996. Sedimentation and shale tectonics of the northwestern Niger Delta front. *Marine and Petroleum Geology*, **13**, 313–328.

Dasgupta, T., Mukherjee, S. Submitted. Sediment Compaction and Applications in Petroleum Geoscience. Springer. Series: Advances in Oil and Gas Exploration & Production. ISSN: 2509-372X. Series Editor: Swenner, R.

Gupta, S.K. 2006. Basin architecture and petroleum system of Krishna Godavari Basin, east coast of India. *The Leading Edge*, **25**, 830–837.

Morley, C.K. and Guerin, G. 1996. Comparison of gravity-driven deformation styles and behavior associated with mobile shales and salt. *Tectonics*, **15**, 1154–1170.

Mourgues, R., Lecomte, E., Vendeville, B. and Raillard, S. 2009. An experimental investigation of gravity-driven shale tectonics in progradational delta. *Tectonophysics*, **474**, 643–656.

Roy, S., Choudhuri, M. and Gupta, P. 2015. Rift grabens and crustal architecture of the offshore North East Coast-Mahanadi Basin, Eastern Continental Margin of India. In: Mukherjee, S. (ed) *Petroleum Geosciences: Indian Contexts*, Springer Geology, pp. 63–86.

35

Growth Fault-Thrust Belts with Multiple Detachments in the Krishna–Godavari Basin, India

Achyuta Ayan Misra

Reliance Industries Ltd, Mumbai, India
Achyuta.Misra@ril.com; achyutaayan@gmail.com

In the previous chapter an example of a linked growth fault-thrust belt (GFTB) in the Krishna–Godavari Basin in eastern India was shown. See Chatterjee *et al.* (2015) and Roy *et al.* (2015) for details of the tectonics of the basin. That was a GFTB with a single detachment. However, in this chapter, the complexity is added to by displaying a profile of another GFTB, this with two or possibly three detachments. This section lies between the Krishna- and the Godavari delta mouths (Figure 35.1). There is an older detachment of Paleocene-Eocene age and one or two younger Miocene-Pliocene detachments (Choudhuri *et al.*, 2010). In the Paleocene-Eocene the depocentre was present in the mouths of both the deltas. It then shifted south of the Krishna delta mouth during the Oligocene and again returned back to its initial location (Choudhuri *et al.*, 2010). This change in depocentre might have produced such multiple detachments. The Paleocene-Eocene growth fault system developed a toe thrust system, which is visible in the section itself. On the other hand, the younger system does not form a shale bulge or a toe thrust visible on the section (Figures 35.2a–35.2d). It may form one further southeast of the section. Thus, the translation zone of the younger system may be larger than the older one. The formation of a shale bulge or toe thrust depends on a number of factors: (i) a sudden spatial decrease in the dip of the detachment, (ii) an increase in the friction of the detachment, (iii) a buttress, for example a basement ridge, (iv) sedimentation and differential compaction in the area and (v) a sudden spatial change in pore pressure in the shale layer. Since the pore pressure may vary, formation of shale bulge or toe thrust can effectively modify the local center of gravity of the rock mass (Mukherjee in press).

Atlas of Structural Geological Interpretation from Seismic Images, First Edition. Edited by Achyuta Ayan Misra and Soumyajit Mukherjee.

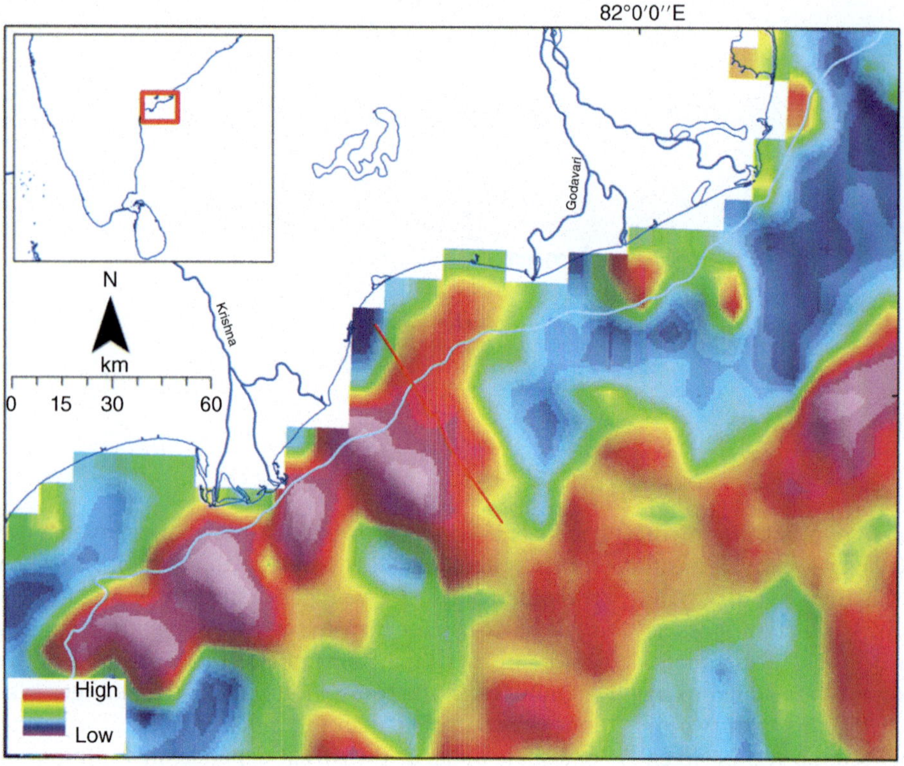

Figure 35.1 Map of the Krishna–Godavari Basin along the ECMI overlaid on hill-shaded, upward-continued (10 km) Bouguer gravity anomaly data (see the colour scale). Blue line: shelf-slope break. Red line: position of seismic section.

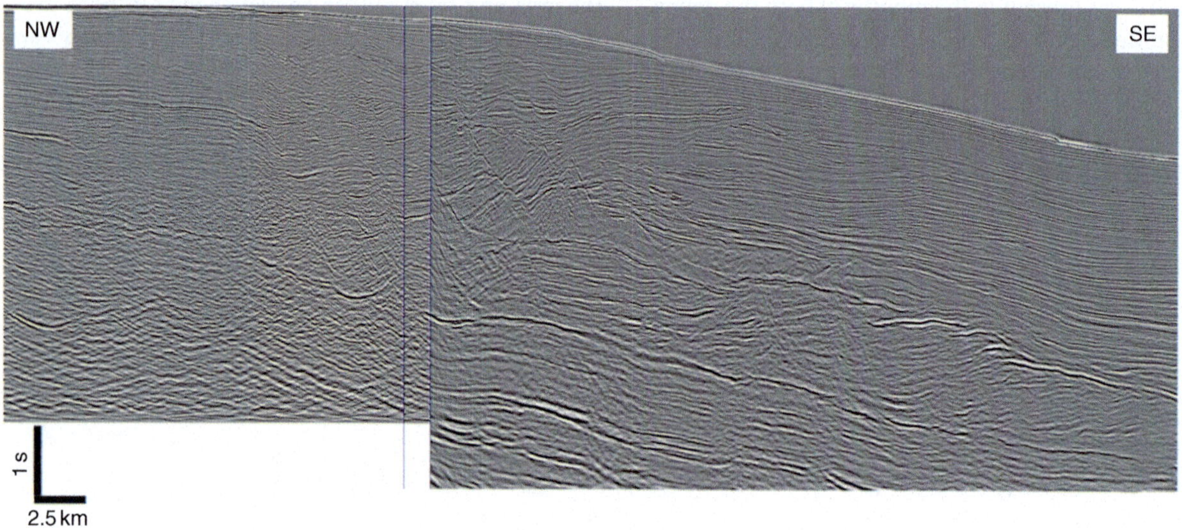

Figure 35.2a Uninterpreted seismic section in the Krishna–Godavari Basin showing a growth fault-thrust belt with multiple detachments. See Figure 35.1 for location. Seismic data courtesy Reliance Industries Ltd. Reproduced with permission from the the Directorate General of Hydrocarbons (DGH), India.

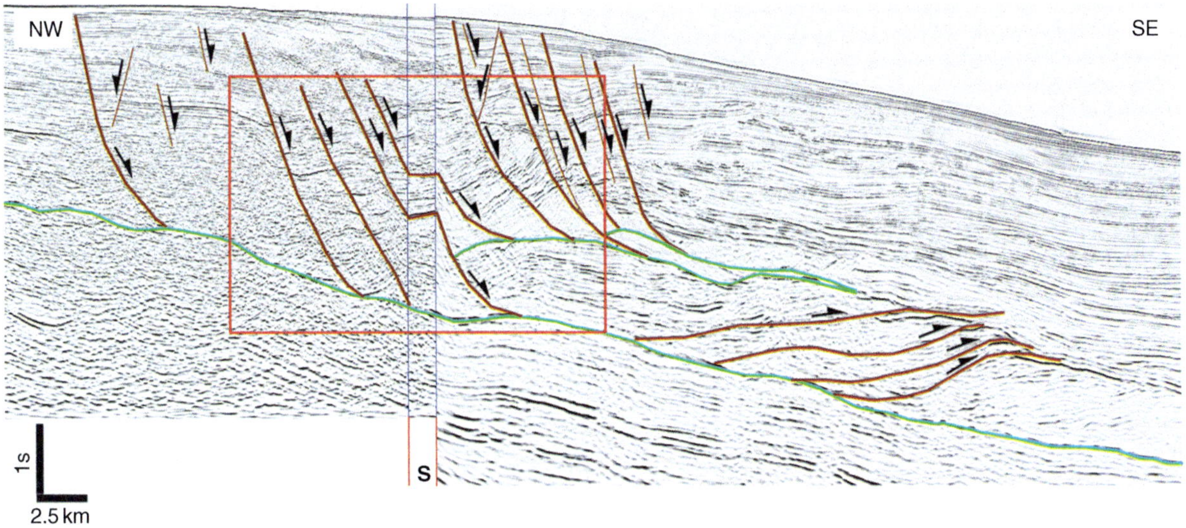

Figure 35.2b Intrepretation of seismic section shown in Figure 35.2a. The translational zone of the older GFTB is smaller than the younger one. The shale bulge or toe thrust of the younger GFTB may be further SE from this section. Green horizon: detachment. Red rectangle: location of sections in (c-d). There is a strike oriented part in this seismic section marked "S". Seismic data courtesy Reliance Industries Ltd. Reproduced with permission from the the Directorate General of Hydrocarbons (DGH), India.

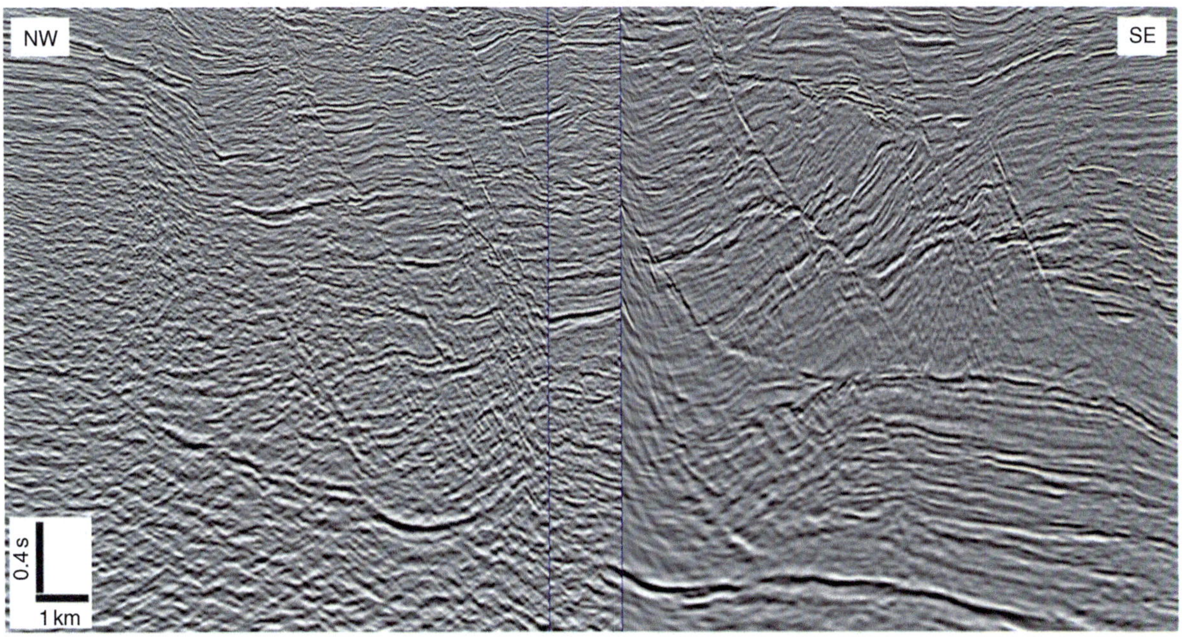

Figure 35.2c Uninterpreted seismic section showing the details of the normal faults in the GFTB. See Figure 35.2b for location of the profile. Note the presence of noise in the lower parts of this seismic section. Seismic data courtesy Reliance Industries Ltd. Reproduced with permission from the the Directorate General of Hydrocarbons (DGH), India.

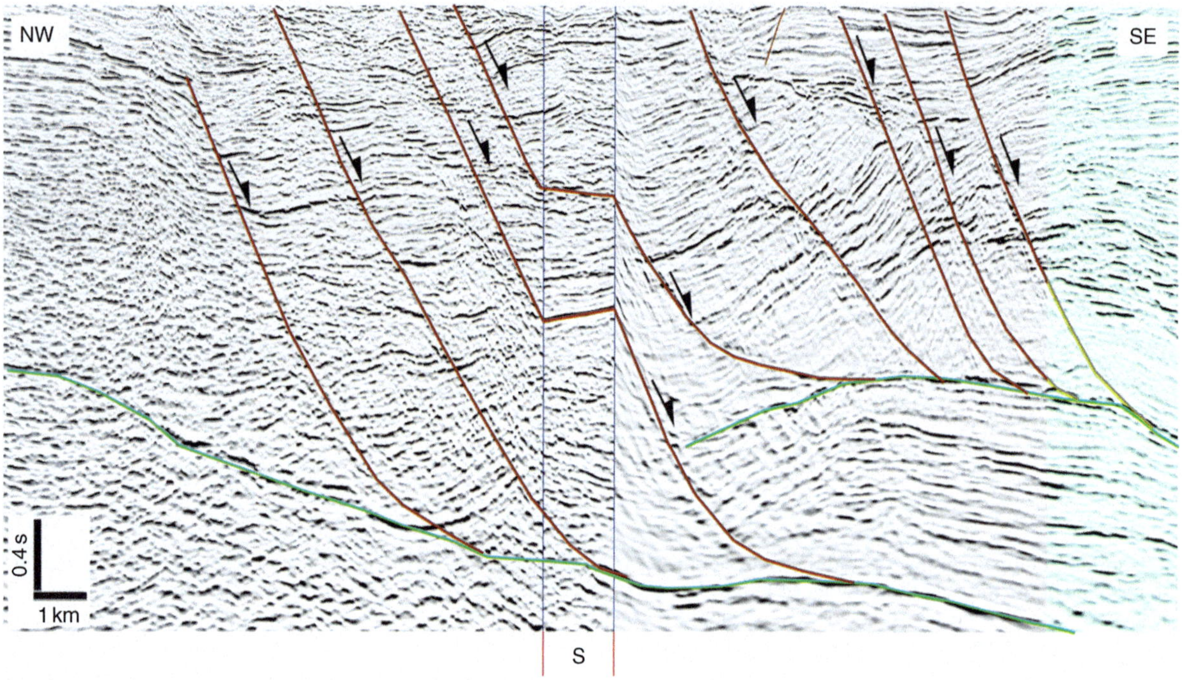

Figure 35.2d Interpretation of seismic section shown in Figure 35.2c. Note the criteria for identification of faults (Chapter 2.3) and the growth of the sediments towards the fault planes. Green horizon: detachment. S: strike-oriented part of the seismic section. Seismic data courtesy Reliance Industries Ltd. Reproduced with permission from the the Directorate General of Hydrocarbons (DGH), India.

References

Chatterjee, R., Paul, S., Singha, D.P. and Mukhopadhyay, M. 2015. Overpressure zones in relation to in situ stress for the Krishna-Godavari Basin, Eastern Continental Margin of India: implications for hydrocarbon prospectivity. In: Mukherjee, S. (ed) *Petroleum Geosciences: Indian Contexts*, Springer Geology, pp. 127–142.

Choudhuri, M., Guha, D., Dutta, A. *et al.* 2010. Spatiotemporal variations and kinematics of shale mobility in the Krishna-Godavari basin, India. In: Wood, L. (ed) *Shale Tectonics*, AAPG Memoir 93, American Association of Petroleum Geologists, pp. 91–109.

Mukherjee, S. (in press) Locating center of gravity in geological contexts. *International Journal of Earth Sciences*. DOI: 10.1007/s00531-017-1560-z

Roy, S., Choudhuri, M. and Gupta, P. 2015. Rift grabens and crustal architecture of the offshore North East Coast-Mahanadi Basin, Eastern Continental Margin of India. In: Mukherjee, S. (ed) *Petroleum Geosciences: Indian Contexts*, Springer Geology, pp. 63–86.

36

Growth Fault-Thrust Belts with a Small Translation Zone in the Mumbai Offshore Basin, India

Achyuta Ayan Misra

Reliance Industries Ltd, Mumbai, India
Achyuta.Misra@ril.com; achyutaayan@gmail.com

In Chapters 35 and 36, examples of linked growth fault thrust belts (GFTBs) from the eastern continental margin of India were shown. Also discussed were translation zones, which are the zones of little to no deformation between the extensional and compressional domains in a GFTB. The locations of the toe thrusts or shale bulges depend on a few factors (e.g. Choudhuri *et al.*, 2010), as mentioned in Chapter 36. In this chapter, yet another example is presented of a GFTB from the Mumbai/Bombay Offshore Basin, West India (Figure 36.1). See Collier *et al.* (2008), Armittage *et al.* (2011), Misra *et al.* (2014, 2015, 2017) and Misra and Mukherjee (2015, 2017) for information about the tectonics of the western continental margin of India. This GFTB has a very small translation zone (Figures 36.2a–36.2d). The normal faults are not clearly visible in the current data (Figures 36.2a and 36.2b). The toe thrusts continue to very shallow levels in the

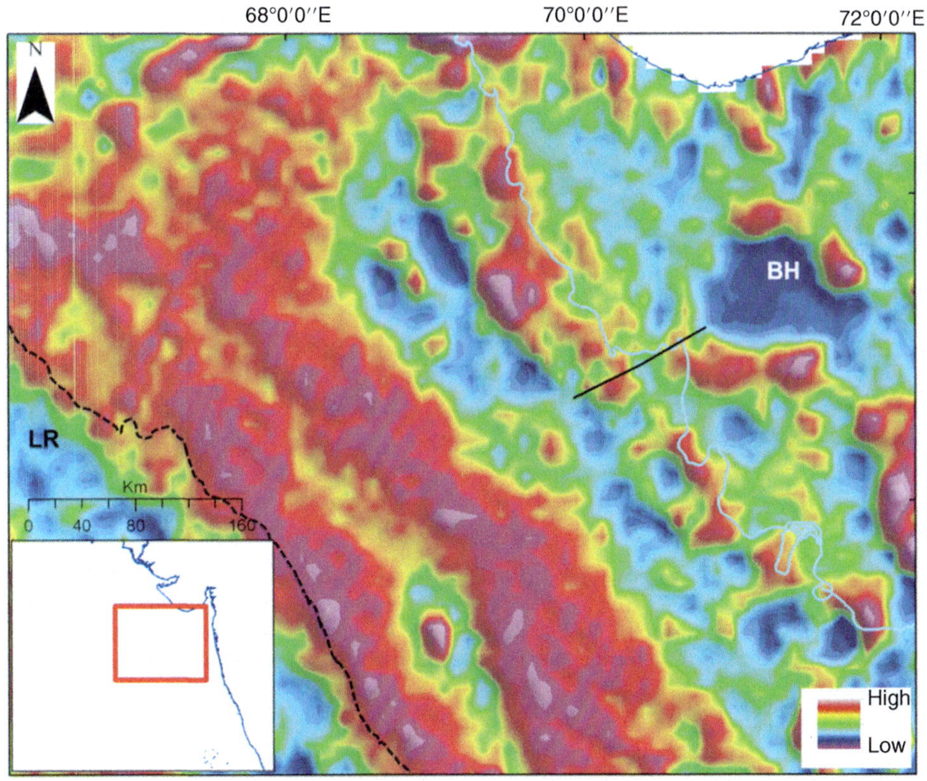

Figure 36.1 Map of the Mumbai Offshore Basin, west India overlaid on hill-shaded, upward-continued (10 km) Bouguer gravity anomaly data (see the colour scale). Blue line: shelf-slope break. Black line: position of seismic section. LR: Laxmi Ridge; BH: Bombay High.

Atlas of Structural Geological Interpretation from Seismic Images, First Edition. Edited by Achyuta Ayan Misra and Soumyajit Mukherjee.

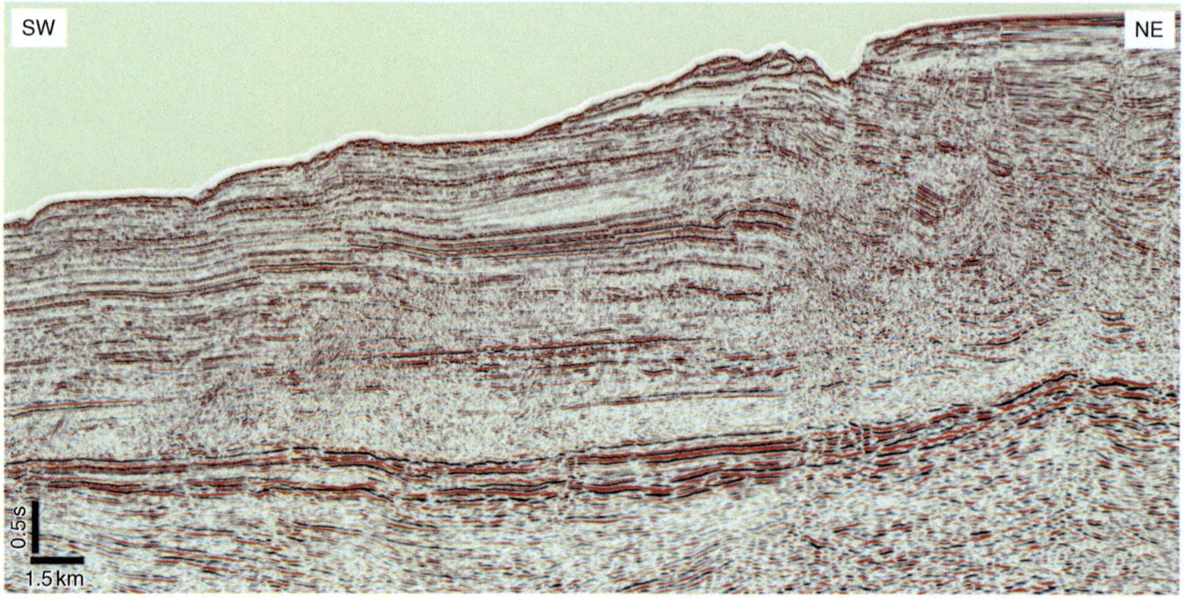

Figure 36.2a Uninterpreted seismic section in the Mumbai Offshore Basin showing a growth fault-thrust belt with a small translation zone. See Figure 36.1 for location. Seismic data courtesy Reliance Industries Ltd. Reproduced with permission from the the Directorate General of Hydrocarbons (DGH), India.

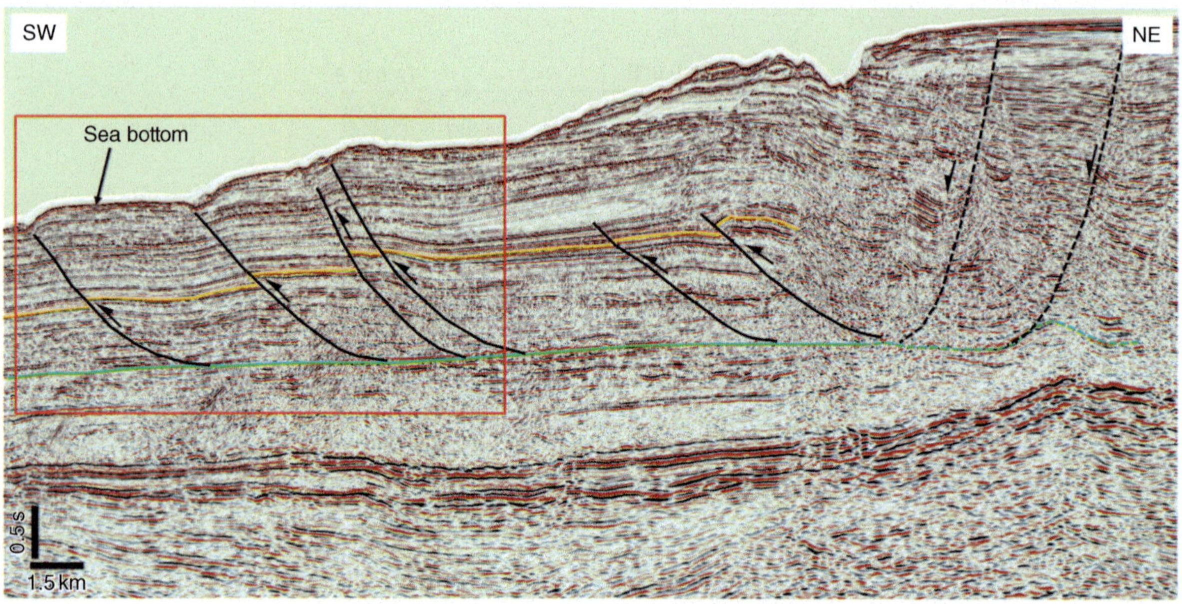

Figure 36.2b Intrepretation of seismic section shown in Figure 36.2a. The translational zone is small here. The toe thrusts are widely spaced and reach up to very shallow levels. Note that the normal faults are difficult to observe. Minor drag related to faults (Mukherjee, 2014) noted. Green horizon: detachment. Red rectangle: location of the the seismic sections in Figures 36.2c and 36.2d. Seismic data courtesy Reliance Industries Ltd. Reproduced with permission from the the Directorate General of Hydrocarbons (DGH), India.

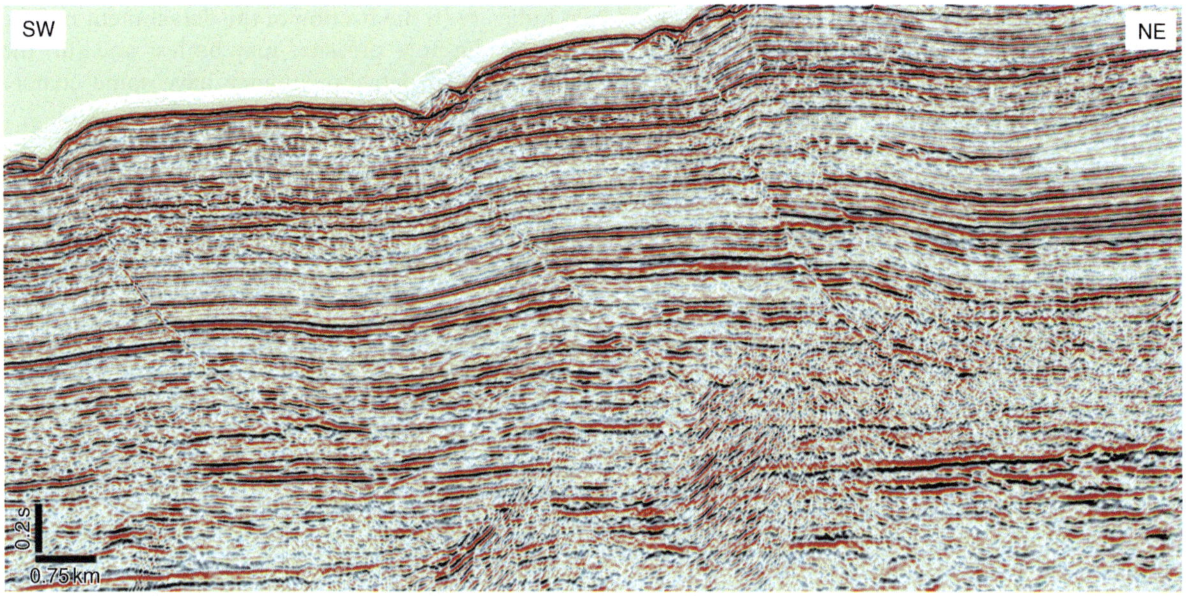

Figure 36.2c Uninterpreted seismic section showing the details of the toe thrusts in the GFTB. See Figure 36.2a for location. Seismic data courtesy Reliance Industries Ltd. Reproduced with permission from the the Directorate General of Hydrocarbons (DGH), India.

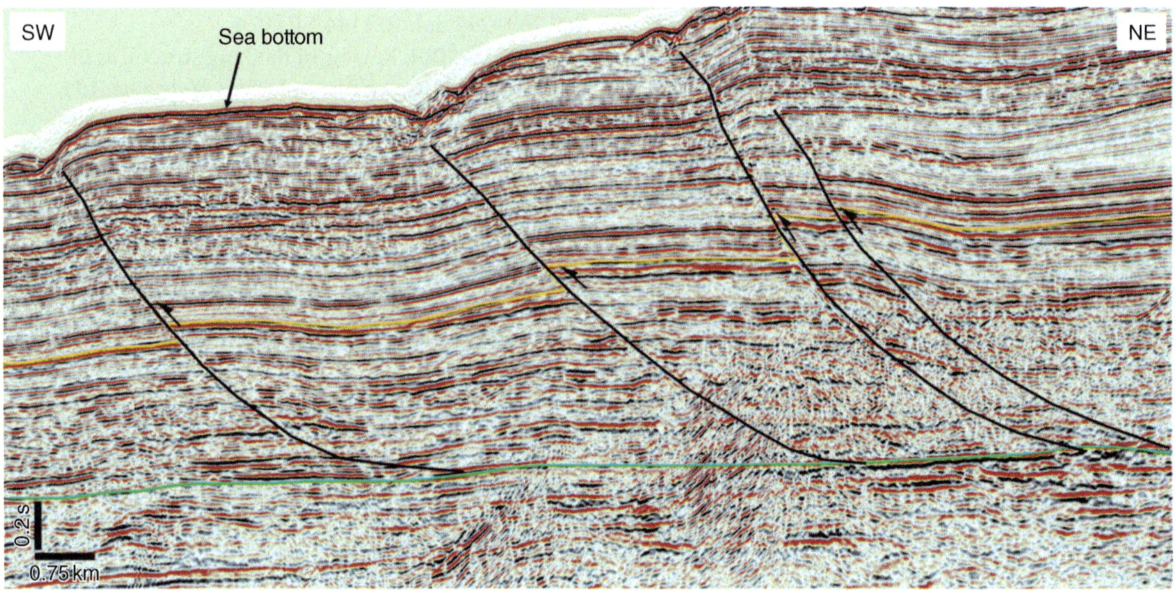

Figure 36.2d Interpretation of seismic section shown in Figure 36.2c. Note that the toe thrusts are widely spaced and reach up to very shallow levels. Green horizon: detachment. Seismic data courtesy Reliance Industries Ltd. Reproduced with permission from the the Directorate General of Hydrocarbons (DGH), India.

section and are very widely spaced. The difference in the geometry is most plausibly due to the nature of the detachment. The most likely differences of the detachment in this seismic section with the earlier ones from east India are (i) the friction of the detachment may be more, (ii) the pore pressure may be less and (iii) the lithology of the detachment may have some coarser clastics mixed with shale.

References

Armitage, J.J., Collier, J.S., Minshull, T.A. and Henstock, T.J. 2011. Thin oceanic crust and flood basalts: India-Seychelles breakup. *Geochemistry, Geophysics, Geosystems*, **12**(5).

Choudhuri, M., Guha, D., Dutta, A. *et al.* 2010. Spatiotemporal variations and kinematics of shale mobility in the Krishna-Godavari basin, India. In: Wood, L. (ed) *Shale Tectonics*, AAPG Memoir 93, American Association of Petroleum Geologists, pp. 91–109.

Collier, J.S., Sansom, V., Ishizuka, O., Taylor, R.N., Minshull, T.A. and Whitmarsh, R.B. 2008. Age of Seychelles–India break-up. Earth and Planetary Science Letters, **272**(1), pp. 264–277.

Misra, A.A., Bhattacharya, G., Mukherjee, S. and Bose, N. 2014. Near N-S paleo-extension in the western Deccan region in India: Does it link strike-slip tectonics with India-Seychelles rifting? *International Journal of Earth Sciences*, **103**, 1645–1680.

Misra, A.A., Sinha, N. and Mukherjee, S. 2015. Repeat ridge jumps and microcontinent separation: insights from NE Arabian Sea. *Marine and Petroleum Geology*, **59**, 406–428.

Misra, A.A., Banerjee, S., Kundu, N. and Mukherjee, B. 2017. Subsidence around oceanic ridges along passive margins: NE Arabian Sea. In: Mukherjee, S., Misra, A.A., Calvés, G. and Nemčok, M. (eds) *Tectonics of the Deccan Large Igneous Province*, Special Publications vol. 445, The Geological Society, London, pp. 119–149. doi: 10.1144/SP445.10.

Misra, A.A. and Mukherjee, S. 2015. *Tectonic Inheritance in Continental Rifts and Passive Margins*. SpringerBriefs in Earth Sciences, Springer International.

Misra, A.A. and Mukherjee, S. 2017. Dyke-brittle shear relationships in the Western Deccan Strike-slip zone around Mumbai (Maharashtra, India). In: Mukherjee, S., Misra, A.A., Calvès, G. and Nemčok, M. (eds) *Tectonics of the Deccan Large Igneous Province*, Special Publications vol. 445, The Geological Society, London, pp. 269–295 doi: 10.1144/SP445.4.

Mukherjee, S. 2014. Review of flanking structures in meso- and micro-scales. *Geological Magazine*, **151**, 957–974.

37

Growth Structures and Unconformities in the Upper Evaporites, Red Sea

Neil C. Mitchell[1] and Marco Ligi[2]*

[1] *University of Manchester, United Kingdom*
[2] *ISMAR-CNR, Bologna, Italy*
* *neil.mitchell@manchester.ac.uk*

Most evaporite (salt) deposits found at continental margins have been deeply buried and remobilized, so the original structures formed during and shortly after their deposition are now obscured. In the Red Sea (Figure 37.1), in contrast, thick evaporites (>1 km) were deposited from the mid-Miocene up to the end of the Miocene (about 5.3 Ma) (Hughes *et al.*, 1992), which is relatively recent compared with other ocean margins. Away from the coasts, the evaporites are covered by hemipelagic sediment of a density similar to that of the evaporates, so diapirism there is not as strongly promoted by sedimentary loading as in the continental marginal settings. An oceanic spreading centre runs down the Red Sea; the tectonic extension associated with this feature and the relief of its axial valley has left many of the evaporites un-buttressed, so that they have flowed into the valley in spectacular flows or 'salt glaciers' observed in multibeam sonar data (Mitchell *et al.*, 2010).

In seismic reflection images (Figure 37.2), the top of the evaporites in the Red Sea is marked by a prominent reflection called 'S' (Ross and Schlee, 1973). Deep-seismic images can show the evaporites to be either massive (transparent reflectivity) or layered (reflective) (Izzeldin, 1987). The former are suspected to represent massive halite deposits, while the latter represent halite interlayered with anhydrite and shale, as found at the deep sea drilling sites in the Red Sea (Stoffers and Kühn, 1974). The base of the evaporites was not imaged by the data, although away from the spreading centre it is expected to lie at generally >0.5 s two-way seismic time below seabed, based on results of a seismic refraction experiment here (Tramontini and Davies, 1969).

In the 2D seismic reflection image presented here (from Mitchell *et al.*, 2017), a reflective Plio-Pleistocene sequence overlies a prominent reflector (S). Beneath S on the left, an anticline can be observed (shot point (SP) 200). Evaporite reflections within and around the anticline vary in dip. This is evidence that the evaporites were mobile during deposition, as well as after deposition (a feature also found farther north in the Red Sea by Ehrhardt *et al.*, 2005). The character of the Plio-Pleistocene sediments overlying S also varies from being conformable with S to discordant. This and disruptions of the stratigraphy are likely the result of the continuing movements of the evaporites.

On the right around SP 400, dipping evaporite reflections beneath S show that the S reflection is locally an unconformity. Although distorted by the halokinetic deformation, this unconformity has little relief. Such flat unconformities in evaporites in the Mediterranean have been interpreted as due to surf erosion when the sea level was depressed and such areas were in the littoral zone, in contrast with more rugged unconformities indicating fluvial erosion (Bache *et al.*, 2009). If this unconformity represents the effects of surf erosion, it marks the local level of the Red Sea at the end of the Miocene. This unconformity lies deeper than other areas (not shown here) where the S reflection is not an unconformity, suggesting that the evaporites here have subsided relative to those other areas. This subsidence has accompanied the broad flowage suggested by the evaporite reflectors and by the multibeam sonar data.

Atlas of Structural Geological Interpretation from Seismic Images, First Edition. Edited by Achyuta Ayan Misra and Soumyajit Mukherjee.

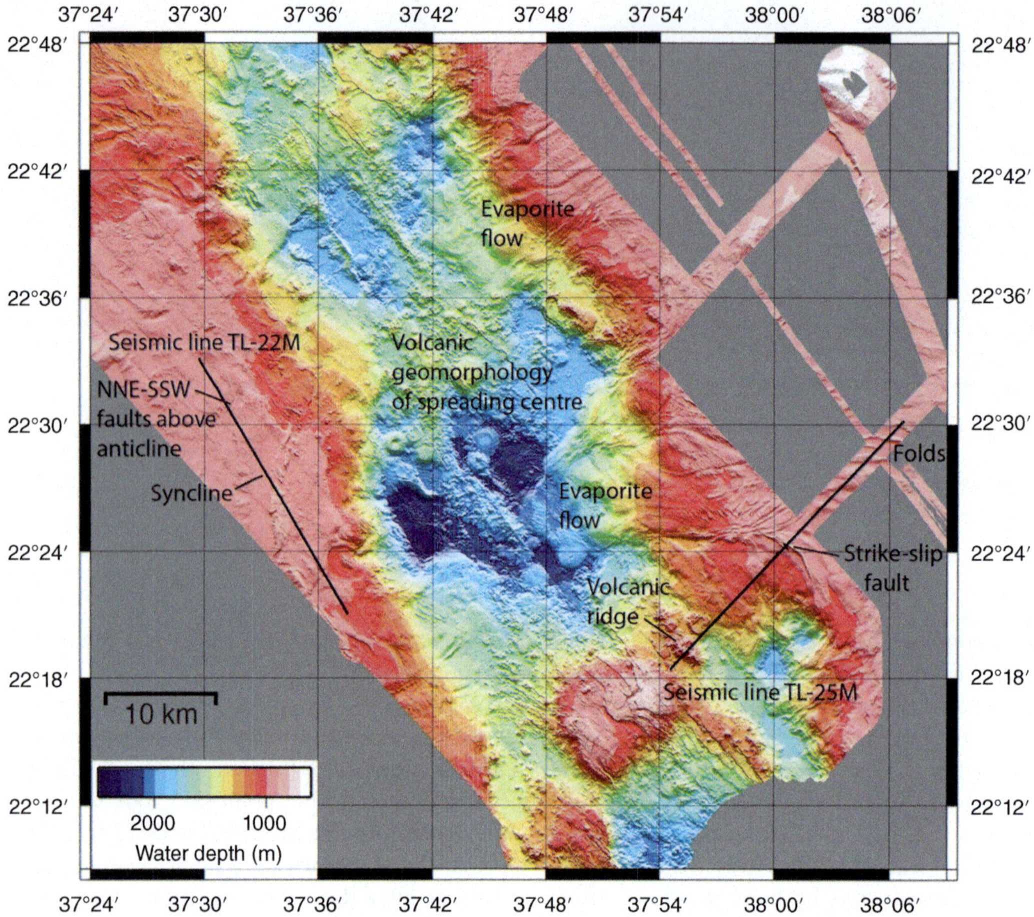

Figure 37.1 Map of the study area showing interpreted multibeam sonar bathymetry with seismic reflection lines marked.

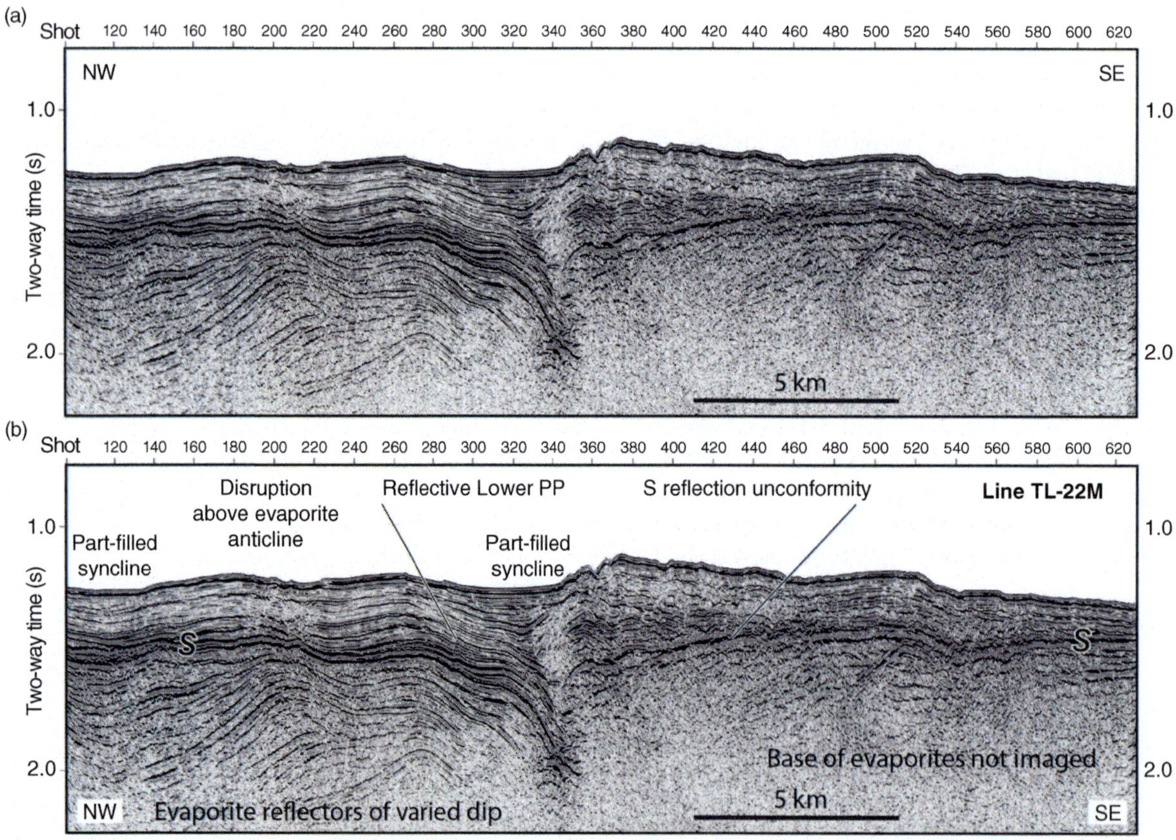

Figure 37.2 (a) Uninterpreted and (b) interpreted reflection seismic section.

References

Ehrhardt, A., Hübscher, C. and Gajewski, D. 2005. Conrad Deep, Northern Red Sea: development of an early stage ocean deep within the axial depression: *Tectonophysics*, **411**, 19–40.

Bache, F., Olivet, J.L., Gorini, C. *et al.* 2009. Messinian erosional and salinity crises: view from the Provence Basin (Gulf of Lions, Western Mediterranean): *Earth and Planetary Science Letters*, **286**, 139–157.

Hughes, G.W. and Beydoun, Z.R. 1992. The Red Sea – Gulf of Aden: biostratigraphy, lithostratigraphy and palaeoenvironments: *Journal of Petroleum Geology*, **15**, 135–156.

Izzeldin, A.Y. 1987. Seismic, gravity and magnetic surveys in the central part of the Red Sea: their interpretation and implications for the structure and evolution of the Red Sea. *Tectonophysics*, **143**, 269–306.

Mitchell, N.C., Ligi, M., Farrante, V. *et al.* 2010. Submarine salt flows in the central Red Sea: Geol. *Soc. Am. Bull.*, **122**, 701–713.

Mitchell, N.C., Ligi, M., Feldens, P. and Hübscher, C. 2017. Deformation of a young salt giant: regional topography of the Red Sea Miocene evaporites. *Basin Research*, **29**, 352–369.

Ross, D.A. and Schlee, J. 1973. Shallow structure and geologic development of the southern Red Sea. *Geol. Soc. Am. Bull.*, **84**, 3827–3848.

Stoffers, P. and Kühn, R. 1974. Red Sea evaporites: a petrographic and geochemical study. In: Whitmarsh, R.B., Weser, O.E., Ross, D.A. *et al.* (eds) Initial Reports of the Deep Sea Drilling Project, vol. 23, US Government Printing Office, Washington, DC, pp. 821–847.

Tramontini, C. and Davies, D. 1969. A seismic refraction survey in the Red Sea. *Geophys. J. R. Astr. Soc.*, **17**, 225–241.

Part IV

Tectonic/Regional Structures

Tectonics deals with structural geology on very large scales or regional scales. In this section examples are given of:

- Precambrian crust (Chapters 39, 40, 41)
- Passive margins (Chapters 42, 46, 47, 48)
- Oceanic crust structures (Chapters 43, 45)
- Subduction zones (Chapters 44)

38

BABEL4: Crustal Structures of the Precambrian Svecofennian Accretionary Orogen in the Fennoscandian Shield

*Pekka Heikkinen and Annakaisa Korja**

University of Helsinki, Finland
**annakaisa.korja@helsinki.fi*

BABEL profiles image the crustal structures of the Precambrian Svecofennian accretionary orogen in the Fennoscandian Shield (BABEL Working Group, 1993; Lahtinen *et al.*, 2005) (Figure 38.1a). BABEL4 profile images, for the first time, a Paleoproterozoic subduction zone suggesting modern plate tectonics (BABEL Working Group, 1993) and collision tectonics (Beaumont and Quinlan, 1994; Korja and Heikkinen, 2005).

The near vertical seismic reflection data are the final stack from BABEL4 (Figure 38.1b). Acquisition parameters and the full processing sequence have been described by the BABEL Working Group (1993). Migrated sections are displayed as instantaneous amplitude sections overlain by automatic line drawings (Korja and Heikkinen, 2005). The section is plotted without normalization, that is, the amplitudes of the different areas in each section are comparable. The large-scale reflectivity changes are pronounced, whereas the details of individual reflections are obscured.

In the geological interpretation it is assumed that reflections image mainly primary igneous or secondary tectonic contacts formed during folding/faulting. Weak reflectivity generally indicates monotonous intrusions and older crustal pieces in which the internal structure has been homogenized in the scale of reflectivity before deforming. Near vertical strike-slip zones are observed indirectly, as they are associated with transparent zones with decreased reflectivity and displacement of continuous reflections.

BABEL4 profile images five major blocks with distinct reflectivity patterns. The lower crust of the northern part is a reflective homogenous block. The upper and lower boundaries are reflective. The block is interpreted as an older continental fragment/crustal indentor with tectonic contacts. The middle crust of the northern part is highly reflective comprising a set of southwest dipping listric reflections, correlated with the volcano-sedimentary sequences of the Savo/Skellefte belt onshore. The lower crust of the southern part is highly reflective and hosts high amplitude, northeast-dipping reflections. The block has seismic P-wave velocities > 6.8 km s^{-1} suggesting mafic granulites. The high amplitude reflections may originate from the interlayering of oceanic crust and metasedimentary rocks. The upper and middle crust of the southern and central parts is weakly reflective. It hosts a package of steeply northeast-dipping reflections bound by low-angle (8°) unreflective bands. The bands are disturbed and reminiscent of large-scale breccias in the central part. The southern and central parts are correlated with the migmatitic and granitoid rocks of the Bothnia Basin and Vaasa complex respectively. The seismic image suggests that the block is separated from the blocks below along a tectonic contact/detachment.

Seismic profile BABEL4 images a frozen paleosubduction zone, where northeast subduction of the lower plate below the upper plate has ceased and collision has started (Beaumont and Quinlan, 1994). A retrowedge comprises remnants of the Savo arc (Savo belt, SB) that have been accreted to the upper plate. The accretionary wedge (Bothnian Basin, BB) is part of the prow edge. Thickening of the crust induced high T – low P regional metamorphism, partial melting and formation of granites (unreflective parts of the upper and middle crust) within the prow edge and uplifted plug (Bothnian Basin and Vaasa complex).

Atlas of Structural Geological Interpretation from Seismic Images, First Edition. Edited by Achyuta Ayan Misra and Soumyajit Mukherjee.

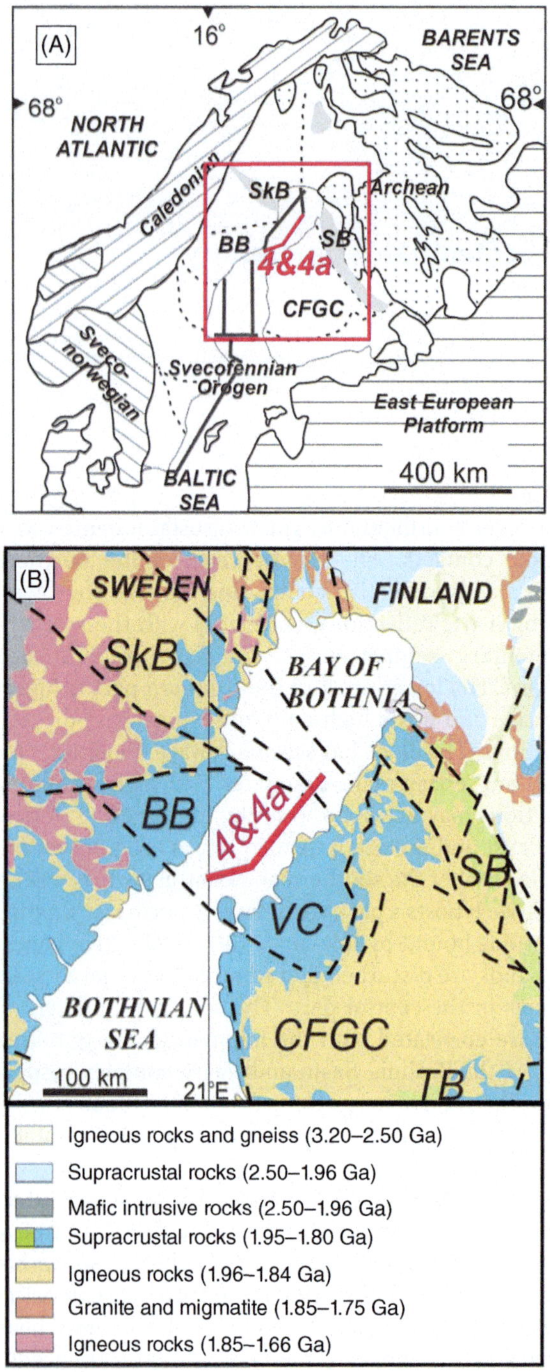

Figure 38.1a A) BABEL4&4a line on a geological index map of Northern Europe. B) BABEL4&4a a lithological map of the study area after Lahtinen *et al.* (2005).

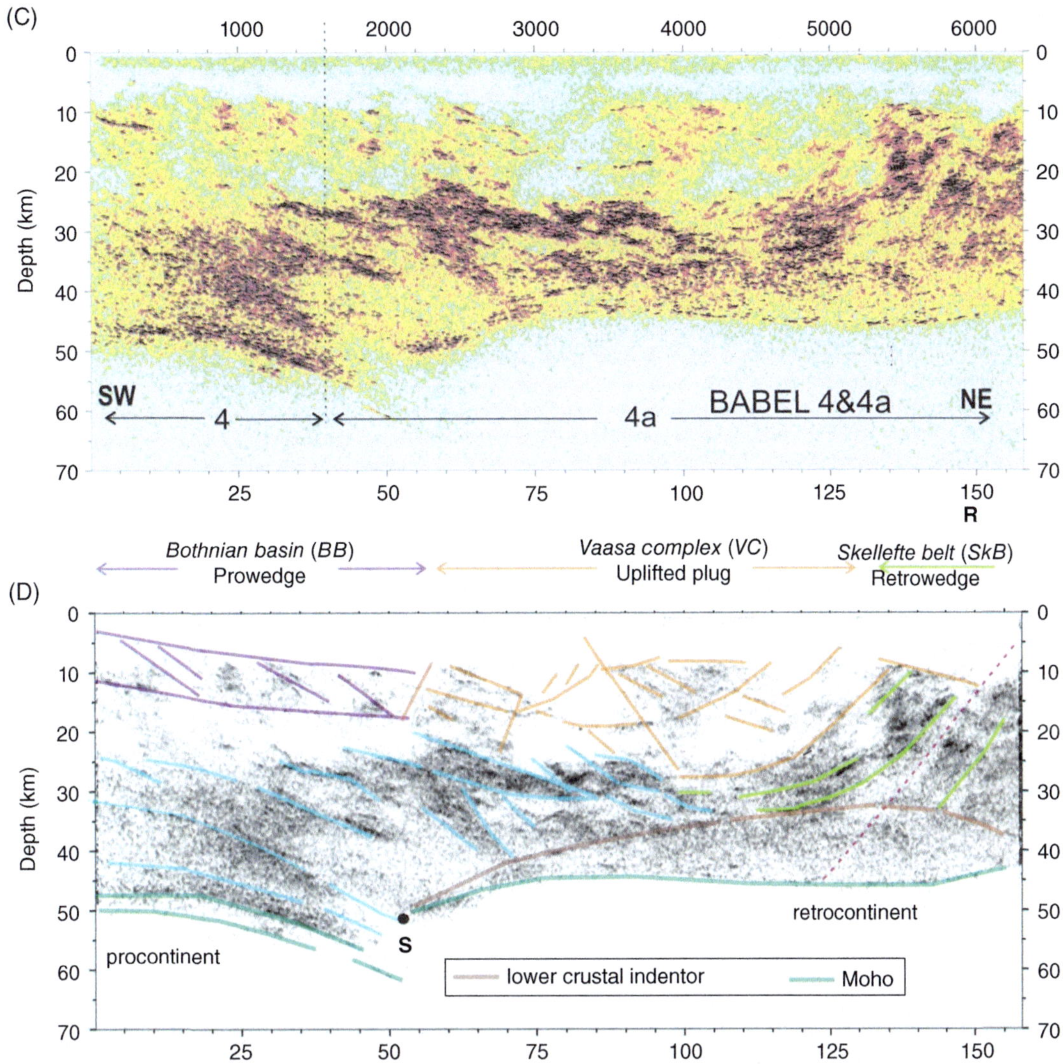

Figure 38.1b C) A vertical seismic section along BABEL4&4a profile shown as an instantaneous amplitude section in colour scale overlain by an automatic line drawing. No vertical exaggeration. D) A schematic geological line interpretation of BABEL4&4a profile on an instantaneous amplitude section in greyscale after Korja and Heikkinen (2005).

References

BABEL Working Group. 1993. Integrated seismic studies of the Baltic Shield using data in the Gulf of Bothnia region. *Geophysical Journal International*, **112**, 305–324.

Beaumont, C. and Quinlan, G. 1994. A geodynamic framework for interpreting crustal-scale seismic-reflectivity patterns in compressional orogens. *Geophysical Journal International*, **116**, 754–783.

Korja, A. and Heikkinen, P. 2005. The Accretionary Svecofennian Orogen – insight from the BABEL profiles. *Precambrian Research*, **136**, 241–268.

Lahtinen, R., Korja, A. and Nironen, M. 2005. Paleoproterozoic tectonic evolution. In: Lehtinen, M., Nurmi, P. and Rämö, O.T. (eds) *Precambrian Geology of Finland – Key to the Evolution of the Fennoscandian Shield*, Elsevier, Amsterdam, pp. 481–532.

39

BABEL7: Crustal-Scale Extension Associated with the Formation of the Baltic Sea Paleorift, Fennoscandian Shield

Annakaisa Korja and Pekka Heikkinen*

University of Helsinki, Finland
** annakaisa.korja@helsinki.fi*

BABEL profiles image the crustal structures of the Fennoscandian Shield (BABEL Working Group, 1993; Korja *et al.*, 2006). BABEL7 (Figure 39.1a) images crustal-scale extension associated with formation of the Baltic Sea paleorift and intrusion of the Åland rapakivi granite batholith (Korja and Heikkinen, 1995; 2005). The Baltic Sea is a Mesoproterozoic failed rift associated with thinning of the crust, voluminous bimodal magmatism (rapakivi granites and gabbro-anorthosites) and continental type sedimentary basins. Rapakivi granite plutons are characteristically multiple intrusions emplaced in an extensional tectonic setting as relatively shallow-level plutons (Kohonen and Rämö, 2005).

The near vertical seismic reflection data are the final stacks from BABEL7 (Figure 39.1b). Acquisition parameters and the full processing sequence have been described by the BABEL Working Group (1993). Migrated sections are displayed as instantaneous amplitude sections overlain by automatic line drawings (Korja and Heikkinen, 2005). The section is plotted without normalization, that is, the amplitudes of the different areas in each section are comparable. The large-scale reflectivity changes are pronounced, whereas the details of individual reflections are obscured.

Geological interpretation presumes that reflections image mainly primary igneous or secondary tectonic contacts formed during deformation. Weak reflectivity generally indicates monotonous intrusions and older crustal pieces in which the internal structure has been homogenized in the scale of reflectivity before deforming.

BABEL7 profile images three major blocks with distinct reflectivity patterns. The upper crust of the eastern part is weakly reflective but it hosts some minor reflections mimicking graben–horst and diamond-shaped structures. The western upper and middle crust is characterized by strong southeast-dipping bands of crustal-scale listric reflections flattening at the middle-lower crust boundary. The lower crust is highly reflective and bulges upwards in the east. The reflection Moho, defined as base of the reflective lower crust, is distinct and can be correlated with the Moho derived from the wide-angle data.

The crustal structure is interpreted to image mostly the latest tectonic extensional event associated with the thinning of the crust and the formation of the Åland and Gävle rapakivi granite batholiths and Bothnian Sea Basin. Some of the extensional normal faults may be inverted stacking surfaces.

The listric reflections are interpreted to image listric normal faults detaching on the mid-lower crust boundary. The offsets of the major reflections can be traced to the surface, where they presumably connect with thin-skin half-grabens of the Bothnian Sea. Thinning of the upper-middle crust counterbalance by bulging and uplifting of the lower crust. Space created by extension is occupied by the Åland batholith: a homogenous and weakly reflective body. It is composed of an upper granitic part (6.15 km s^{-1}, 2625 kg m^{-3}) and a lower diamond-shaped gabbro-anorthositic part (6.50 km s^{-1}, 2870 kg m^{-3}). A reflective graben and horst-structure within the batholith outlines the contact between the granitic and gabbro-anorthositic parts. Breaks in the reflections are interpreted as normal and reverse faults.

Atlas of Structural Geological Interpretation from Seismic Images, First Edition. Edited by Achyuta Ayan Misra and Soumyajit Mukherjee.

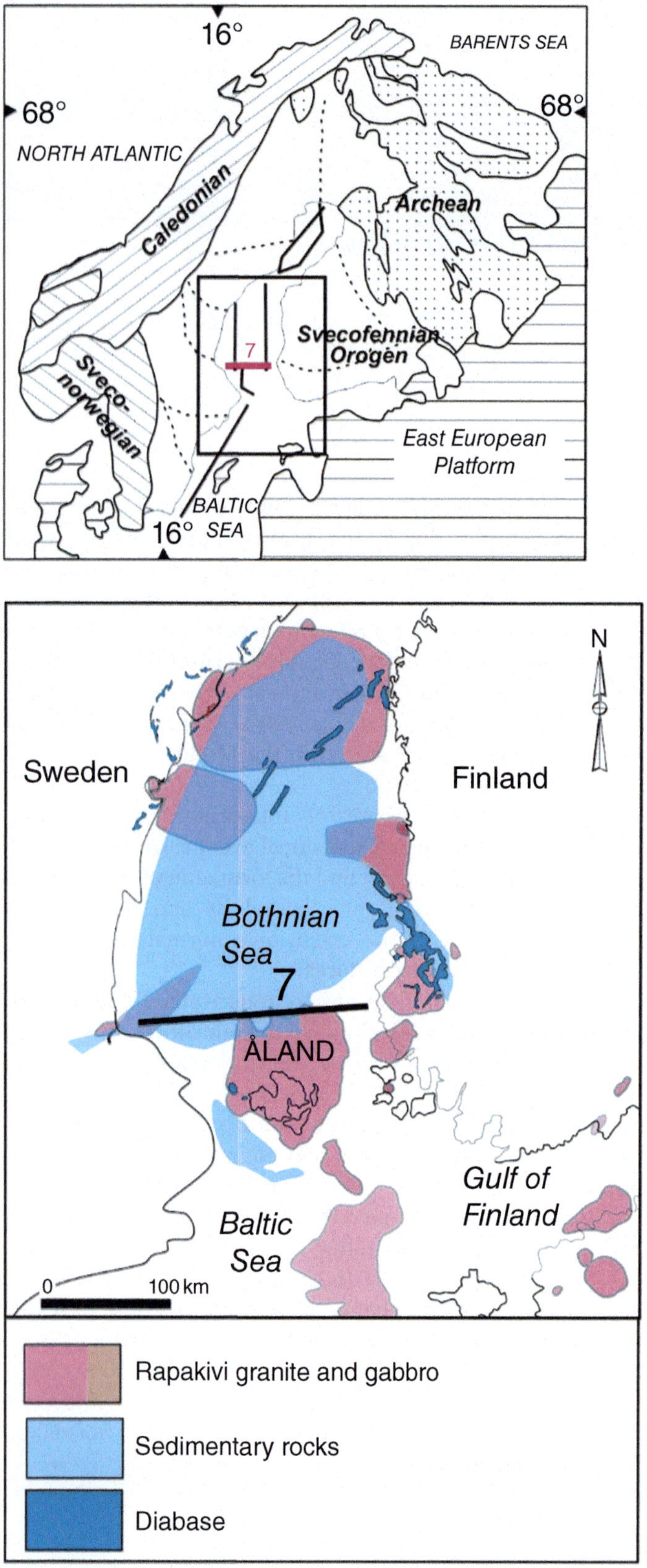

Figure 39.1a (top) BABEL7 line on a geological index map of Northern Europe. (bottom) BABEL7a lithological map of the study area after Kohonen and Rämö (2005).

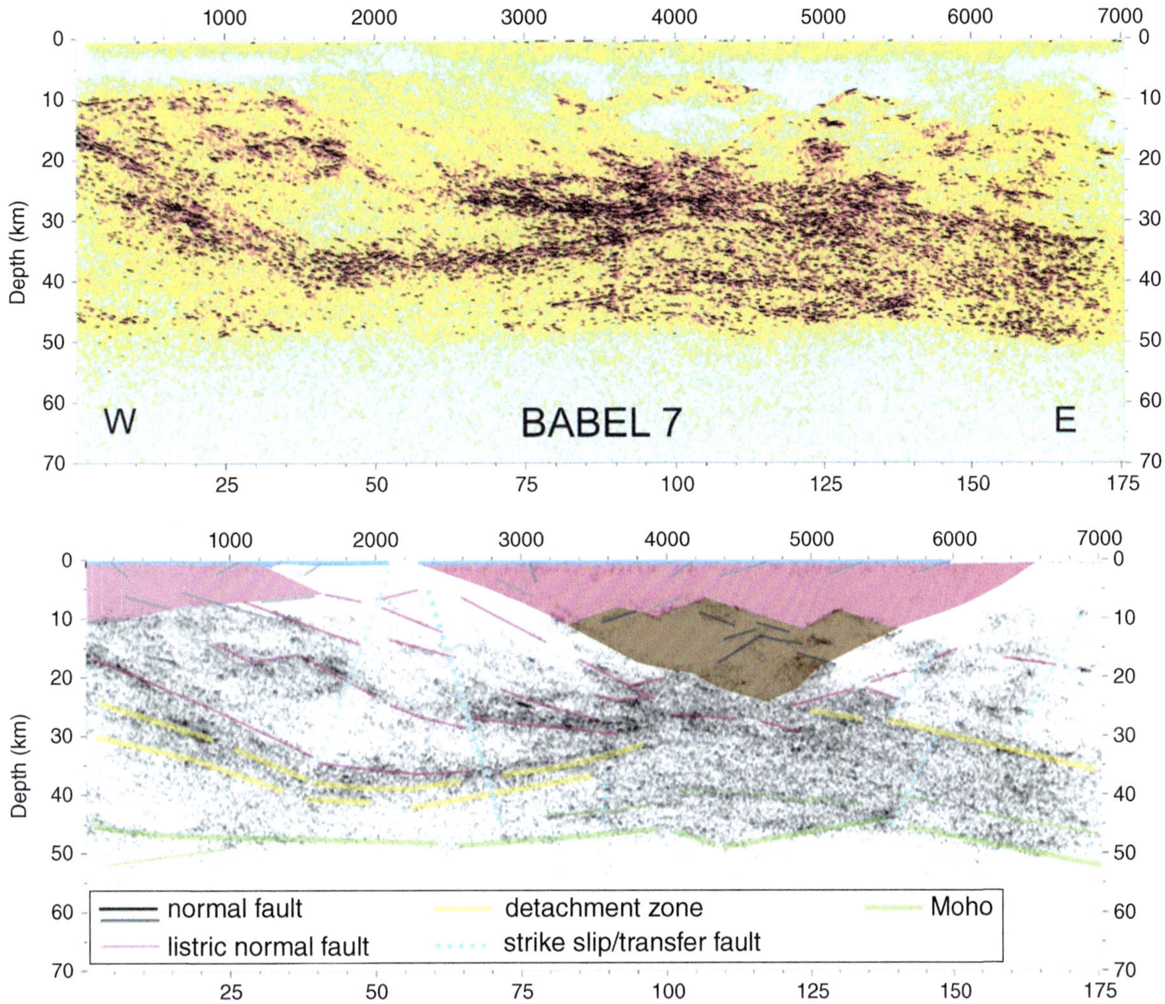

Figure 39.1b (top) A vertical seismic section along BABEL7 profile shown as an instantaneous amplitude section in colour scale overlain by an automatic line drawing. No vertical exaggeration. (bottom) A schematic geological line interpretation of BABEL7 profile on an instantaneous amplitude section in greyscale modified after Korja and Heikkinen (1995; 2005).

References

BABEL Working Group. 1993. Integrated seismic studies of the Baltic Shield using data in the Gulf of Bothnia region. *Geophysical Journal International*, **112**, 305–324.

Korja, A. and Heikkinen, P.J. 1995. Proterozoic extensional tectonics of the central Fennoscandian Shield: Results from the Baltic and Bothnian Echoes from the Lithosphere experiment. *Tectonics*, **14**, 504–517.

Korja, A. and Heikkinen, P. 2005. The Accretionary Svecofennian Orogen – insight from the BABEL profiles. *Precambrian Research*, **136**, 241–268.

Korja, A., Lahtinen, R. and Nironen, M. 2006. The Svecofennian orogen: a collage of microcontinents and island arcs. In: Gee, D.G. and Stephenson, R.A. (eds) *European Lithosphere Dynamics*. Memoirs vol. 32, The Geological Society, London, pp.561–578.

Kohonen, J. and Rämö, O.T. 2005. Sedimentary rocks, diabases, and late Cratonic evolution. In: Lehtinen, M., Nurmi, P. and Rämö, O.T. (eds) *Precambrian Geology of Finland – Key to the Evolution of the Fennoscandian Shield*. Elsevier, Amsterdam, pp. 563–603.

40

FIRE3&3a: Crustal-Scale Structures of the Precambrian Svecofennian Accretionary Orogen in the Fennoscandian Shield

Annakaisa Korja[1], Pekka Heikkinen[1], Timo Tiira[1] and Kaisa Nikkilä[1,2]*

[1] *University of Helsinki, Finland*
[2] *Åbo Academi University, Finland*
* *annakaisa.korja@helsinki.fi*

A series of seismic deep reflection profiles in Finland – FIRE profiles – image the crustal-scale structures of the Precambrian Svecofennian accretionary orogen in the Fennoscandian Shield (Figure 40.1a) (Lahtinen *et al.*, 2005; Kukkonen *et al.*, 2006). The exposed surface hosts granitoid intrusions and low P–high T metamorphic rocks (Korsman *et al.*, 1999), typical for large hot orogens.

The near vertical reflection data are the crustal-scale migrated NMO-stacks from FIRE3. They are displayed as instantaneous amplitude sections, averaged both horizontally and vertically and plotted as greyscale intensities (technical detail in Kukkonen *et al.*, 2006). The section is plotted without normalization, that is, the amplitudes of the different areas in each section are comparable. In the sections, large-scale reflectivity changes are pronounced, whereas the details of individual reflections are obscured.

In the geological interpretation it is assumed that reflections image mainly primary igneous or secondary tectonic contacts formed during folding/faulting. Weak reflectivity generally indicates monotonous intrusions and older crustal pieces in which the internal structure has been homogenized in the scale of reflectivity prior to deformation. Near vertical strike-slip zones are observed indirectly as they are associated with transparent zones with decreased reflectivity and displacement of continuous reflections.

FIRE3 profile displays a heterogenous structure where reflectivity patterns change both vertically and laterally (Figure 40.1b). Comparison of the reflection section with the tomographic velocity model (Figure 40.1b, top) (Hyvönen *et al.*, 2007) suggests a three-layer structure with distinct reflection and velocity properties in each crustal layer. The upper crust is associated with 5.8–6.2 km s^{-1} velocity and *Vp*/*Vs* ratios of 1.68–1.70; the middle crust with 6.3–6.6 km s^{-1} velocity and *Vp*/*Vs* ratios of 1.71–1.74; and the lower crust with 6.8–7.6 km s^{-1} velocity and *Vp*/*Vs* ratios of 1.74–1.76. Subcontinuos subhorizontal reflective boundaries separate these layers. The profile images also a lateral structural change from eastern Archean parts through a boundary zone into the western Paleoproterozoic part.

Common to all the parts is that the lower crust has a subdued and patchy subhorizontal reflectivity that dies out at the Moho boundary marked by a small amplitude decrease in background reflectivity coinciding with the wide-angle reflection Moho. In the western to central parts, the middle crust has two components: large blocks of poorly reflective background material boarded by high amplitude, shallow-to-steeply dipping crustal-scale reflections. Listric to anticlinal reflections sole out at the middle and lower crust boundary. In the eastern parts, the middle crust has large blocks of both poorly and well-reflective material (Ma). In the central parts, the upper crust displays fine-scale structures with listric reflections flattening at a highly reflective upper-middle crustal boundary. In the eastern parts, the upper crust displays small-scale thrusts and normal faults.

FIRE3 profile transects the core of the Svecofennian orogeny. It has been interpreted to image accretion by stacking and thrusting (Figure 40.1c, top) (Kukkonen *et al.*, 2006; Sorjonen-Ward, 2006) or by stacking and thrusting followed by inversion to normal faults during lateral spreading and gravitational collapse and formation of super-infrastructure (Figure 40.1c) (Korja and Heikkinen, 2008; Nikkilä *et al.*, 2016). Note that thrusting is one mechanism of stacking taking place at different scales (Mukherjee, 2014). Large-scale crustal structures along FIRE3 might result from gravitational collapse of the thickened and thermally matured hot Svecofennian orogenic crust (Figure 40.1c, middle and bottom) (Nikkilä *et al.*, 2015).

Atlas of Structural Geological Interpretation from Seismic Images, First Edition. Edited by Achyuta Ayan Misra and Soumyajit Mukherjee.

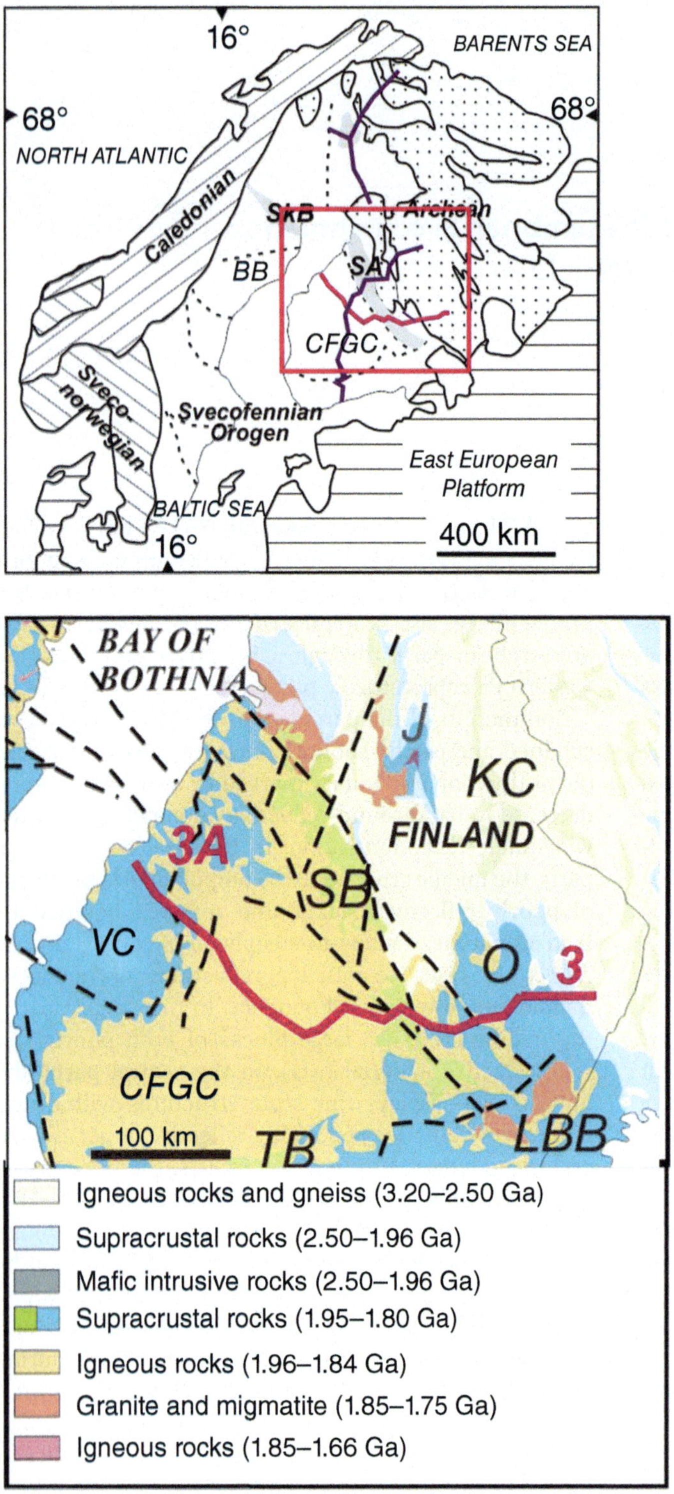

Figure 40.1a (top) FIRE3&3a line on a geological index map of Northern Europe. (bottom) FIRE3&3a a lithological map of the study area. Adapted from Lahtinen *et al.* (2005).

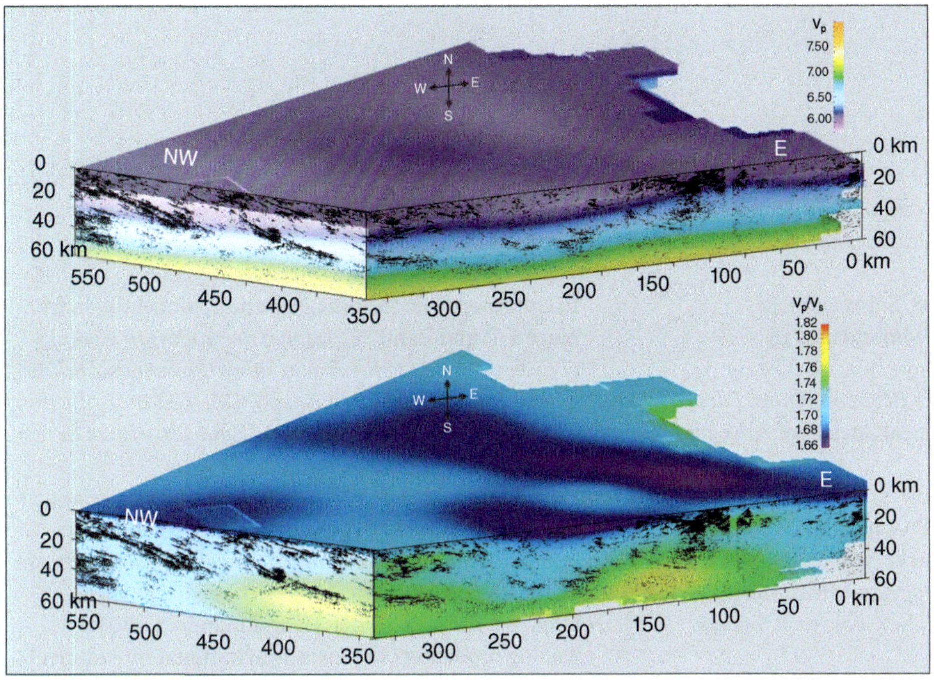

Figure 40.1b (top) Three-dimensional block diagrams of the seismic P-wave velocity and *Vp/Vs* ratio models (Hyvönen *et al.*, 2007) overlain by automatic line drawings along FIRE3&3a. A view to N. (bottom). A vertical seismic section along FIRE3&3a profile shown as an instantaneous amplitude section in greyscale. No vertical exaggeration.

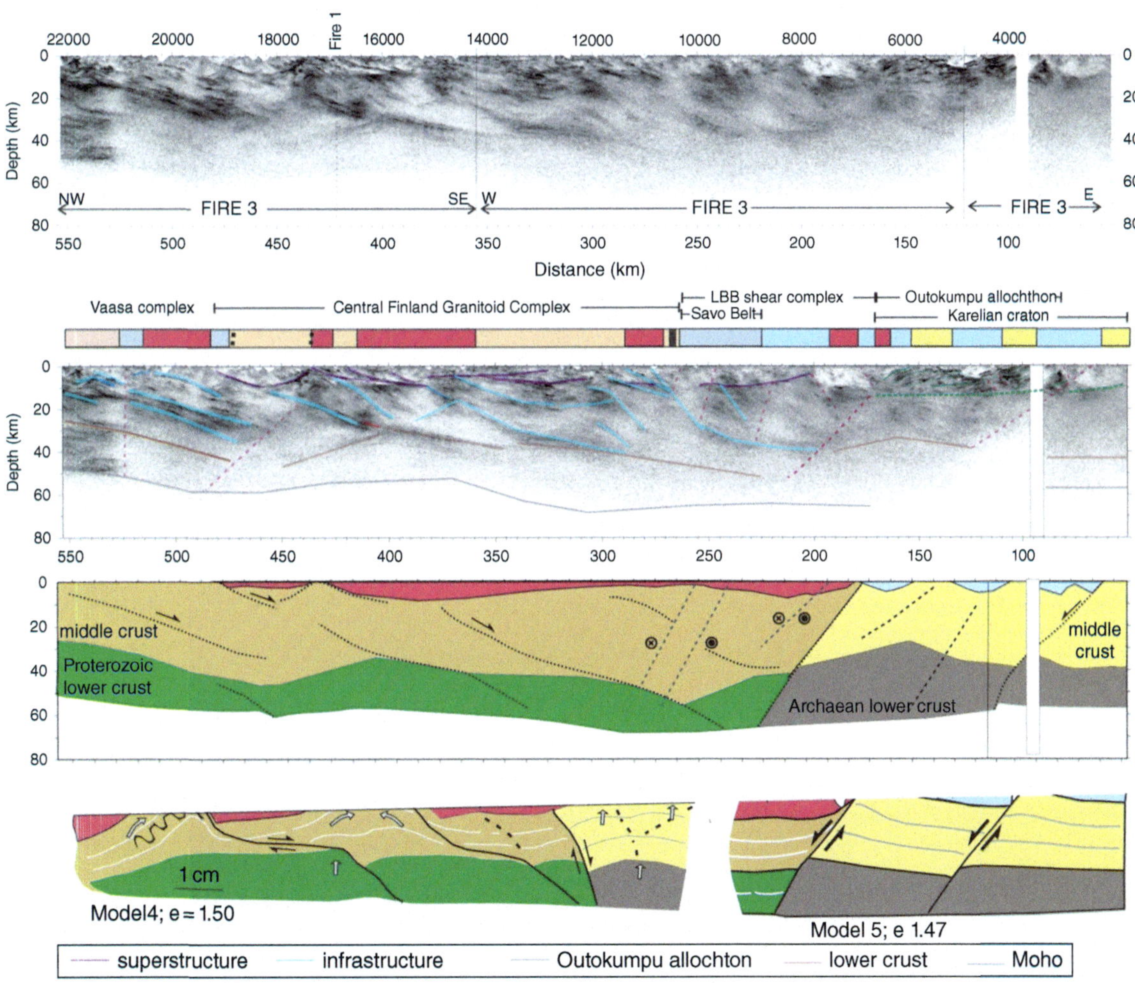

Figure 40.1c (top) A schematic geological line interpretation of FIRE3&3a profile on an instantaneous amplitude section in greyscale modified from Korja and Heikkinen (2008) and Korja *et al.* (2009). (middle) A simplified three-layer interpretation of FIRE3&3a after Nikkilä *et al.* 2015. (bottom) A comparison of analogue models of gravitational collapse of a thick three layer crust and FIRE3&3a profile. Model 4 and Model 5 by Nikkilä *et al.* (2015) are compared to the western Paleoproterozoic part and to the eastern Archean part, respectively.

References

Hyvönen, T., Tiira, T., Korja, A. *et al.* 2007. A tomographic crustal velocity model of the central Fennoscandian Shield. *Geophysical Journal International*, **168**, 1210–1226.

Korja, A. and Heikkinen, P.J. 2008. Seismic images of Paleoproterozoic microplate boundaries in Fennoscandian Shield. In: Condie, K.C. and Pease, V. (eds) *When Did Plate Tectonics Begin on Planet Earth?* Special Paper 440, The Geological Society of America, pp. 229–248.

Korja, A., Kosunen, P. and Heikkinen, P.J. 2009. A case study of lateral spreading the Precambrian Svecofennian Orogen In: Ring, U. and Wernicke, B. (eds) *Extending a Continent: Architecture, Rheology and Heat Budget*, Special Publications vol. 321, The Geological Society, London, pp. 225–251.

Kukkonen, I.T., Heikkinen, P., Ekdahl, E. *et al.* 2006. Acquisition and geophysical characteristics of reflection seismic data on FIRE transects, Fennoscandian Shield. In: Kukkonen, I.T. and Lahtinen, R. (eds) *Finnish Reflection Experiment FIRE 2001–2005*, Special Paper 43, The Geological Survey of Finland, pp. 13–43.

Korsman, K., Korja, T., Pajunen, M. *et al.* 1999. The GGT/SVEKA Transect – Structure and evolution of the continental crust in the Palaeoproterozoic Svecofennian Orogen in Finland. *International Geology Review*, **41**, 287–333.

Lahtinen, R., Korja, A. and Nironen, M. 2005. Paleoproterozoic tectonic evolution. In: Lehtinen, M., Nurmi, P. and Rämö, O.T. (eds) *Precambrian Geology of Finland – Key to the Evolution of the Fennoscandian Shield.*,Elsevier, Amsterdam, pp. 481–532.

Mukherjee, S. 2014. *Atlas of Shear Zone Structures in Meso-scale*, Springer.

Nikkilä, K., Mänttäri, I., Nironen, M. *et al.* 2016. Three stages to form a large batholith after terrane accretion – An example from the Svecofennian orogen. *Precambrian Res.* **281**, 618–638.

Nikkilä, K., Korja, A., Koyi, H. and Eklund, O. 2015. Analog modeling of oneway gravitational spreading of hot orogens – A case study from the Svecofennian orogen, Fennoscandian Shield. *Precambrian Research*, **268**,135–152.

Sorjonen-Ward, P. 2006. Geological and structural framework and preliminary interpretation of the FIRE 3 and FIRE3A reflection seismic profiles, central Finland. *Proceedings of the Finnish Reflection Experiment 2001–2005*, pp.105–159.

41

Magma-Poor Rifting in the Indian Eastern Continental Margin

Achyuta Ayan Misra

Reliance Industries Ltd, Mumbai, India
Achyuta.Misra@ril.com; achyutaayan@gmail.com

The Indian eastern continental margin formed when Antarctica separated from India in the Late Jurassic-Early Cretaceous (Rao, 2001). The margin exhibits most features of a typical 'magma-poor' rift, for example stretching, thinning and exhumation domains besides genesis of an oceanic crust of normal thickness (Manatschal, 2004; Manatschal *et al.*, 2007; Manatschal and Müntener, 2009; Péron-Pinvidic and Manatschal, 2009; Franke, 2013; Masini *et al.*, 2013; Nemčok *et al.*, 2013). Structures of the stretching domain are normal fault bound grabens with conspicuous growth packages (Chapter 4). In this chapter, acquired by Ion-GXT, the thinning, exhumation and oceanic domains are shown (Figures 41.1, 41.2a and 41.2b). Tectonic detail of specific zones on the Indian east coast can also be found elsewhere (Chatterjee *et al.*, 2015; Roy *et al.*, 2015).

The continental Moho and the oceanic Moho are imaged very clearly in this seismic, due to the acquisition parameters (Nemčok *et al.*, 2013). The continental Moho rises gently towards the oceanic domain while the oceanic Moho is subhorizontal (Figures 41.2a and 41.2b). The Moho appears as one or a group of high amplitude reflectors due to impedance contrast between the continental granitic-granodioritic rocks of about 2.67 $g\,cm^{-3}$ density or oceanic basalts of 2.7–2.9 $g\,cm^{-3}$ and peridotitic mantle rocks of about 3.0–3.3 $g\,cm^{-3}$ density. These two crustal domains can be identified based on the character of the Moho. Additionally, the continental crust is brittle faulted and sheared in the upper part (Figures 41.2c–41.2f). In contrast, the oceanic crust, presumably brittle, is layered due to the basaltic flows. Oceanic crust is usually unfaulted, although it may be rugged and possess volcanic pods and/or small volcanoes/volcanic mounds (Figure 41.2g).

The thinned continental crust also shows rifted grabens with prominent synrift growth sedimentary packages and a top-synrift/breakup unconformity, geometrically resembling the stretching domain (Chapter 4). However, the rifts and unconformities become younger towards the thinning domain, as the deformation progresses towards the oceanic domain during rifting.

The proto-oceanic crust (POC) or alternately zone of exhumed continental mantle (ZECM) lies between the continental and the oceanic crust. The mantle exhumes at the POC due to unroofing of the continental crust. Peridotites are serpentinized in contact with sea water. The densities of the POC serpentinites and mantle rocks are similar. So, the Moho reflection is subdued or invisible at the POC/ZECM. The seismic interpretation presented has been supported by gravity-magnetic studies and several well data (Nemčok *et al.*, 2013; Sinha *et al.*, 2015).

Atlas of Structural Geological Interpretation from Seismic Images, First Edition. Edited by Achyuta Ayan Misra and Soumyajit Mukherjee.

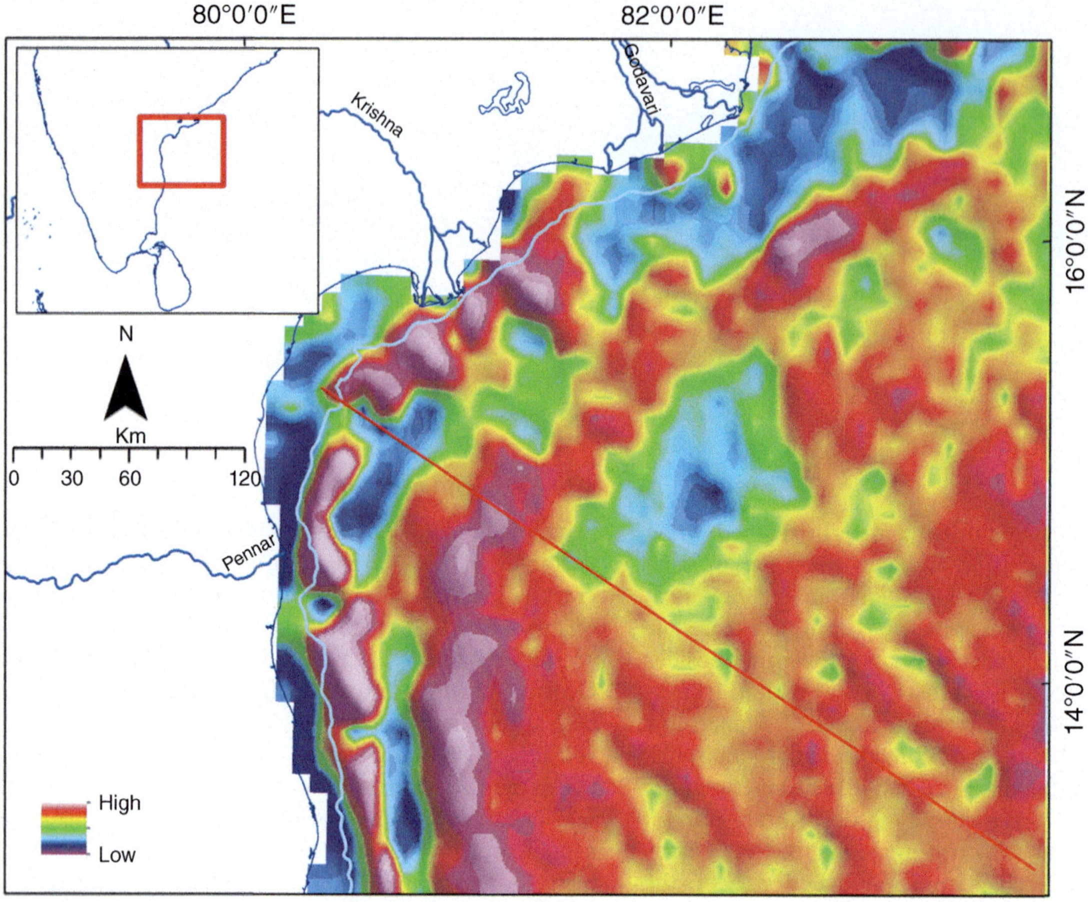

Figure 41.1 Map of the Krishna–Godavari Basin along the ECMI overlaid on hill-shaded, upward-continued (10 km) Bouguer gravity anomaly data (see the colour scale). Blue line: shelf-slope break. Red line: position of seismic section.

Figure 41.2a Uninterpreted regional seismic section in the Krishna–Godavari Basin, India. See Figure 41.1 for location of the seismic section. Seismic data courtesy Reliance Industries Ltd. Reproduced with permission from the Directorate General of Hydrocarbons (DGH), India.

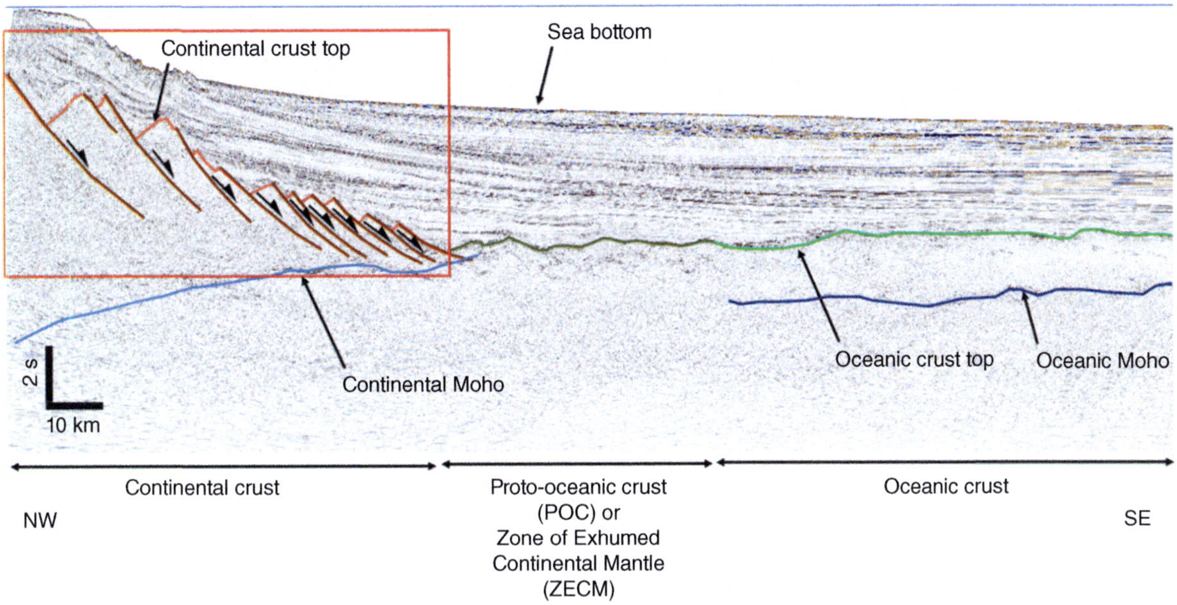

Figure 41.2b Interpreted regional seismic section from Figure 41.2a showing the tectonic elements of a magma-poor passive margin. Note the differences in Moho character beneath the thinned continental crust, oceanic crust and proto-oceanic crust. Red rectangle: extent of seismic section in Figures 41.2c and 41.2d). Seismic data courtesy Reliance Industries Ltd. Reproduced with permission from the Directorate General of Hydrocarbons (DGH), India.

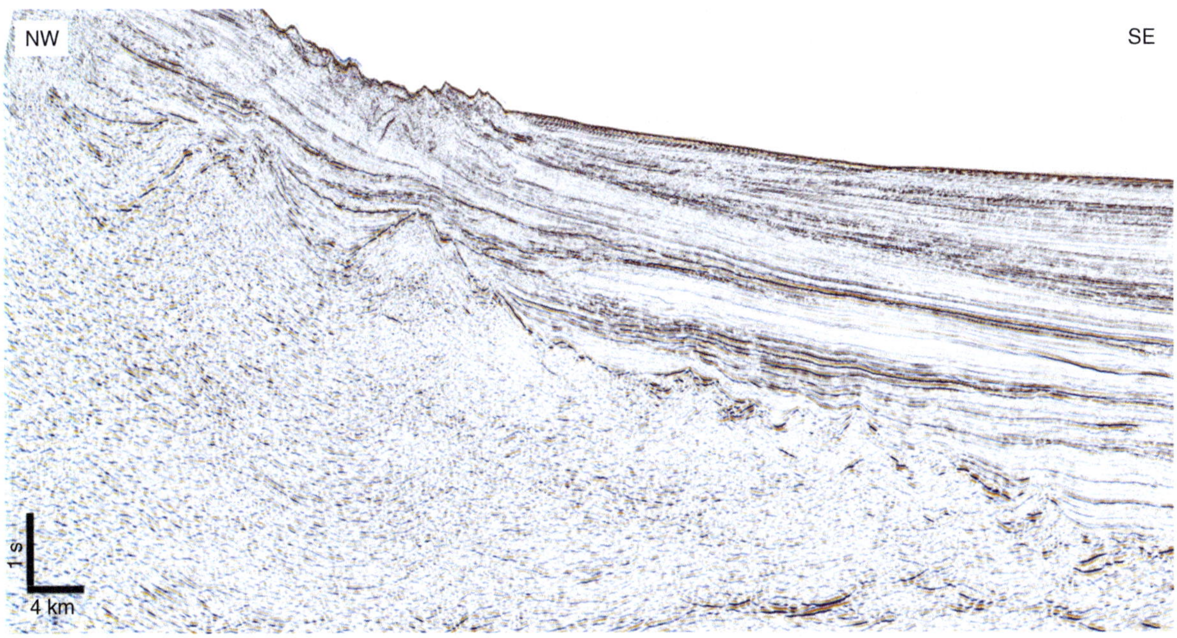

Figure 41.2c Uninterpreted seismic section showing the thinned continental crust. See Figure 41.2b for the location of the seismic section. Seismic data courtesy Reliance Industries Ltd. Reproduced with permission from the Directorate General of Hydrocarbons (DGH), India.

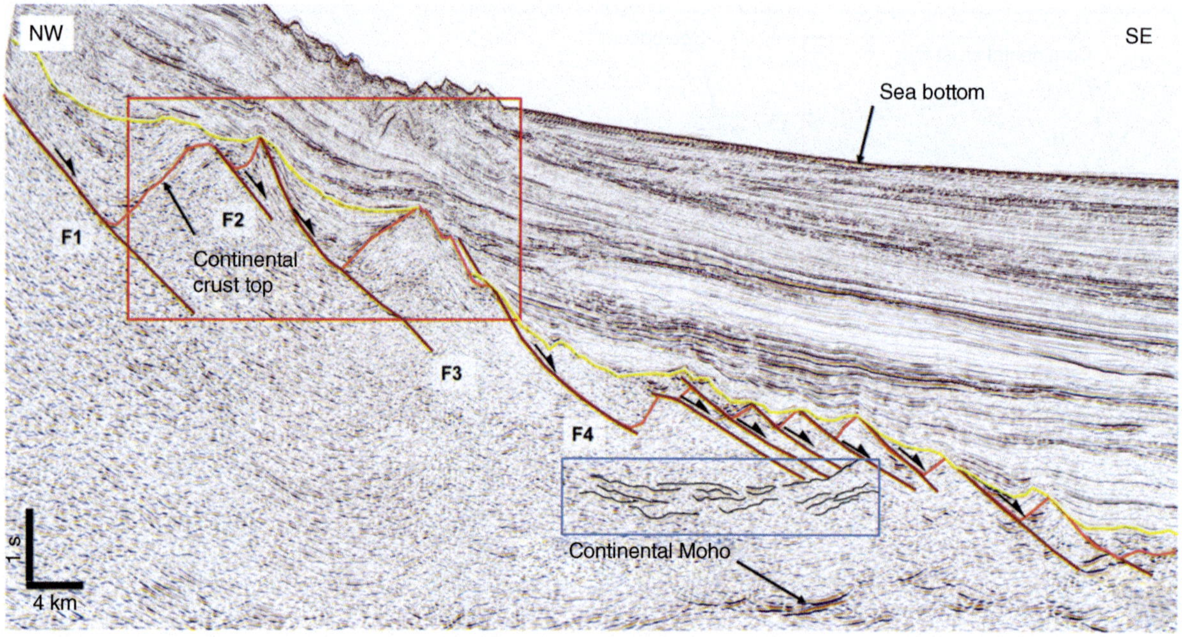

Figure 41.2d Interpreted seismic section of the thinned continental crust. Blue rectangle: subhorizontal reflection that may indicate the boundary between the upper and lower continental crust, also known as the brittle-ductile transition zone. Yellow line: breakup unconformity. Note the continental Moho. Red rectangle: extent of seismic section in Figures 41.2e and 41.2f. Seismic data courtesy Reliance Industries Ltd. Reproduced with permission from the Directorate General of Hydrocarbons (DGH), India.

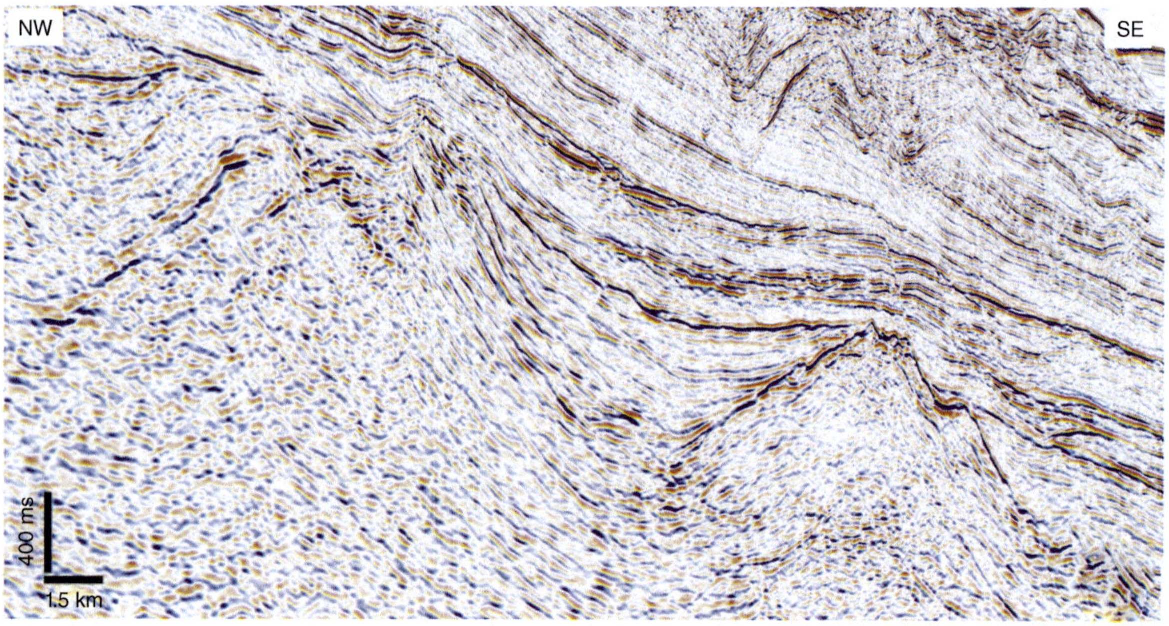

Figure 41.2e Uninterpreted seismic section showing rifted graben in the thinned continental crust. See Figure 41.2d for the location of the seismic section. Seismic data courtesy Reliance Industries Ltd. Reproduced with permission from the Directorate General of Hydrocarbons (DGH), India.

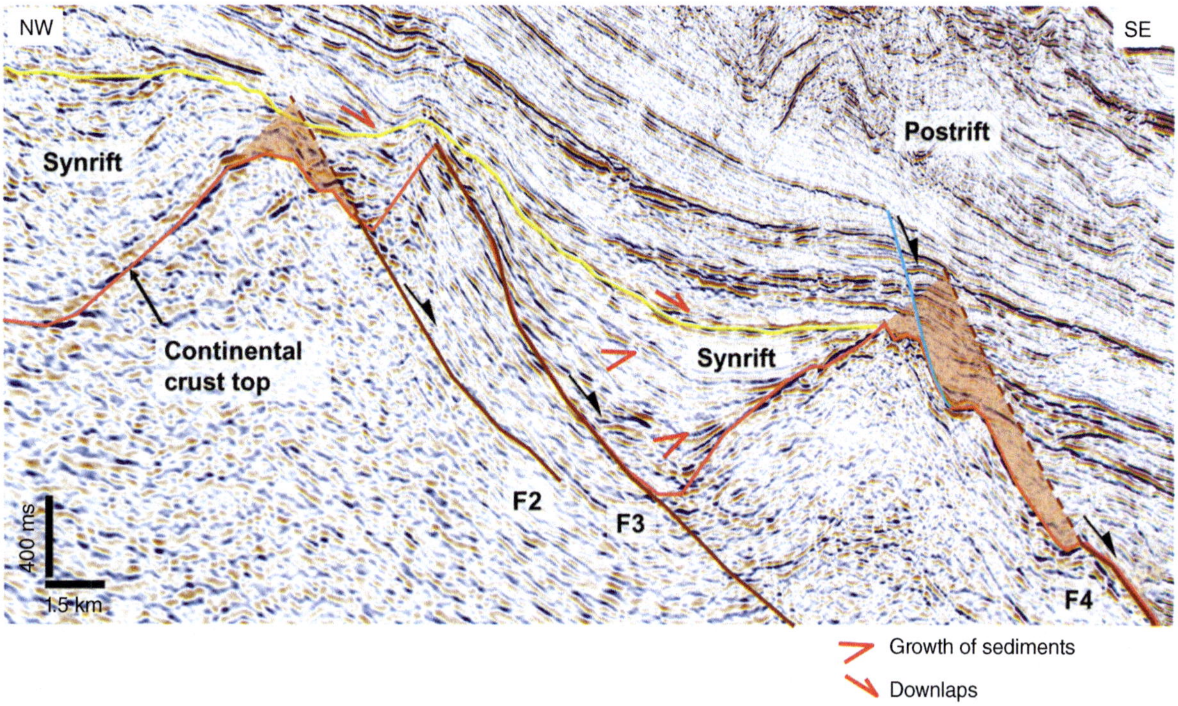

Figure 41.2f Interpreted seismic section showing rifted graben in the thinned continental crust. Footwall erosion is also evident along faults F2 and F4 here similar to Figure 41.2d. Yellow line: breakup unconformity. Seismic data courtesy Reliance Industries Ltd. Reproduced with permission from the Directorate General of Hydrocarbons (DGH), India.

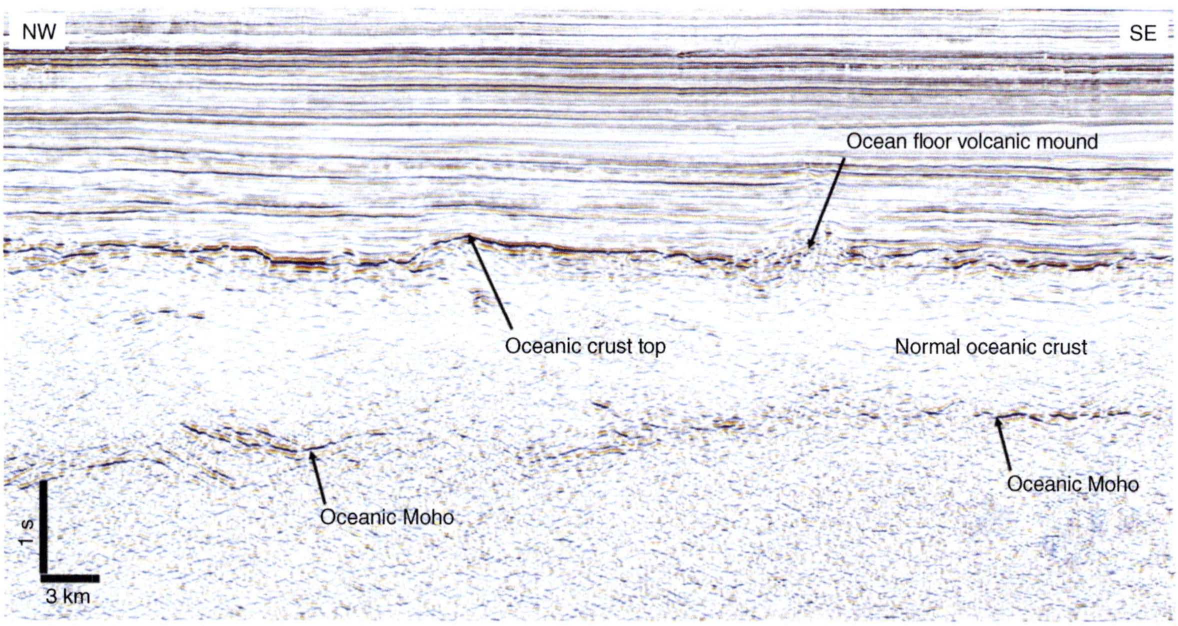

Figure 41.2g Interpreted seismic section showing the oceanig crust top and oceanic Moho. Note the layered nature of the oceanic crust. Seismic data courtesy Reliance Industries Ltd. Reproduced with permission from the Directorate General of Hydrocarbons (DGH), India.

References

Chatterjee, R., Paul, S., Singha, D.K. and Mukhopadhyay, M. 2015. Overpressure zones in relation to in situ stress for the Krishna-Godavari Basin, Eastern Continental Margin of India: implications for hydrocarbon prospectivity. In: Mukherjee, S. (ed) *Petroleum Geosciences: Indian Contexts*, Springer, pp. 127–142.

Franke, D. 2013. Rifting, lithosphere breakup and volcanism: Comparison of magma-poor and volcanic rifted margins. *Marine and Petroleum Geology*, **43**, 63–87.

Manatschal, G. 2004. New models for evolution of magma-poor rifted margins based on a review of data and concepts from West Iberia and the Alps. *International Journal of Earth Sciences*, **93**, 432–466.

Manatschal, G. and Müntener, O. 2009. A type sequence across an ancient magma-poor ocean–continent transition: the example of the western Alpine Tethys ophiolites. *Tectonophysics*, **473**, 4–19.

Manatschal, G., Müntener, O., Lavier, L.L. *et al.* 2007. Observations from the Alpine Tethys and Iberia–Newfoundland margins pertinent to the interpretation of continental breakup. In: Karner, G.D., Manatschal, G. and Pinheiro, L.M. (eds) *Imaging, Mapping and Modelling Continental Lithosphere Extension and Breakup*, Special Publications vol. 282, The Geological Society, London, pp. 291–324.

Masini, E., Manatschal, G. and Mohn, G., 2013. The Alpine Tethys rifted margins: Reconciling old and new ideas to understand the stratigraphic architecture of magma-poor rifted margins. *Sedimentology*, **60**, 174–196.

Nemčok, M., Sinha, S.T., Stuart, C.J. *et al.* 2013. East Indian margin evolution and crustal architecture: integration of deep reflection seismic interpretation and gravity modelling. In: Mohriak, W.U., Danforth, A., Post, P.J. *et al.* (eds) Conjugate Divergent Margins, Special Publications vol. 369, The Geological Society, London, pp. 477–496.

Péron-Pinvidic, G. and Manatschal, G. 2009. The final rifting evolution at deep magma-poor passive margins from Iberia-Newfoundland: a new point of view. *International Journal of Earth Sciences*, **98**, 1581–1597.

Rao, G.N. 2001. Sedimentation, stratigraphy, and petroleum potential of Krishna-Godavari basin, East Coast of India. *AAPG Bulletin*, **85**, 1623–1643.

Roy, S., Choudhury, M. and Gupta, P. 2015. Rift grabens and crustal architecture of the offshore North East Coast-Mahanadi Basin, Eastern Continental Margin of India. In: Mukherjee, S. (ed) *Petroleum Geosciences: Indian Contexts*, Springer, pp. 63–86.

Sinha, S.T., Nemčok, M., Choudhuri, M. *et al.* 2015. The role of break-up localization in microcontinent separation along a strike-slip margin: the East India–Elan Bank case study. In: Nemčok, M., Rybár, S., Sinha, S.T. *et al.* (eds) *Transform Margins: Development, Controls and Petroleum Systems*, Special Publications vol. 431, The Geological Society, London, pp. 95–123.

42

The Ratnagiri Fracture Zone: A Paleo Oceanic-Fracture-Zone in the Mumbai-Ratnagiri Offshore Region, West India

Achyuta Ayan Misra[1], Neeraj Sinha[1], and Soumyajit Mukherjee[2]*

[1] *Reliance Industries Ltd, Mumbai, India*
[2] *Indian Institute of Technology Bombay, Mumbai, India*
* *Achyuta.Misra@ril.com; achyutaayan@gmail.com*

Oceanic fracture zones are large deformation zones with strong strike-slip movement. They separate segments of the spreading centres or mid-oceanic ridges (Searle, 2013). The seismic section presented here from the Mumbai/Bombay offshore region, in the western continental margin of India (WCMI) shows a paleo-oceanic fracture zone (Figure 42.1). The oceanic fracture zone is characterized by approximately 12 faults (Figures 42.2a–42.2d). The well BRDW-4-1 drilled a clay section from the sea bottom to the basaltic basement (NELP VII, 2007). The faults interpreted here are based on transparent fault shadows, bending of horizons and displacement of horizons (Chapter 2.3). The faults dip approximately 50–60°, as expected from oceanic fracture zones (Searle, 2013). Fracture zones also resemble aborted rifts. However, fundamental differences exist! Unlike aborted rifts, here the faults dip in the same direction. Secondly, in the transparent sediment pack, which appears similar to a sag basin deposit, the thickest section is higher than the thinner flanks: just opposite to that in aborted rifts. Thirdly, there are no growth sequences in these seemingly half grabens. Fourthly, the well shows a complete clay section, while in aborted rifts sand sections are expected. Chapter 4 shows rifted graben geometry with respect to sedimentation and deformation.

The geometry of the fracture zones connotes differential thermal subsidence on either sides of the fracture zone (Watts, 2001). Figure 42.2e schematically presents a Mid-Oceanic Ridge (MOR) and fracture zone-transform fault (in thick brown) set up. The part of the fracture zone between the spreading ridges is called a transform fault. Note, although the MOR segments appear to be displaced dextrally, the transform fault slipped sinistrally. Green arrows: spreading directions of the ridges. On the schematic section (Figure 42.2f) the plate at right is older with more subsidence than the plate at left. To maintain the elevation difference 'h' due to the initial thermal age variation, the higher and lower edges of the plate flex, causing the reverse drag (similar to Mukherjee, 2014a; also see Mukherjee and Koyi 2009; Mukherjee, 2013, 2014b) appearance on the fracture zones. The trends of the fracture zones (Figure 42.1) indicate the spreading direction. Here, it is NNW–SSE directed plate movement of the Seychelles microcontinent (Misra *et al.*, 2015; Misra and Mukherjee 2015).

Atlas of Structural Geological Interpretation from Seismic Images, First Edition. Edited by Achyuta Ayan Misra and Soumyajit Mukherjee.

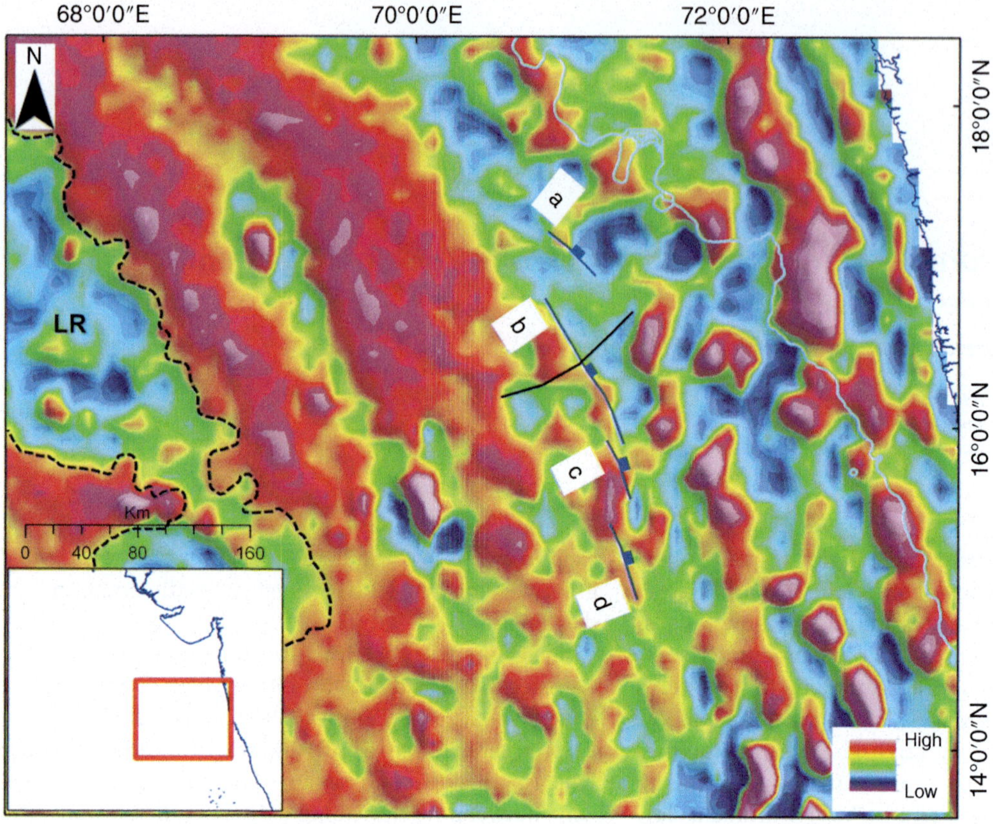

Figure 42.1 Map of Mumbai Offshore Basin, west India showing the Laxmi Ridge (LR) and adoining areas overlain on a hill-shaded, upward-continued (10 km) Bouguer gravity anomaly data (see the colour scale). The Ratnagiri Fracture Zone (RFZ) is indicated by normal faults in blue. a–d: four segments of the RFZ. Blue line: Shelf-slope break. Black line: position of seismic section.

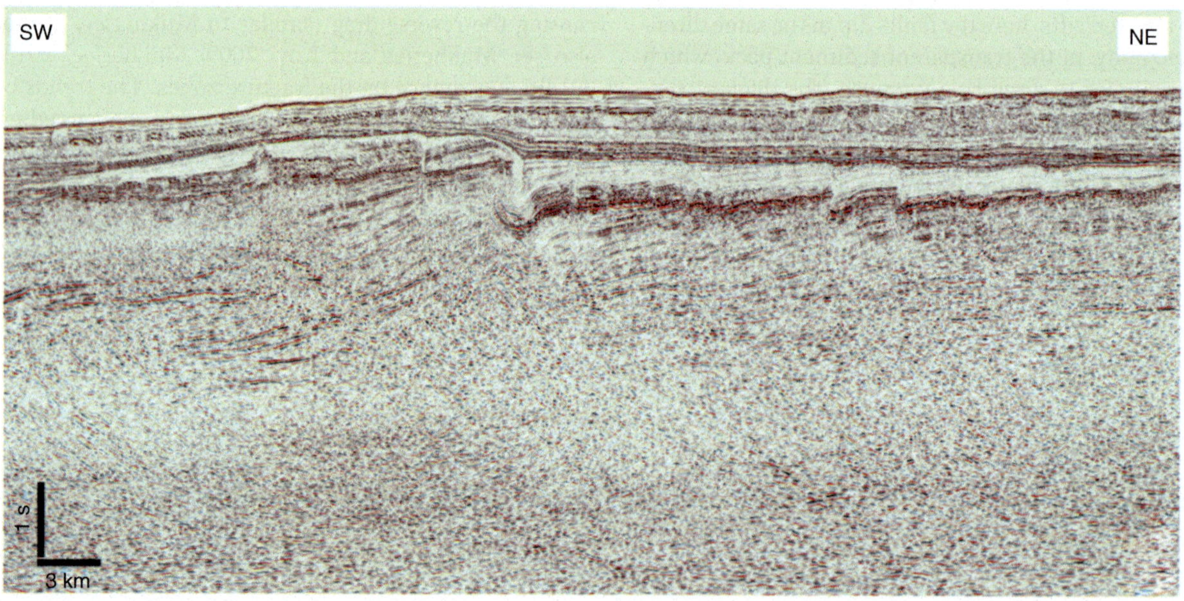

Figure 42.2a Uninterpreted seismic section from the WCMI showing the Ratnagiri Fracture Zone. Seismic data courtesy Reliance Industries Ltd. Reproduced with permission from Directorate General of Hydrocarbons (DGH), India.

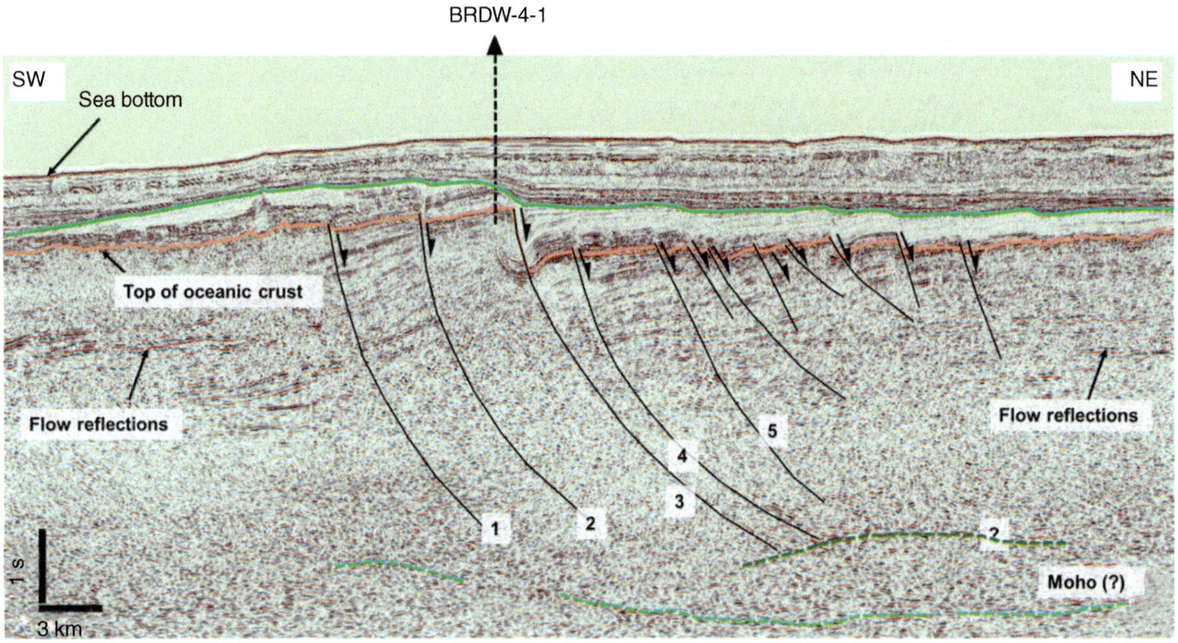

Figure 42.2b Interpretation of seismic section shown in Figure 42.2a. Note the marked faults are actually fracture zones and the slip marked by the half arrow on fault number 3, that is the fault with highest throw is not true. This is not a slipped marker, rather forms due to difference in age of the plates between the two sides of the fracture zone. The marker, that is the top of the oceanic crust, is required to be of the same age to be correlated across a fault. On fault 3, it is not (see Figure 42.2e–f). The age of the marker across the other faults in this section may be similar, which indicates that they can be actual fault planes. Note the layered nature of the oceanic crust due to the volcanic flows. Note the possible Moho reflections. There is another reflection above the Moho reflections where the faults/fracture zones seem to detach. It may be the brittle ductile transition, which needs further studies to confirm. Seismic data courtesy Reliance Industries Ltd. Reproduced with permission from Directorate General of Hydrocarbons (DGH), India.

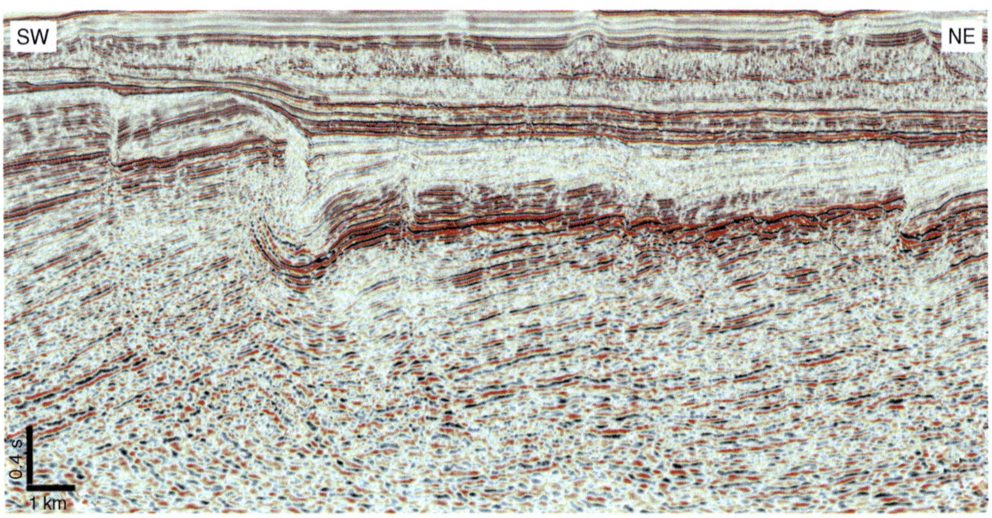

Figure 42.2c Uninterpreted seismic section showing the details of the fault with largest throw. Seismic data courtesy Reliance Industries Ltd. Reproduced with permission from Directorate General of Hydrocarbons (DGH), India.

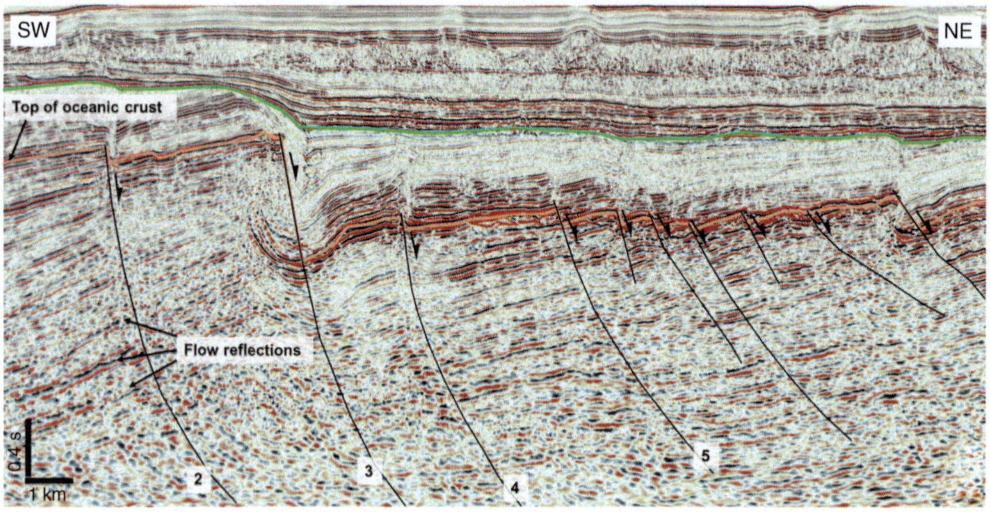

Figure 42.2d Interpretation of seismic section shown in Figure 42.2c. The fault shadows, slips of the top of the oceanic crust and abrupt changes in seismic facies, are the main criteria to identify the faults/fracture zones. Note the bend of the top of the transparent facies (green) above the oceanic crust, the top of which also subsides in response to the subsidence of the older crust on the NE of fault 3. Seismic data courtesy Reliance Industries Ltd. Reproduced with permission from Directorate General of Hydrocarbons (DGH), India.

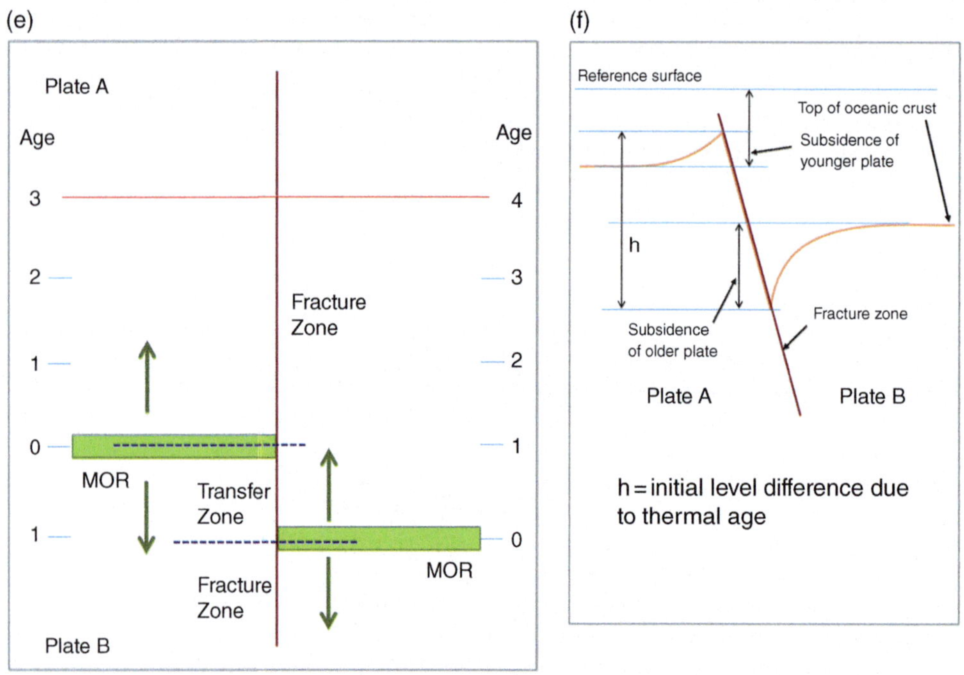

Figure 42.2e–f (e) Schematic map showing two segments of a mid-oceanic ridge (MOR) as green rectangles separated by a fracture zone-transfer zone. Indicative ages mentioned along the sides. Red line: location of the schematic section in 42.2f. 42.2f: Cause of the level difference and drag of the oceanic crust across a fracture zone. The difference in level 'h' is due to the age difference of the plates and the top of the oceanic crust flexes to keep 'h' constant (see Watts, 2001 for details).

References

Misra, A.A. and Mukherjee, S. 2015. *Tectonic Inheritance in Rifting and Continental Passive Margins*. SpringerBriefs, Springer International.

Misra, A.A., Sinha, N. and Mukherjee, S. 2015. Repeat ridge jumps and microcontinent separation: insights from NE Arabian Sea. *Marine and Petroleum Geology*, **59**, 406–428.

Mukherjee, S. 2013. *Deformation Microstructures in Rocks*. Springer.

Mukherjee, S. 2014a. Review of flanking structures in meso- and micro-scales. *Geological Magazine*, **151**, 957–974.

Mukherjee, S. 2014b. *Atlas of Shear Zone Structures in Meso-scale*. Springer.

Mukherjee, S. and Koyi, H.A. 2009. Flanking Microstructures. *Geological Magazine*, **146**, 517–526.

NELP VII. 2007. Mumbai Offshore Basin. http://www.infraline.com/ong/upstream/nelp-vii/mumbai-offshore_basin_nelp_vii.pdf (last accessed on 26 September 2017).

Searle, R. 2013. *Mid-Ocean Ridges*. Cambridge University Press, Cambridge.

Watts, A.B. 2001. *Isostasy and Flexure of the Lithosphere*. Cambridge University Press, Cambridge, pp. 137, 360–366, 458.

43

Normal Faults on the Downgoing Plate in an Arc–Continent Collision at East Timor, N Timor Sea

Achyuta Ayan Misra and Abhimanyu Maitra*

Reliance Industries Ltd, Mumbai, India
** Achyuta.Misra@ril.com; achyutaayan@gmail.com*

Orogenic belts resulting from collision of a continental crust with an island arc are arc–continent collision zones. They are smaller and less extensive than continent–continent collision zones. Arc–continent collisions produce new land masses and expose rocks lying in deep oceanic basins for analyses (Frisch *et al.*, 2010; Harris, 2011; Keary *et al.*, 2013).

The two sections shown here are from the Banda arc–continent collision zone near Timor in the N Timor Sea (Figure 43.1). It is an incipient arc–continent collision where the earliest stages of a compressional margin can be studied (Kaneko *et al.*, 2007; Shulgin *et al.*, 2009; Spakman and Hall, 2010). The accretionary complex/wedge, the downgoing slab and the trench fill are very well imaged (Figures 43.2 and 43.3). The sedimentary sequence of the continental crust is remarkably uniform and individual horizons are traceable throughout the southeast continental part of the seismic section. Normal faults are the only deformation observable on the continental crust (Figures 43.2c–43.2f). These normal faults form as the crust bent by the load accretionary complex. The faults can be identified by offset stratigraphic markers and fault shadows (Chapter 2.3). There are normal faults visible in the subducting plate beneath the accretionary complex also (Figures 43.2c and 43.2d). The top of the continental crust is possibly imaged in the second section (Figures 43.3a–43.3d). It is probably a high amplitude reflector dippling towards the Timor Trough. Further studies would confirm it. The second section also shows the same normal faulted geometry over the continental crust. The trench, the Timor Trough, is sediment starved in the first section (Figure 43.2). However, the second section (Figure 43.3) consists of an approximately 1 km thick turbidite sequence. The accretionary complex, in both the sections, has a chaotic reflection pattern, in contrast to the downgoing plate. This is possibly due to numerous imbricate complexes present, which cannot be imaged well in these seismic sections.

Atlas of Structural Geological Interpretation from Seismic Images, First Edition. Edited by Achyuta Ayan Misra and Soumyajit Mukherjee.

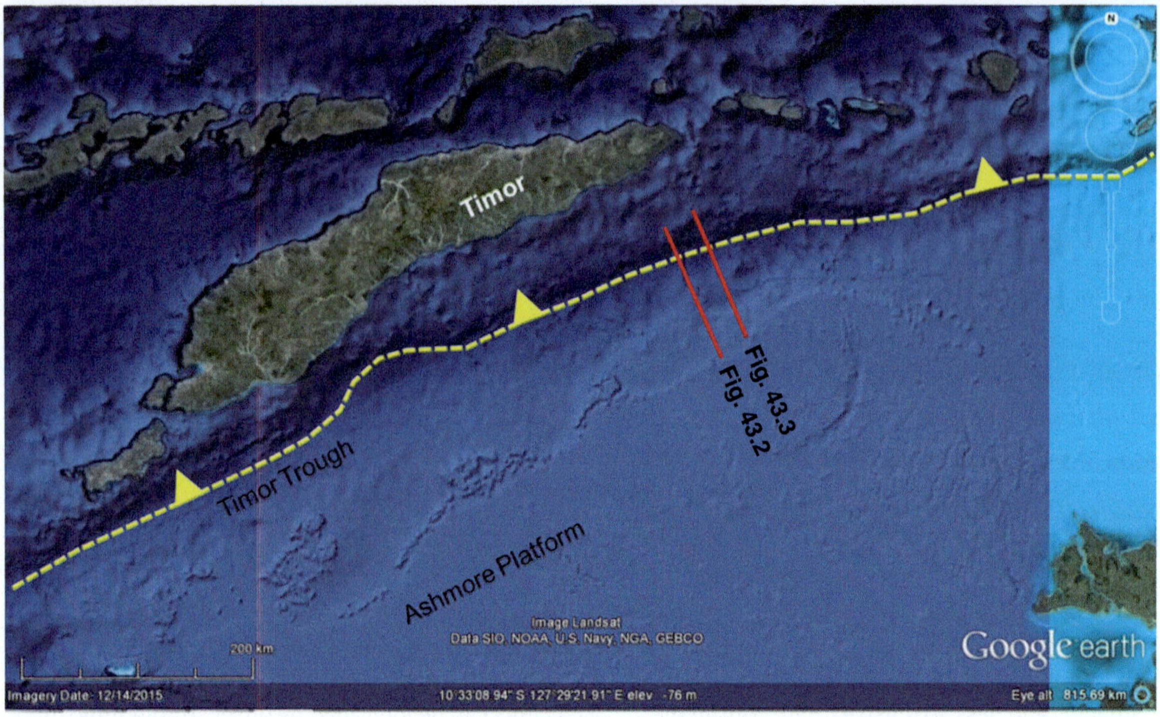

Figure 43.1 Map of the N Timor Sea showing the Timor Island, Timor Trough and Ashmore Platform. Red lines: locations of the seismic sections in this section.

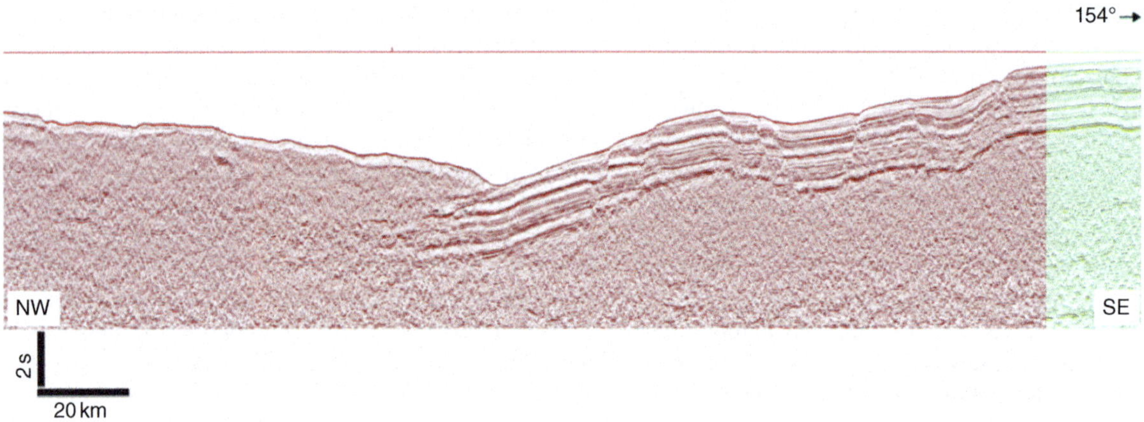

Figure 43.2a Uninterpreted regional seismic section. See Figure 43.1 for location of the seismic section. Seismic data courtesy Reliance Industries Ltd.

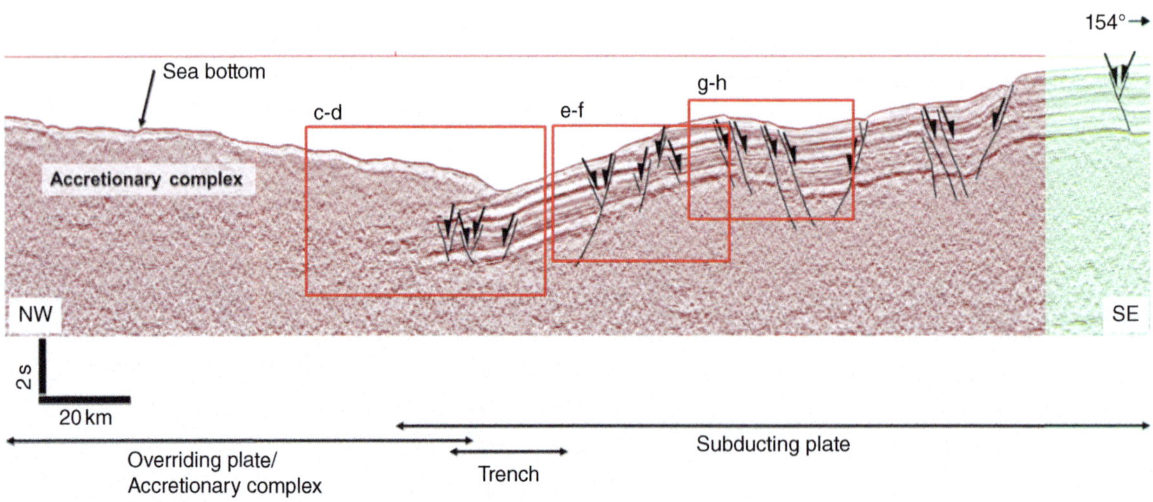

Figure 43.2b Interpretation of the section shown in Figure 43.2a. Note the continuity of the seismic reflections on the subducting plate and the chaotic reflection of the accretionary complex. The bend of the subducting plate is facilitated by NW and SE dipping normal faults. Red rectangles: details shown in Figures 43.2c–43.2h. Seismic data courtesy Reliance Industries Ltd.

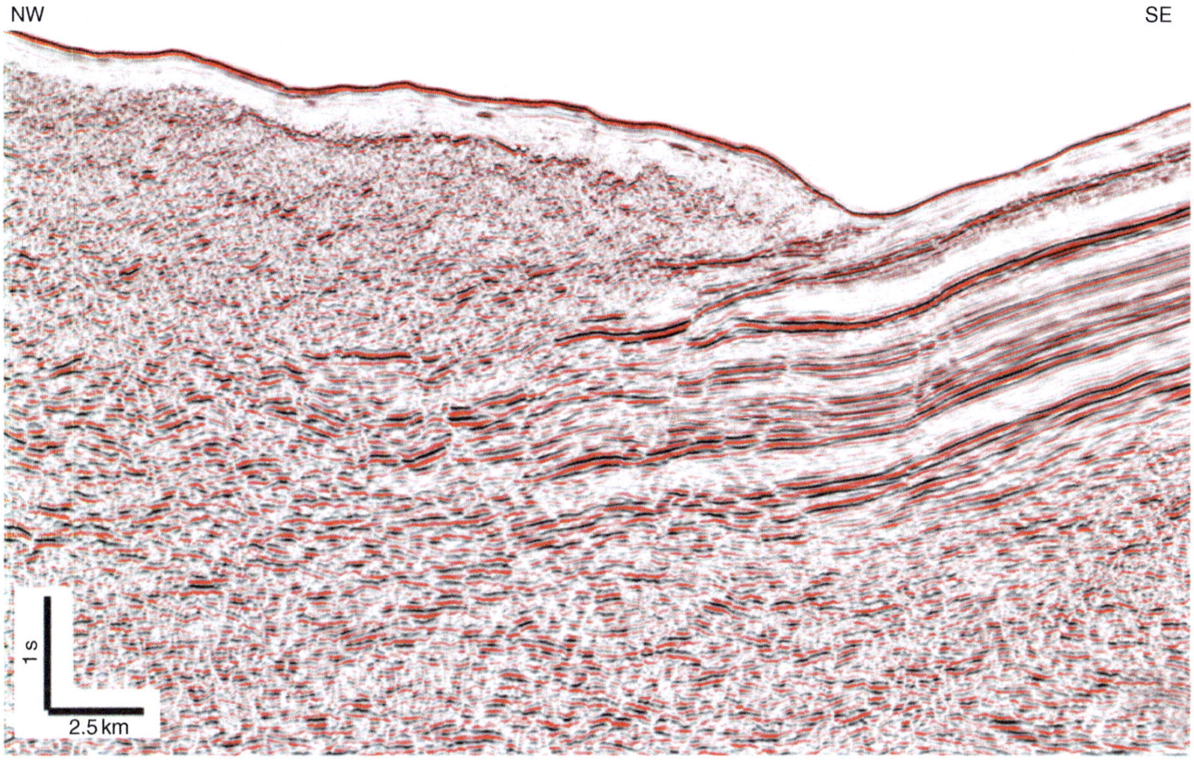

Figure 43.2c Uninterpreted section showing the trench, subducting plate and accretionary complex. See Figure 43.2b for location of the profile. Seismic data courtesy Reliance Industries Ltd.

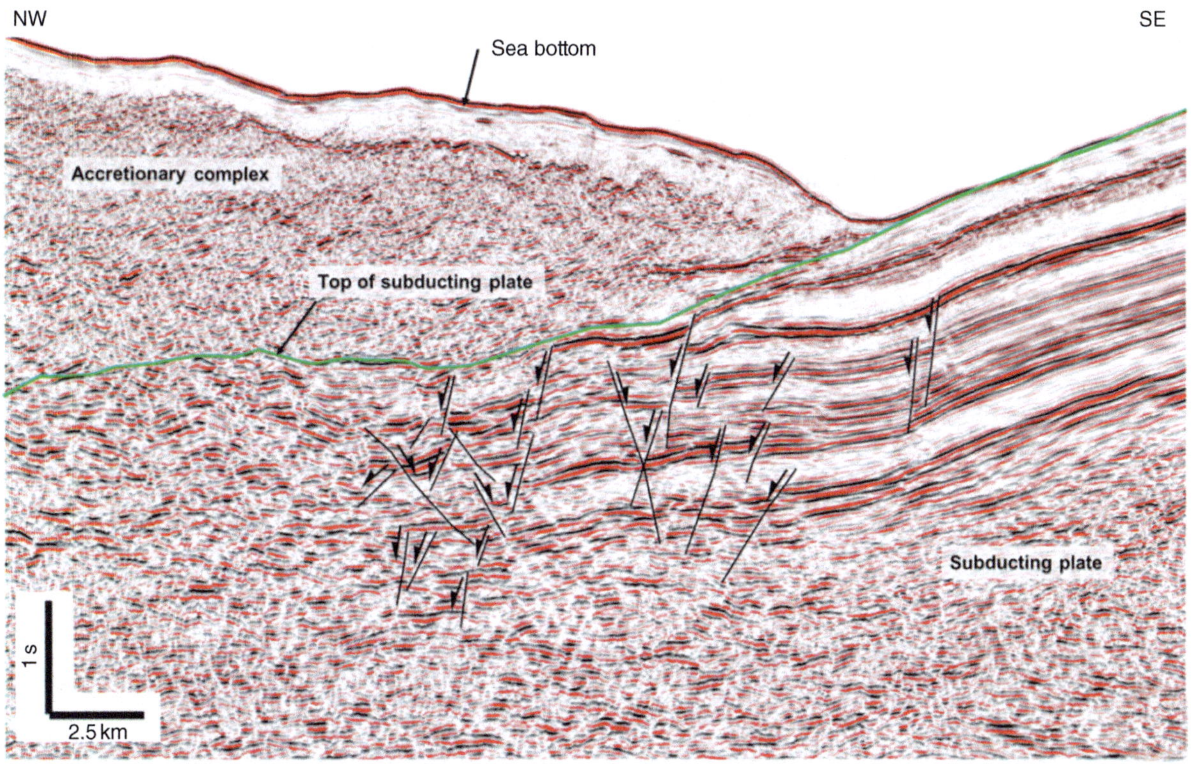

Figure 43.2d Interpretation of seismic section shown in Figure 43.2c showing the trench, subducting plate and accretionary complex. Note the continuity of the seismic reflections of the sedimentary section on the subducting plate and the chaotic reflection of the accretionary complex. The normal faults beneath the accretionary complex help load adjustment. See Figure 43.2b for location of the profile. Seismic data courtesy Reliance Industries Ltd.

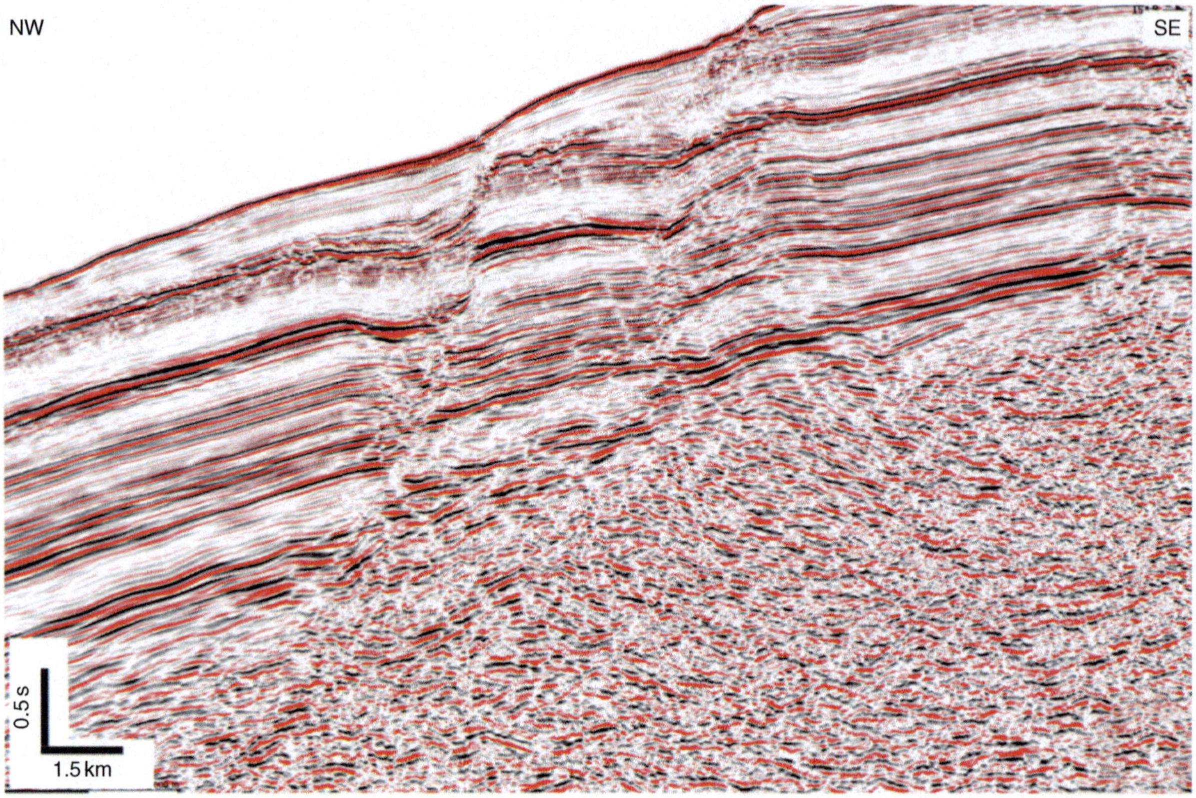

Figure 43.2e Uninterpreted section showing the sedimentary section atop the subducting plate. See Figure 43.2b for location of the profile. Seismic data courtesy Reliance Industries Ltd.

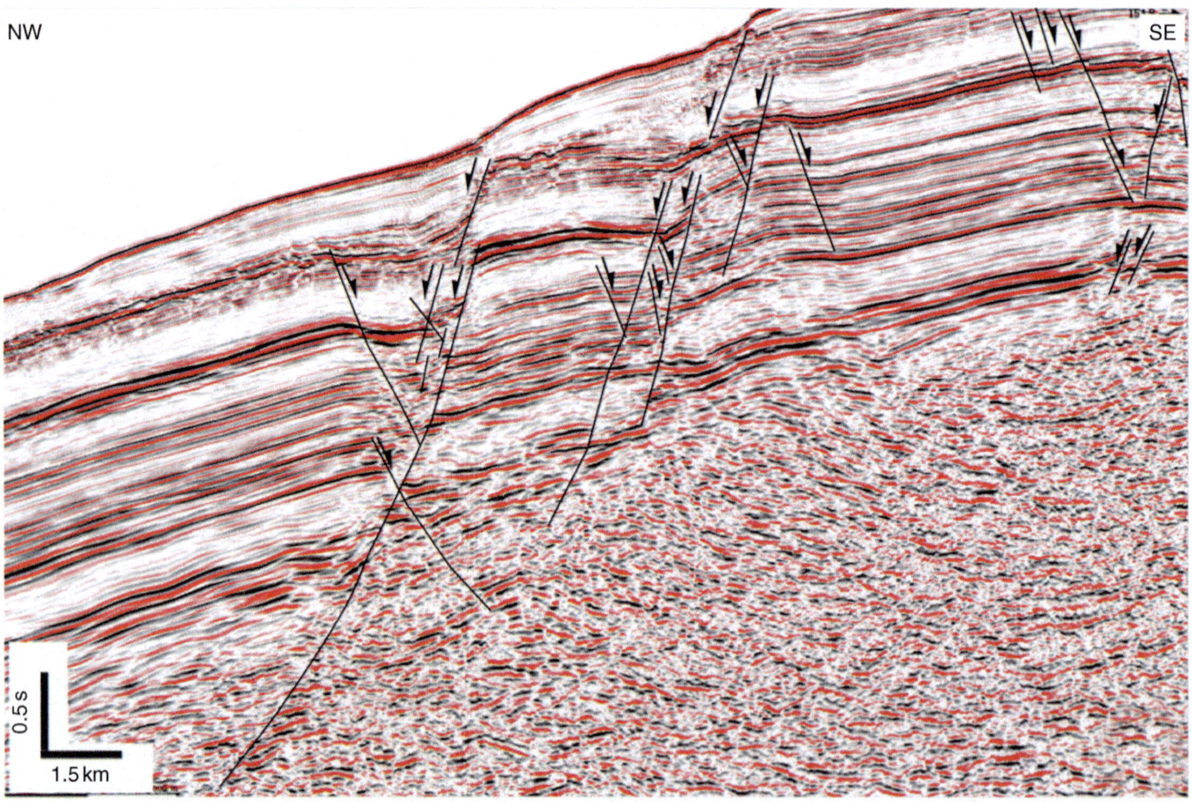

Figure 43.2f Interpretation of seismic section shown in Figure 43.2e showing the sedimentary section atop the subducting plate. Note the remarkable continuity of the seismic reflections in the sedimentary section throughout the profile. The continental crust beneath the sedimentary sequence shows chaotic seismic facies, though the top of the continental crust is not imaged. See Figure 43.2b for location of the profile. Seismic data courtesy Reliance Industries Ltd.

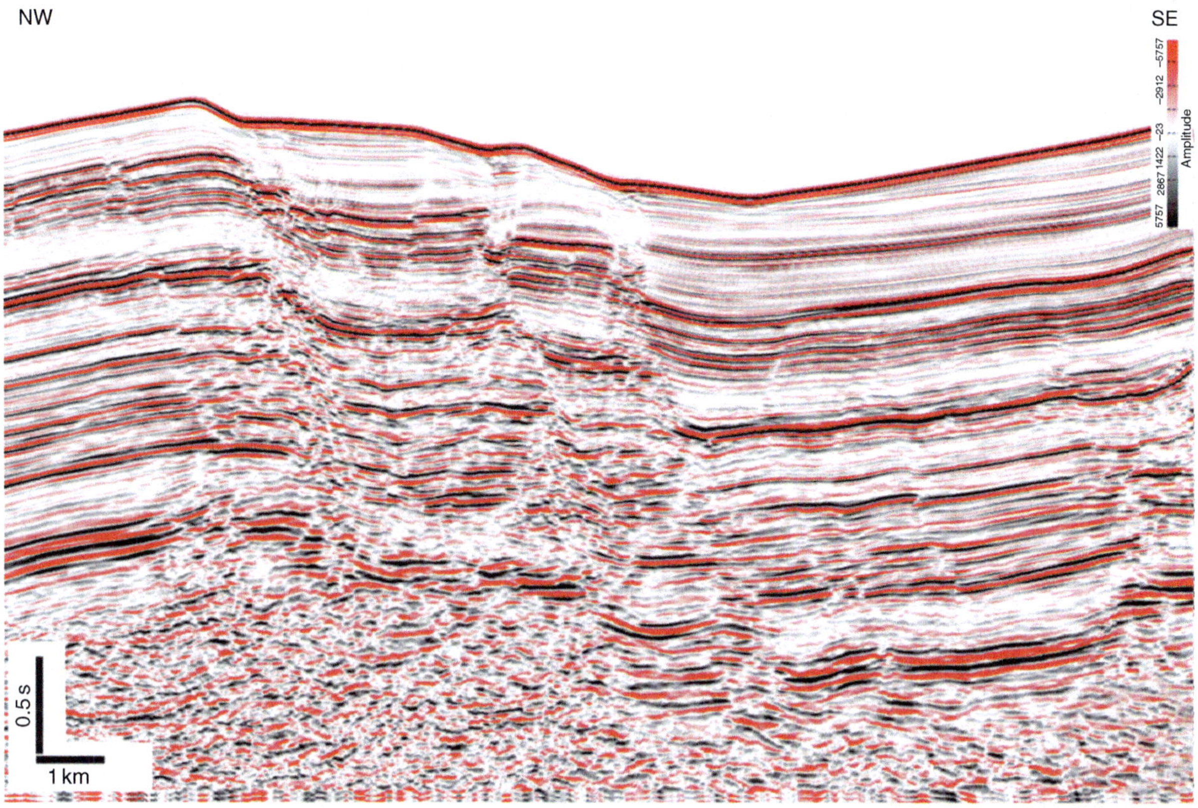

Figure 43.2g Uninterpreted section showing a low on the continental crust. See Figure 43.2b for location of the profile. Seismic data courtesy Reliance Industries Ltd.

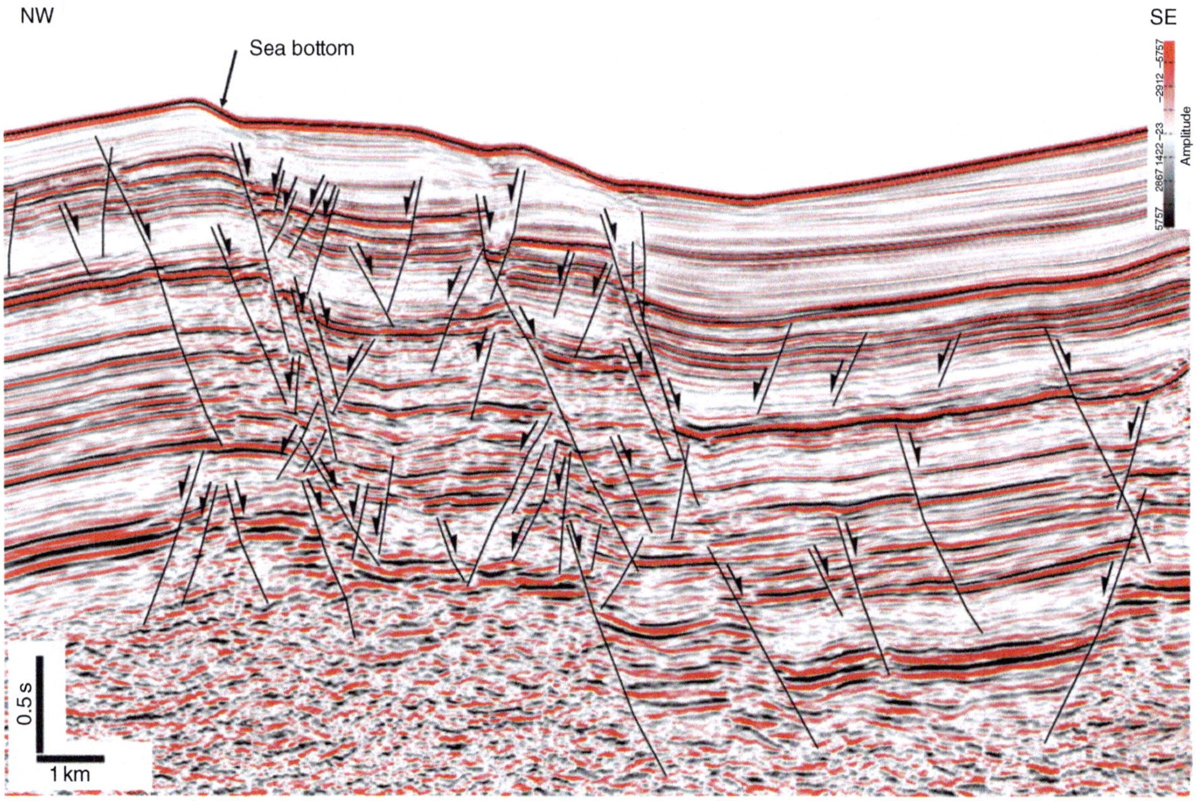

Figure 43.2h Interpretation of seismic section shown in Figure 43.2g. Note the normal faults facilitating the bends on the subducting plate. Seismic data courtesy Reliance Industries Ltd.

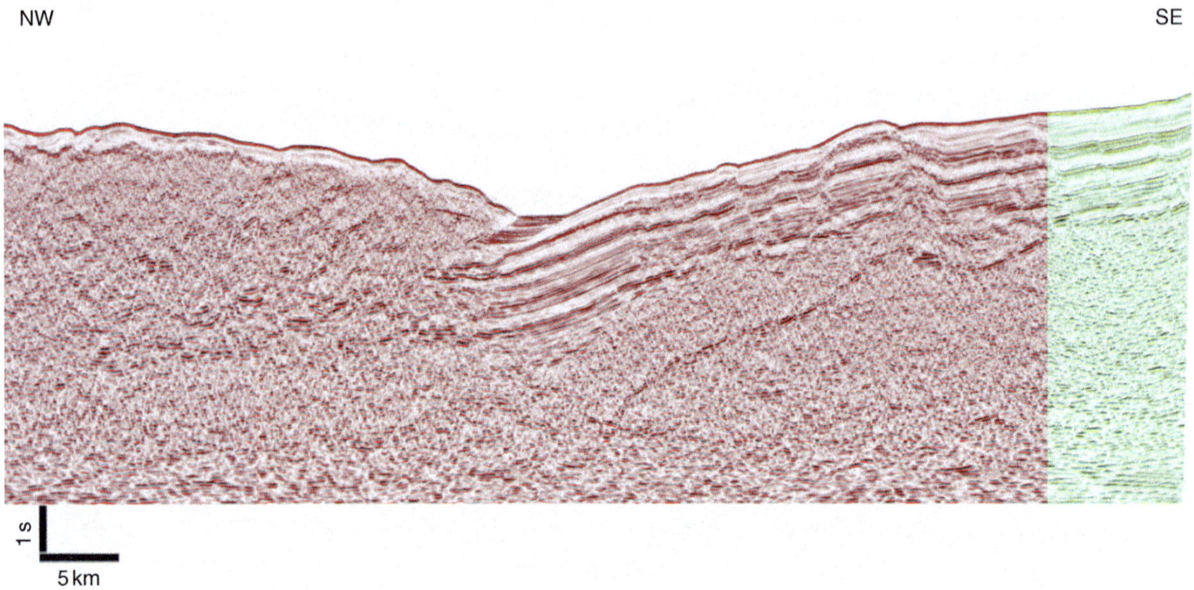

Figure 43.3a Uninterpreted regional seismic section. See Figure 43.1 for location of the seismic section. Seismic data courtesy Reliance Industries Ltd.

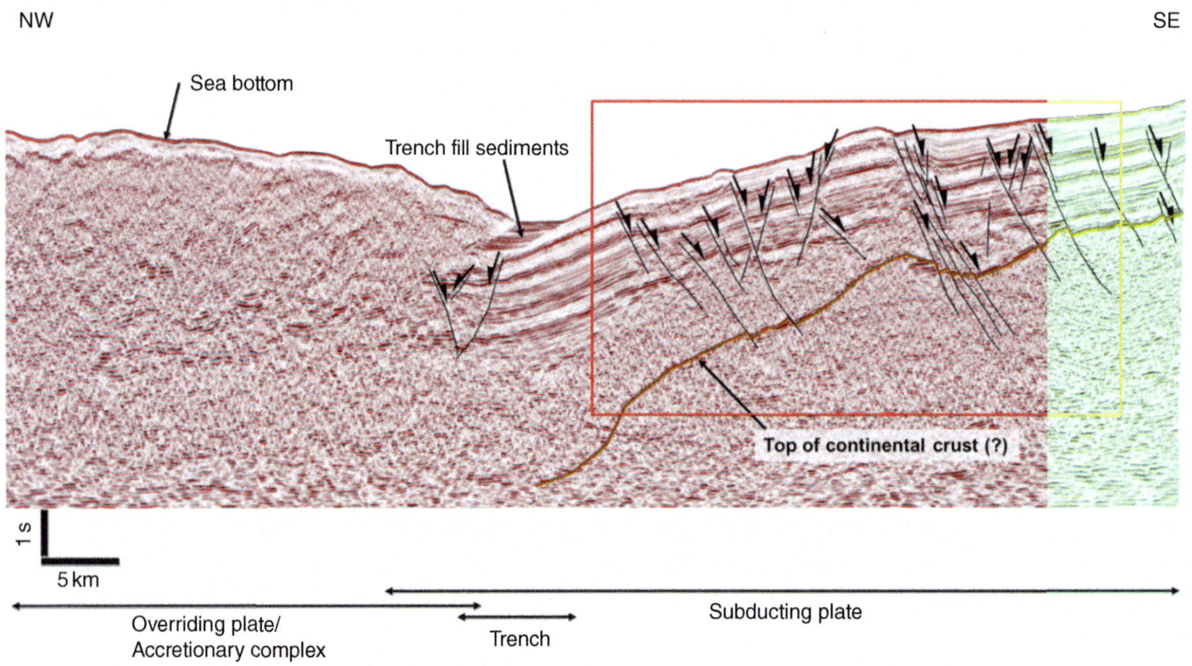

Figure 43.3b Interpretation of the section shown in Figure 43.3a. Similar to Figure 43.2, note the continuity of the seismic reflections on the subducting plate and the chaotic reflection of the accretionary complex. Trench fill sediments, usually turbidites, are observed. Top of the continental crust has been imaged in this section as a continuous, high amplitude reflection dipping towards the trench. Red rectangle: details shown in Figures 43.3c–43.3d. Seismic data courtesy Reliance Industries Ltd.

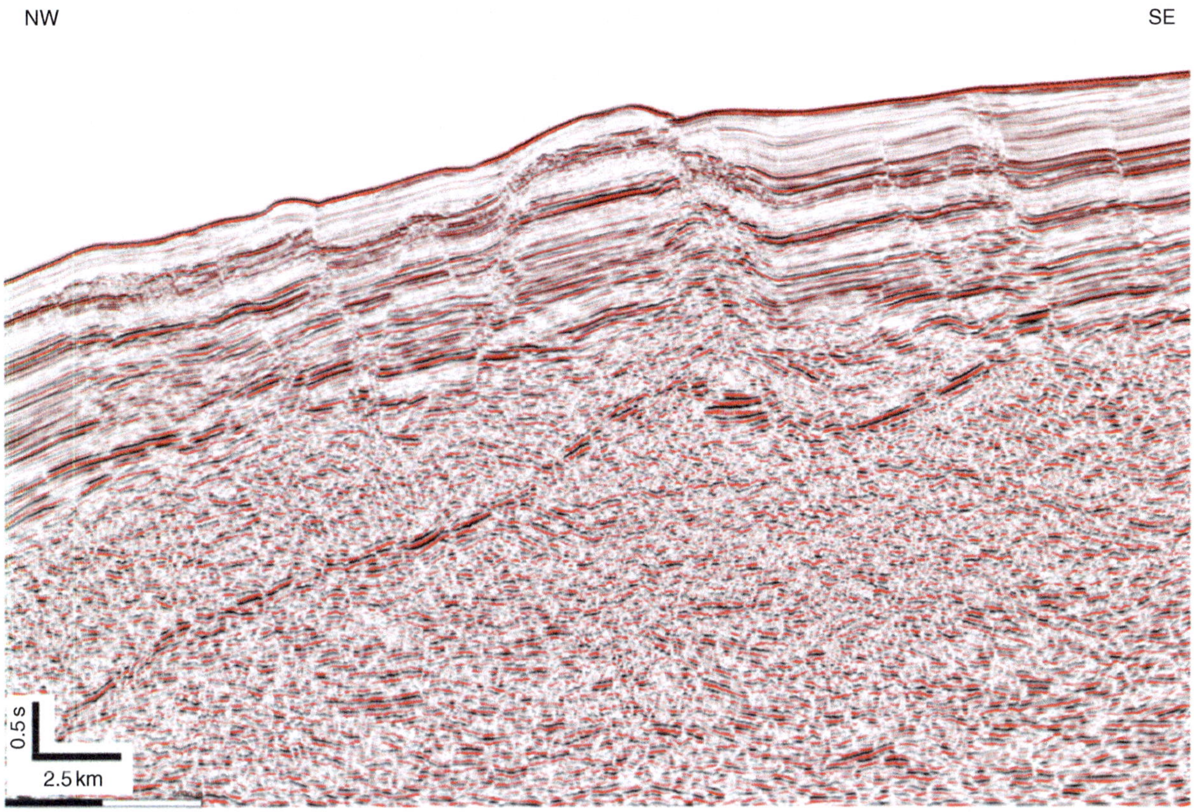

Figure 43.3c Uninterpreted section showing the sedimentary section atop the subducting plate. See Figure 43.3b for location of the profile. Seismic data courtesy Reliance Industries Ltd.

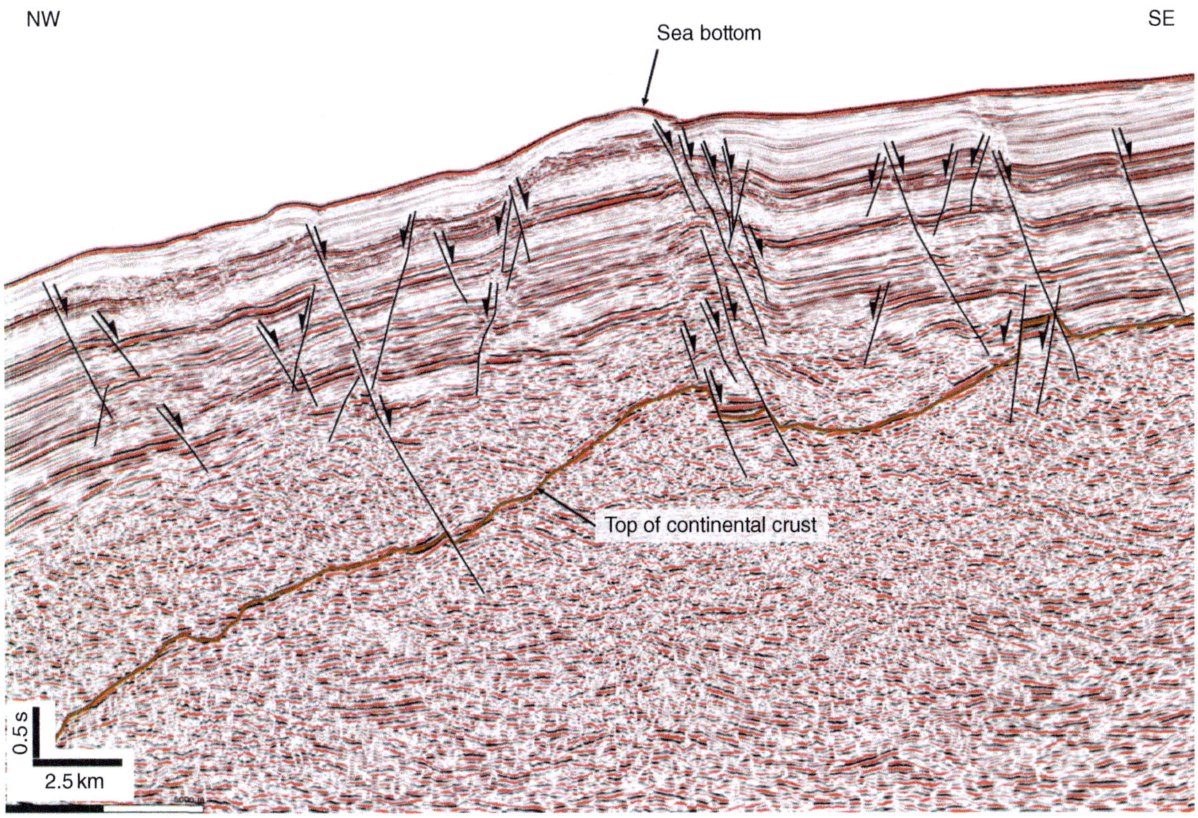

Figure 43.3d Interpretation of Figure 43.3c. Note the normal faults in the sedimentary section that also continue down to the top of the continental crust. Seismic data courtesy Reliance Industries Ltd.

References

Frisch, W., Meschede, M. and Blakey, R.C. 2010. *Plate Tectonics: Continental Drift and Mountain Building*. Springer-Verlag, Berlin/Heidelberg.

Harris, R. 2011. The nature of the Banda Arc–continent collision in the Timor region. In: Brown, D. and Ryan, P.D. (eds) *Arc–Continent Collision*, Springer, Berlin/Heidelberg, pp. 163–211.

Kaneko, Y., Maruyama, S., Kadarusman, A. *et al.* 2007. On-going orogeny in the outer-arc of the Timor–Tanimbar region, eastern Indonesia. *Gondwana Research*, **11**, 218–233.

Kearey, P., Klepeis, K.A. and Vine, F.J. 2013. *Global Tectonics*. John Wiley & Sons Ltd, Chichester.

Shulgin, A., Kopp, H., Mueller, C. *et al.* 2009. Sunda-Banda arc transition: Incipient continent-island arc collision (northwest Australia). *Geophysical Research Letters*, **36**, L10304.

Spakman, W. and Hall, R. 2010. Surface deformation and slab–mantle interaction during Banda arc subduction rollback. *Nature Geoscience*, **3**, 562–566.

44

The Gop Rift: A Paleo Slow Spreading Centre, Offshore Gujarat, India

Achyuta Ayan Misra[1], Neeraj Sinha[1], and Soumyajit Mukherjee[2]*

[1] *Reliance Industries Ltd, Mumbai, India*
[2] *Indian Institute of Technology Bombay, Mumbai, India*
* *Achyuta.Misra@ril.com; achyutaayan@gmail.com*

The Gop Rift, an oceanic spreading centre, has been studied and reviewed recently (Yatheesh *et al.*, 2009; Armitage *et al.*, 2010; Bhattacharya and Yatheesh 2015; Misra *et al.*, 2015; Misra and Mukherjee 2015; Nemčok and Rybar, 2017). The Gop Rift occurs at the SE end on the Saurashtra Volcanic Platform and is underlain by ~18 km thick oceanic crust (Corfield *et al.*, 2010, Figure 3). Presented here is a deep crustal seismic section, shot by Ion-GX Technology (Figure 44.1). (The seismic acquisition parameters can be found in Misra *et al.*, 2015). Figure 44.2 shows the Gop Rift in a seismic section. This is clearly a rift because of the normal faults dip oppositely and face each other. Rifts are either continental (Chapter 4) or oceanic (i.e. mid-oceanic rift ridge systems). Continental rift systems characteristically possess syn-rift sediments with sediment packages (Chapter 4). Oceanic rift systems, that is mid oceanic ridges, especially slow ones, also display a structural set up with normal faults, resembling continental rifts (Standish and Sims, 2010; Searle, 2013). However, the typical sedimentary signature is absent (Figure 44.2). There are normal faults with a central graben with maximum slip. There are high amplitude, low frequency reflections beneath the central graben (Figure 44.2). Moho is imaged ~18–20 km below the fault system (Figure 44.2b). The oceanic crust is anomalously thick, possibly due to underplating (Misra *et al.*, 2015). There are subparallel reflections in the top 5–7 km of the oceanic crust, related to basalt flow. There are reflections that rise up towards the central graben. Their geologic significance needs more study. There are compaction-related faults in the sedimentary sequence (Figures 44.2c–44.2f). There are also channel-levee complexes in the sedimentary section (Figures 44.2c–44.2f) (Carmichael *et al.*, 2009).

Atlas of Structural Geological Interpretation from Seismic Images, First Edition. Edited by Achyuta Ayan Misra and Soumyajit Mukherjee.

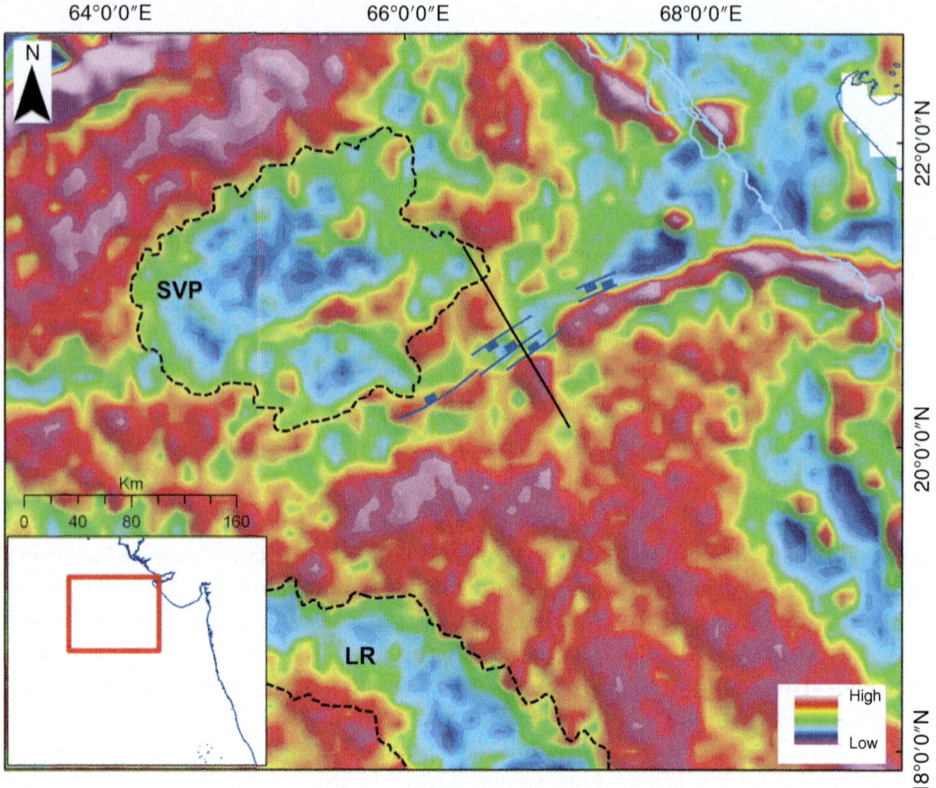

Figure 44.1 Map of offshore Gujarat, India showing the area with Saurashtra Volcanic Platform (SVP) and Laxmi Ridge (LR) overlain on a hill-shaded, upward-continued (10 km) Bouguer gravity anomaly data (see the colour scale). Gop Rift is indicated by normal faults in blue. Blue line: shelf-slope break. Black line: position of seismic section.

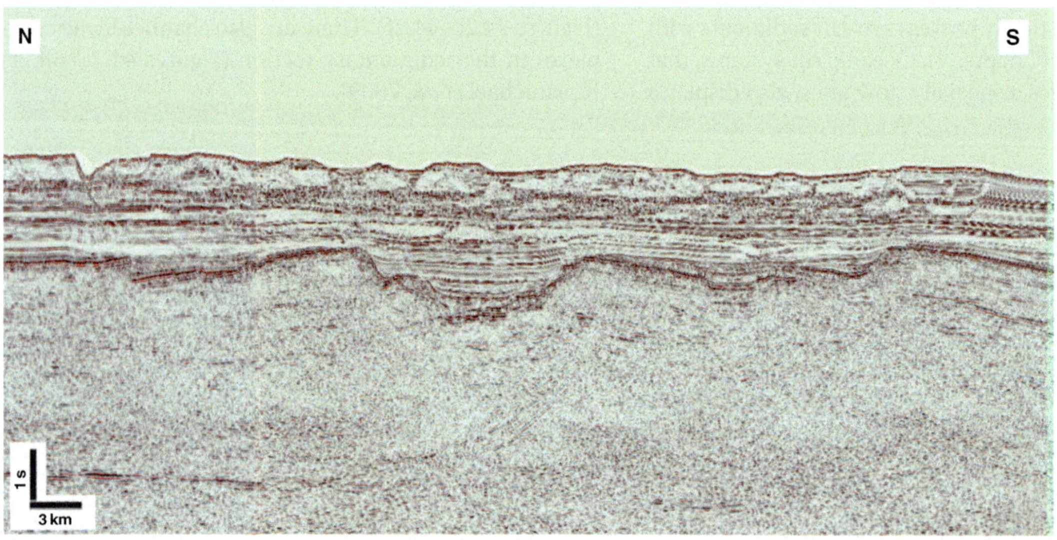

Figure 44.2a Uninterpreted seismic section of the Gop Rift region. See Fig 44.1 for location of the seismic section.Seismic data courtesy Reliance Industries Ltd. Reproduced with permission from the Directorate General of Hydrocarbons (DGH), India.

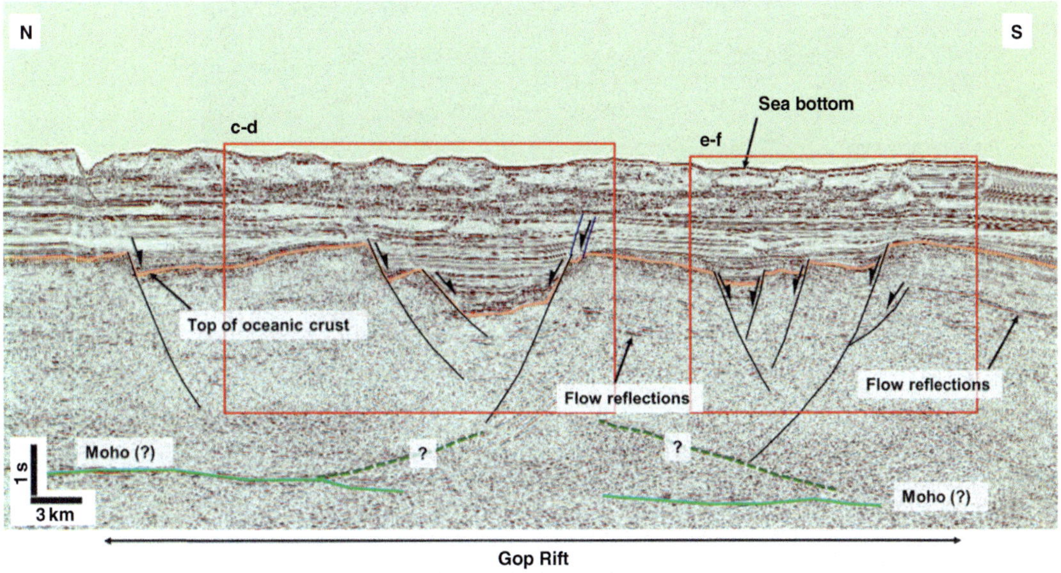

Figure 44.2b Interpretation of seismic section shown in Figure 44.2a. Note the normal faulted nature of the Gop Rift. The faults are identified based on the offset of the marker, that is the top of the oceanic crust. In the deeper parts, there are broken reflections of the fault planes, for example Figure 44.2e. The possible Moho reflection is prominent. The sloping reflections above the Moho can be well observed, but it is difficult to ascertain their geological nature without further studies. Seismic data courtesy Reliance Industries Ltd. Reproduced with permission from Directorate General of Hydrocarbons (DGH), India.

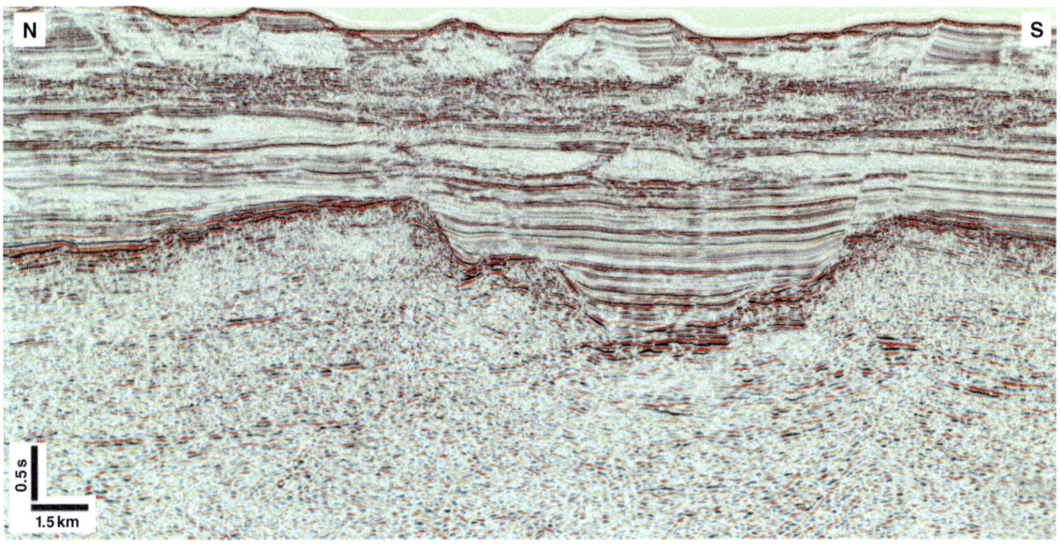

Figure 44.2c Uninterpreted seismic section of the central graben in the Gop Rift. See Figure 44.2b for location of the seismic section. Seismic data courtesy Reliance Industries Ltd. Reproduced with permission from Directorate General of Hydrocarbons (DGH), India.

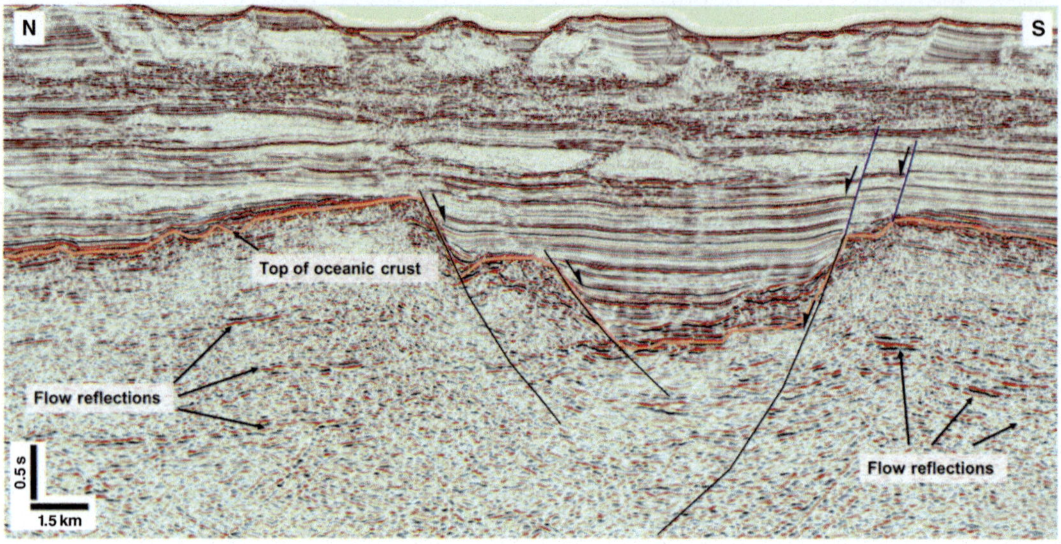

Figure 44.2d Interpretation of Figure 44.2c Note the subparallel flow reflections related to the basaltic flows and also the high amplitude, low frequency reflection in the central graben. Seismic data courtesy Reliance Industries Ltd. Reproduced with permission from Directorate General of Hydrocarbons (DGH), India.

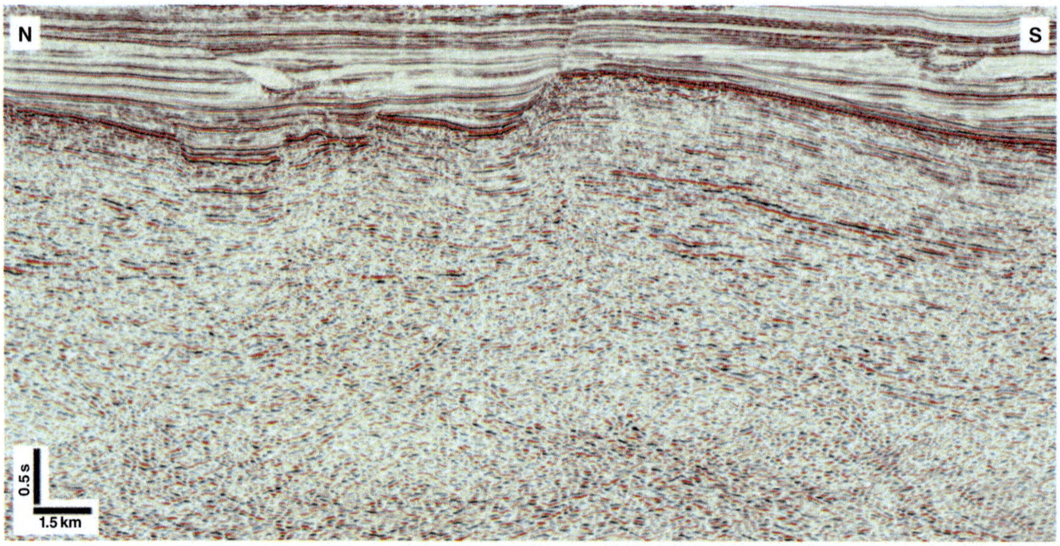

Figure 44.2e Uninterpreted seismic section of the southern part of the Gop Rift. See Figure 44.2b for location of the seismic section. Seismic data courtesy Reliance Industries Ltd. Reproduced with permission from Directorate General of Hydrocarbons (DGH), India.

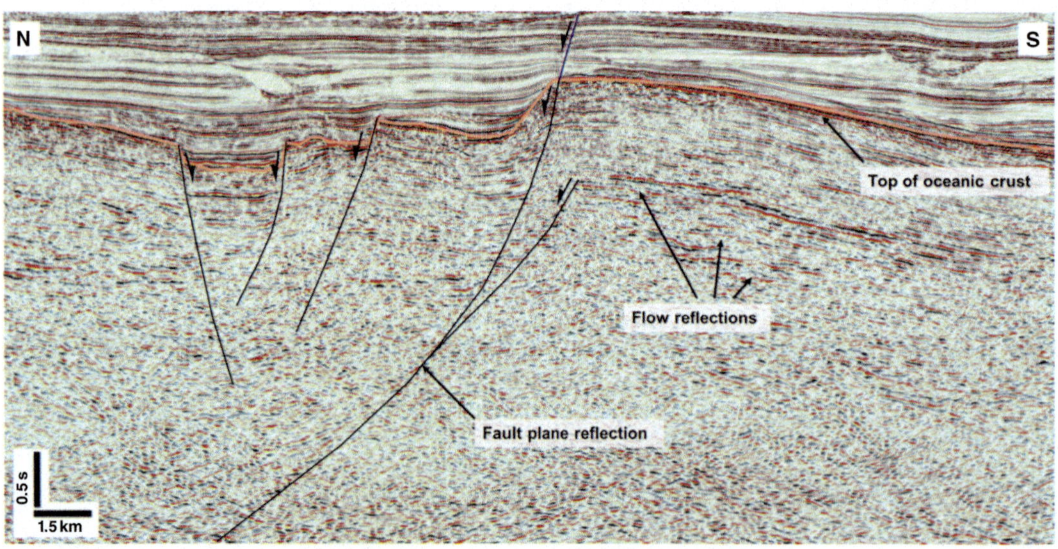

Figure 44.2f Interpretation of seismic section shown in Figure 44.2e. Note the subparallel reflections, which are characteristic of oceanic crust tops and also the fault plane reflections. Fault-related drag (Mukherjee 2014) noted. Seismic data courtesy Reliance Industries Ltd. Reproduced with permission from Directorate General of Hydrocarbons (DGH), India

References

Armitage, J.J., Collier, J.S. and Minshull, T.A. 2010. The importance of rift history for volcanic margin formation, *Nature*, **465**, 913–917.

Bhattacharya, G.C. and Yatheesh, V. 2015. Plate-tectonic evolution of the deep ocean basins adjoining the western continental margin of India – a proposed model for the early opening scenario. In: Mukherjee, S. (ed) *Petroleum Geosciences: Indian Contexts*, Springer, pp. 1–62.

Carmichael, S.M., Akhter, S., Bennett, J.K. *et al.* 2009. Geology and hydrocarbon potential of the offshore Indus Basin, Pakistan. *Petroleum Geoscience*, **15**, 107–116.

Corfield, R.I., Carmichael, S., Bennett, J. *et al.* 2010. Variability in the crustal structure of the West Indian Continental Margin in the Northern Arabian Sea. *Petroleum Geoscience*, **16**(3), 257–265.

Misra, A.A. and Mukherjee, S. 2015. *Tectonic Inheritance in Continental Rifts and Passive Margins*. Springer.

Misra, A.A., Sinha, N. and Mukherjee, S. 2015. Repeat ridge jumps and microcontinent separation: insights from NE Arabian Sea. *Marine and Petroleum Geology*, **59**, 406–428.

Mukherjee, S. 2014. Review of flanking structures in meso- and micro-scales. *Geological Magazine*, **151**, 957–974.

Nemčok, M. and Rybár, S. 2017. Rift–drift transition in a magma-rich system: the Gop Rift–Laxmi Basin case study, West India. In: Mukherjee, S., Misra, A.A., Calvès, G. and Nemčok, M. (eds) *Tectonics of the Deccan Large Igneous Province*, Special Publications vol. 445, The Geological Society, London, pp. 95–117. doi: 10.1144/SP445.5.

Standish, J.J. and Sims, K.W. 2010. Young off-axis volcanism along the ultraslow-spreading Southwest Indian Ridge. *Nature Geoscience*, **3**, 286–292.

Searle, R. 2013. *Mid-Ocean Ridges*. Cambridge University Press, Cambridge.

Yatheesh, V., Bhattacharya, G.C. and Dyment, J. 2009. Early oceanic opening off western India–Pakistan margin: The Gop Basin revisited. *Earth and Planetary Science Letters*, **284**, 399–408.

45

Transform Margin in the Cauvery Basin, Indian East Coast Passive Margin

Achyuta Ayan Misra and Swagato Dasgupta*

Reliance Industries Ltd, Mumbai, India
** Achyuta.Misra@ril.com; achyutaayan@gmail.com*

Transform/sheared margins (Misra *et al.* 2014, 2015; Misra and Mukherjee 2015) form when tectonic extension is at low angle (<30°) to the margin. Mazumder *et al.* (2018) presents a recent selected review on tectonics of this basin. The Cauvery Basin developed as a transform margin as the Elan Bank microcontinent rifted from India (Nemčok *et al.*, 2013; Sinha *et al.*, 2015). A typical seismic section from the Cauvery Basin is presented here (Figure 45.1).The transform passive margins have steep continental slopes due to steeply dipping strike-slip faults that thinned the lithosphere (Figures 45.2a and 45.2b). The continental Moho rises steeply towards the proto-oceanic crust (POC: Chapter 42). The oceanic Moho is flat and the oceanic crust depicts its normal nature in thickness. The POC also shows its typical nature with a subdued Moho reflection and chaotic seismic facies in the top parts. The synrift grabens are smaller than those in orthogonal rift margins (Chapter 42). Footwall erosion is also evident (Figures 45.2c and 45.2d).

Additionally, compactional polygonal faulting is shown in the postrift section (Figures 45.2e and 45.2f). Polygonal faults form due to fluid escape from thick shale layers (Stuevold *et al.*, 2003; Gay *et al.*, 2004, 2006). They are normal faults bound between sediment layers – mainly identified by offset markers and fault shadows (Chapter 2.3).

Atlas of Structural Geological Interpretation from Seismic Images, First Edition. Edited by Achyuta Ayan Misra and Soumyajit Mukherjee.

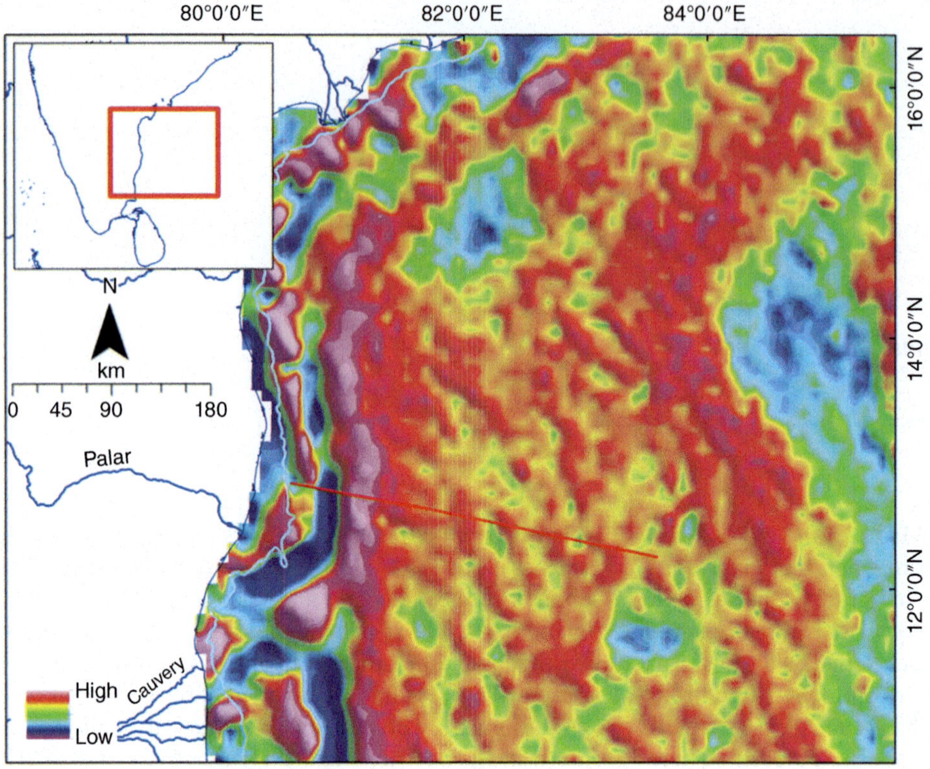

Figure 45.1 Map of the Cauvery Basin along the ECMI overlaid on hill-shaded, upward-continued (10 km) Bouguer gravity anomaly data (see the colour scale). Blue line: shelf-slope break. Red line: position of seismic section.

Figure 45.2a Uninterpreted regional seismic section in the Cauvery Basin, India. See Figure 45.1 for location of the seismic section. Seismic data courtesy Reliance Industries Ltd. Reproduced with permission from Directorate General of Hydrocarbons (DGH), India.

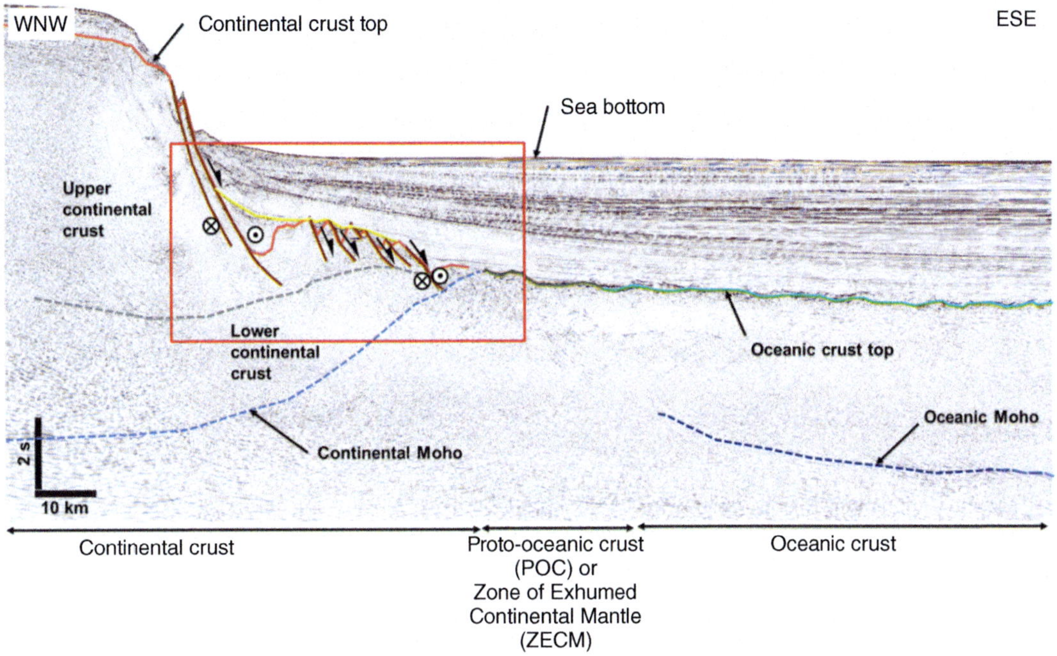

Figure 45.2b Interpreted regional seismic section shown in Figure 45.2a showing the major tectonic elements of a transform/sheared passive margin. Note the differences in Moho character beneath the thinned continental crust, oceanic crust and proto-oceanic crust. The steep fault with throw of more than 4 km is the Coromaldal Strike-slip fault (Nemčok *et al.*, 2013). The boundary of the upper and lower continental crust is determined by the deepest level where faults can be interpreted. Red rectangle: extent of seismic section shown in in Figure 45.2c and 45.2d). Seismic data courtesy Reliance Industries Ltd. Reproduced with permission from Directorate General of Hydrocarbons (DGH), India.

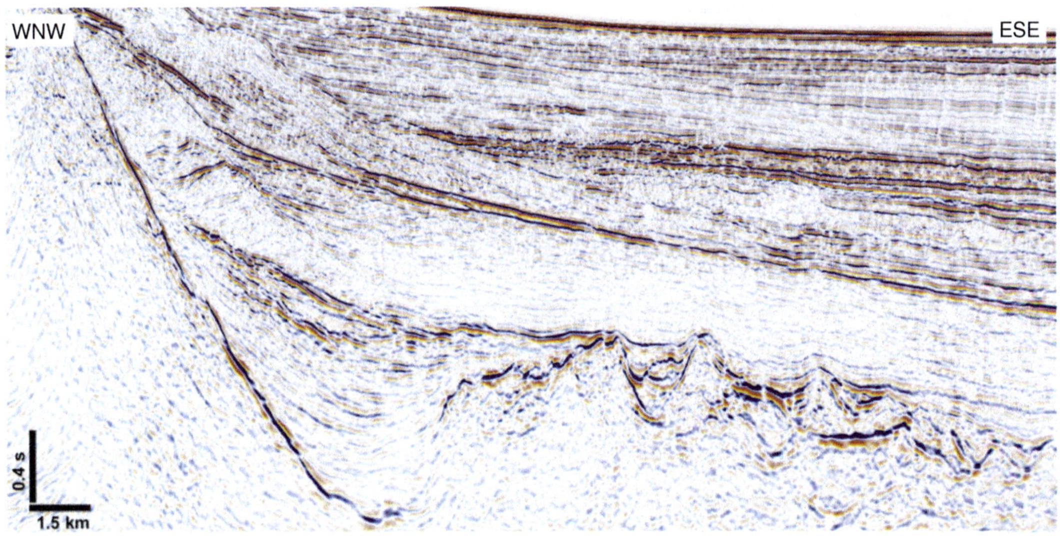

Figure 45.2c Uninterpreted seismic section showing the thinned continental crust in a transform margin. See Figure 45.2b for the location of the seismic section. Seismic data courtesy Reliance Industries Ltd. Reproduced with permission from Directorate General of Hydrocarbons (DGH), India.

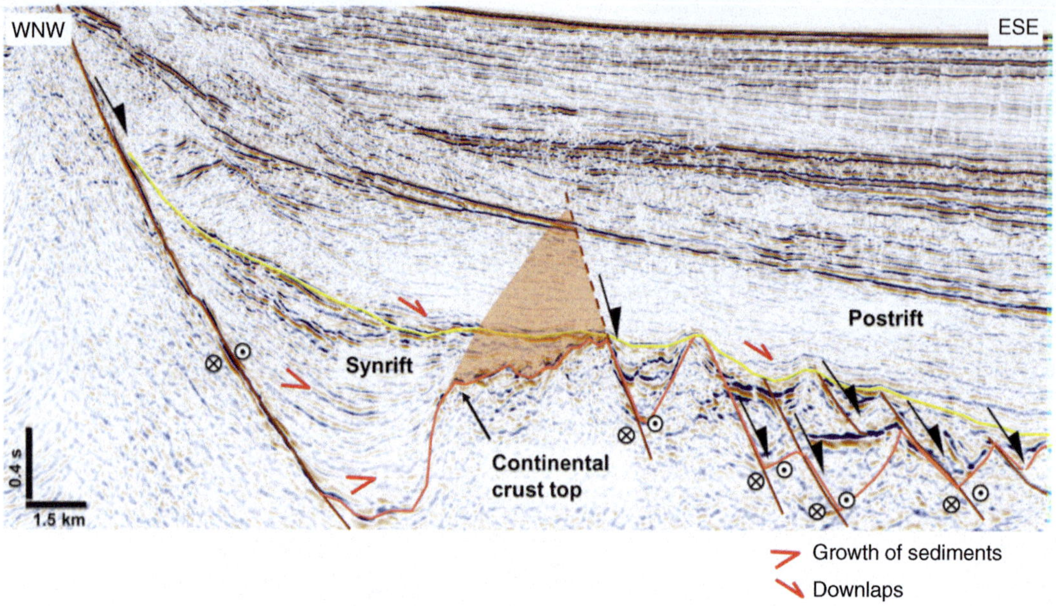

Figure 45.2d Interpreted seismic section of the thinned continental crust in the sheared margin. Note the numerous reflections beneath the basement top, which may be mineralized shear zones in the basement. Yellow line: breakup unconformity. Brown transparent polygon: possible original basement configuration. Seismic data courtesy Reliance Industries Ltd. Reproduced with permission from Directorate General of Hydrocarbons (DGH), India.

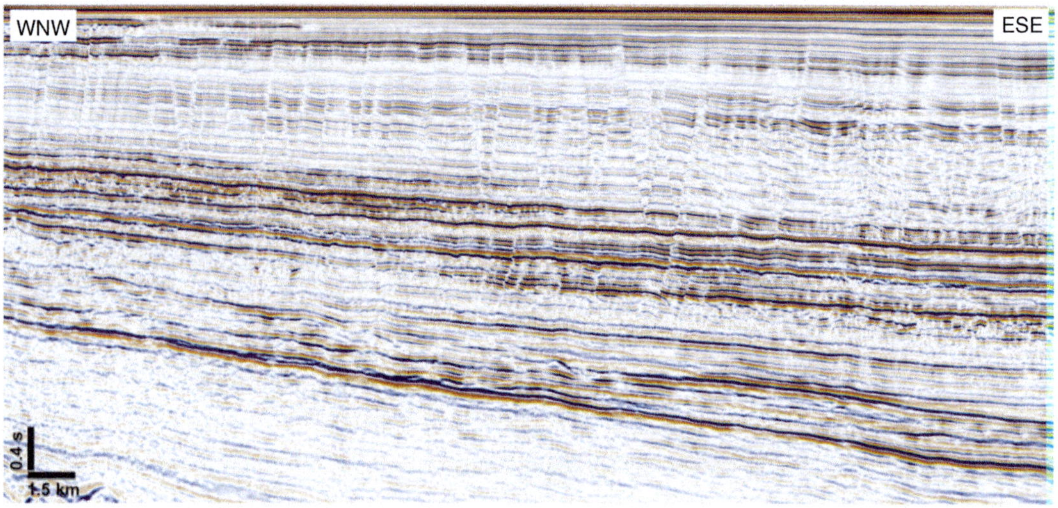

Figure 45.2e Uninterpreted seismic section showing the postrift sediments. Seismic data courtesy Reliance Industries Ltd. Reproduced with permission from Directorate General of Hydrocarbons (DGH), India.

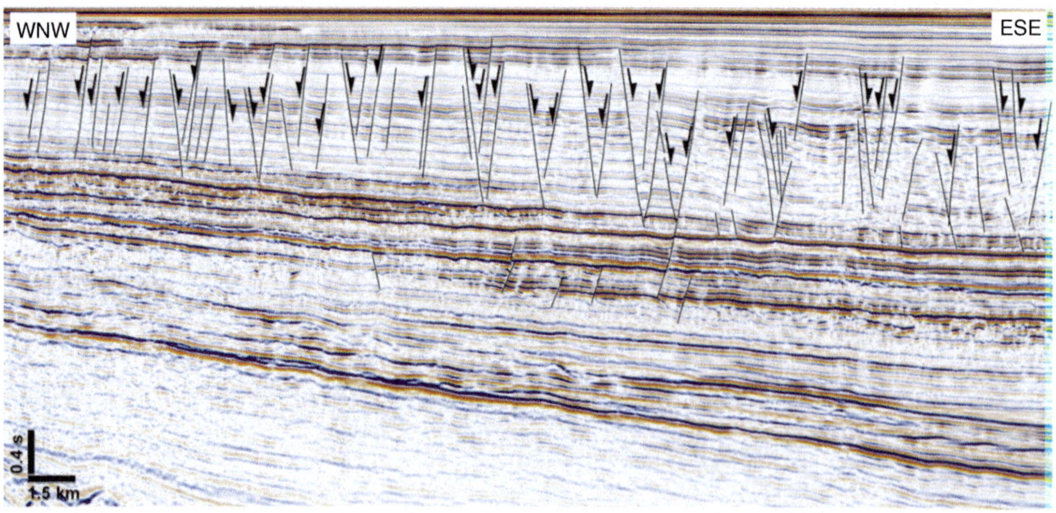

Figure 45.2f Interpretation of seismic section shown in Figure 45.2e. Note the remarkably planar faults. These faults are bound between two sedimentary horizons. The lower one is shale rich layer, from which fluids escaped to form these faults. Seismic data courtesy Reliance Industries Ltd. Reproduced with permission from Directorate General of Hydrocarbons (DGH), India.

References

Gay, A., Lopez, M., Cochonat, P. and Sermondadaz, G. 2004. Polygonal faults-furrows system related to early stages of compaction–upper Miocene to recent sediments of the Lower Congo Basin. *Basin Research,* **16,** 101–116.

Gay, A., Lopez, M., Cochonat, P. *et al.* 2006. Isolated seafloor pockmarks linked to BSRs, fluid chimneys, polygonal faults and stacked Oligocene–Miocene turbiditic palaeochannels in the Lower Congo Basin. *Marine Geology,* **226,** 25–40.

Mazumder, S., Tep, B., Pangtey, KKS., *et al.* 2018. Basement Tectonics and Shear Zones in Cauvery Basin: Implications in Hydrocarbon Exploration. In: Mukherjee, S. (Ed) Tectonics and Structural Geology: Indian Context. Springer International Publishing AG, Cham.

Misra, A.A. and Mukherjee S. 2015. *Tectonic Inheritance in Continental Rifts and Passive Margins.* SpringerBriefs in Earth Sciences, Springer International

Misra, A.A., Bhattacharya, G., Mukherjee, S. and Bose N. 2014. Near N-S paleo-extension in the western Deccan region in India: Does it link strike-slip tectonics with India-Seychelles rifting? *International Journal of Earth Sciences,* **103,** 1645–1680.

Misra, A.A., Sinha, N. and Mukherjee, S. 2015. Repeat ridge jumps and microcontinent separation: insights from NE Arabian Sea. *Marine and Petroleum Geology,* **59,** 406–428.

Nemčok, M., Sinha, S.T., Stuart, C.J. *et al.* 2013. East Indian margin evolution and crustal architecture: integration of deep reflection seismic interpretation and gravity modelling. In: Mohriak, W.U., Danforth, A.L.; Post, P.J. *et al.* (eds) *Conjugate Divergent Margins,* Special Publications vol. 369, The Geological Society, London, pp. 477–496.

Sinha, S.T., Nemčok, M., Choudhuri, M. *et al.* 2015. The role of break-up localization in microcontinent separation along a strike-slip margin: the East India–Elan Bank case study. In: Nemčok, M., Rybár, S., Sinha, S.T. *et al.* (eds) *Transform Margins: Development, Controls and Petroleum Systems,* Special Publications vol. 431, The Geological Society, London, pp. 95–123.

Stuevold, L.M., Faerseth, R.B., Arnesen, L. *et al.* 2003. Polygonal faults in the Ormen Lange field, Møre basin, offshore mid Norway. In: van Rensbergen, P., Hillis, R.R., Maltman, A.J. and Morley, C.K. (eds) *Subsurface Sediment Mobilization,* Special Publications vol. 216, The Geological Society, London, pp. 263–281.

46

Proximal and Distal Margin Examples from the Eastern Continental Passive Margin of India

M. Nemčok[1,2], C. Welker[1], S. Sinha[3], M. Choudhuri[3], S. Sharma[3], S. Venkatraman[4], and N. Sinha[3]*

[1] *EGI, University of Utah, Salt Lake City, UT, USA*
[2] *EGI Laboratory at SAV, Bratislava, Slovakia*
[3] *Reliance Industries Ltd, Mumbai, India*
[4] *ION Geophysical, Houston, TX, USA*
* *mnemcok@egi.utah.edu*

Examples are given here of different structures from proximal and distal margins in the East Indian passive margin. Details of the structures are provided in the captions to the figures. Further information can be found elsewhere (Nemčok *et al.*, 2013).

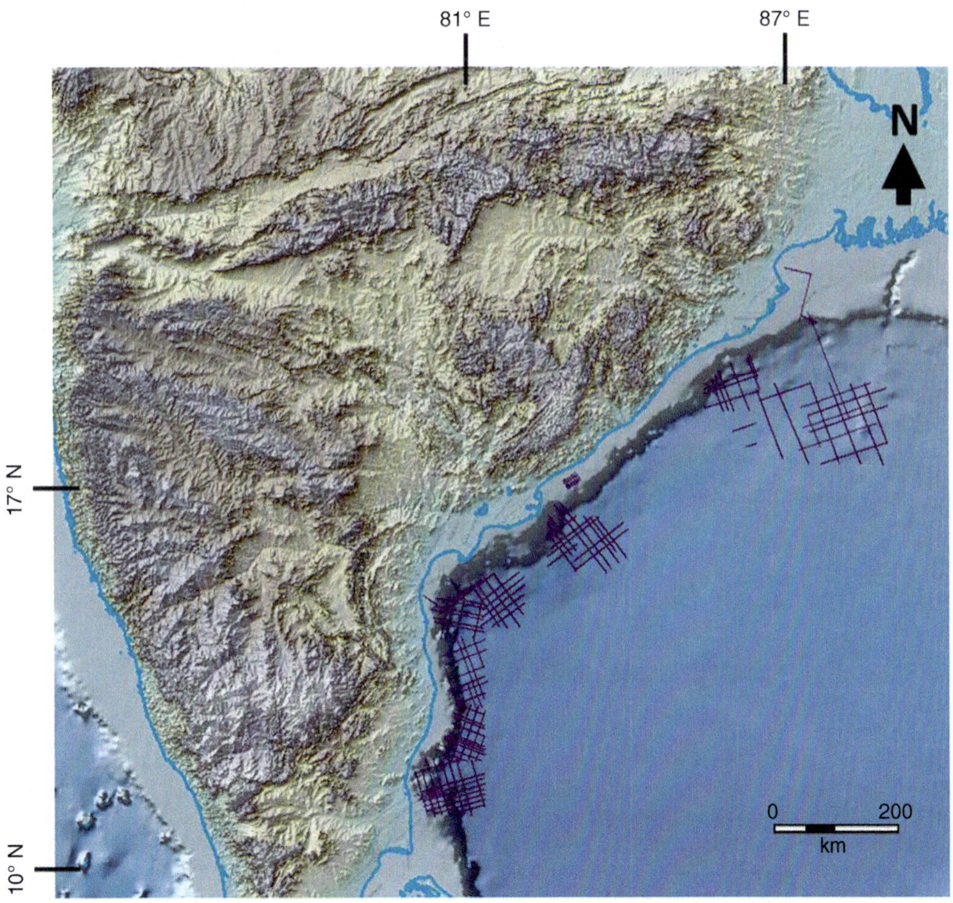

Figure 46.1 Map of the eastern continental margin of India on hill-shaded topography and bathymetry. Purple lines in offshore: seismic lines for the examples showed here.

Atlas of Structural Geological Interpretation from Seismic Images, First Edition. Edited by Achyuta Ayan Misra and Soumyajit Mukherjee.

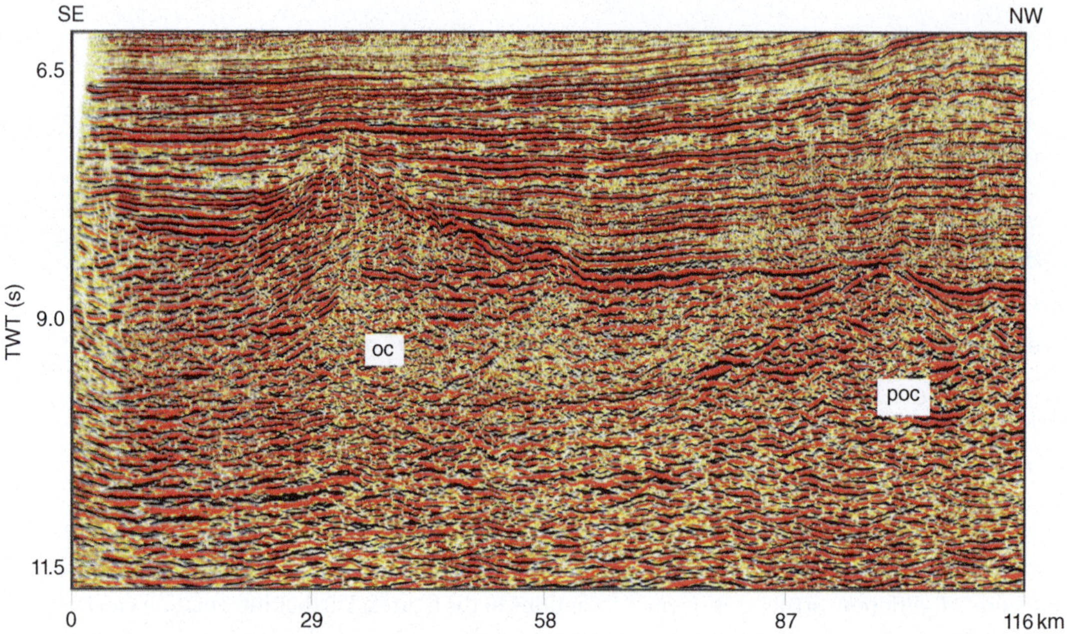

Figure 46.2a Uninterpreted dip-oriented reflection seismic profile through the transition between oceanic crust and proto-oceanic crust (exhumed continental lithospheric mantle) in the East Indian offshore.

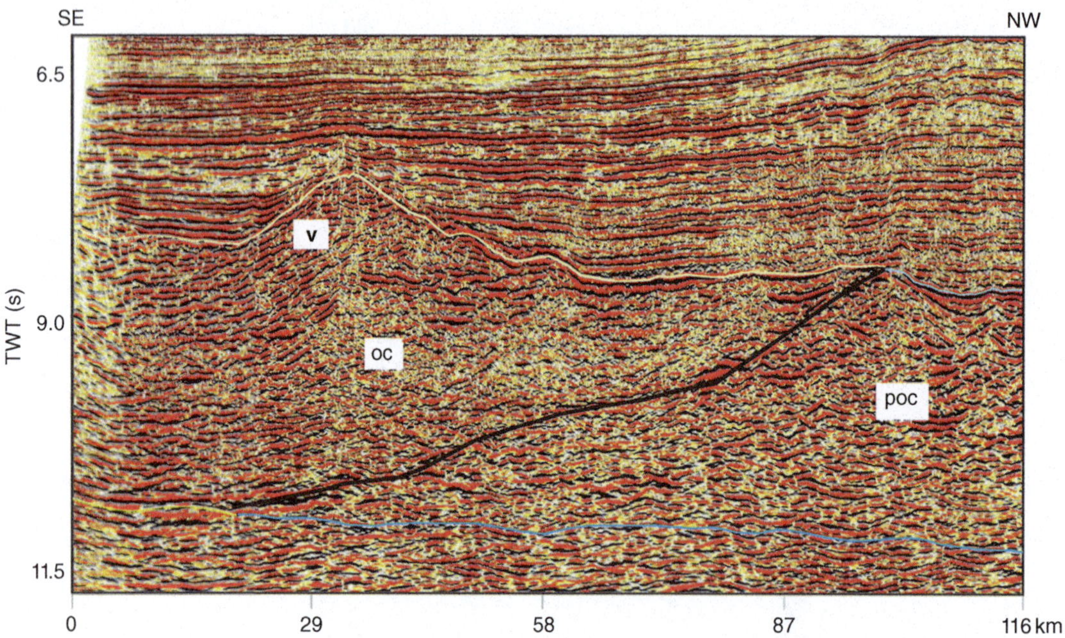

Figure 46.2b Interpretation of seismic section shown in Figure 46.2a. (Adapted from Nemčok *et al.*, 2013. Seismic data courtesy Reliance Industries Ltd.) The breakup process was delivered in the crust first-mantle second (magma-poor) scenario. Abbreviations: oc – oceanic crust, poc – proto-oceanic crust, v – volcano. Horizons: light blue – top poc, dark blue – poc Moho, light yellow – top oc, dark yellow – oc Moho. Interpretation indicates that the poc breakup was accommodated by the upward-convex fault with normal fault displacement sense. Profile location is not given for confidentiality reasons. Reliance Industries Ltd is gratefully acknowledged for allowing the seismic image publishing in the original article.

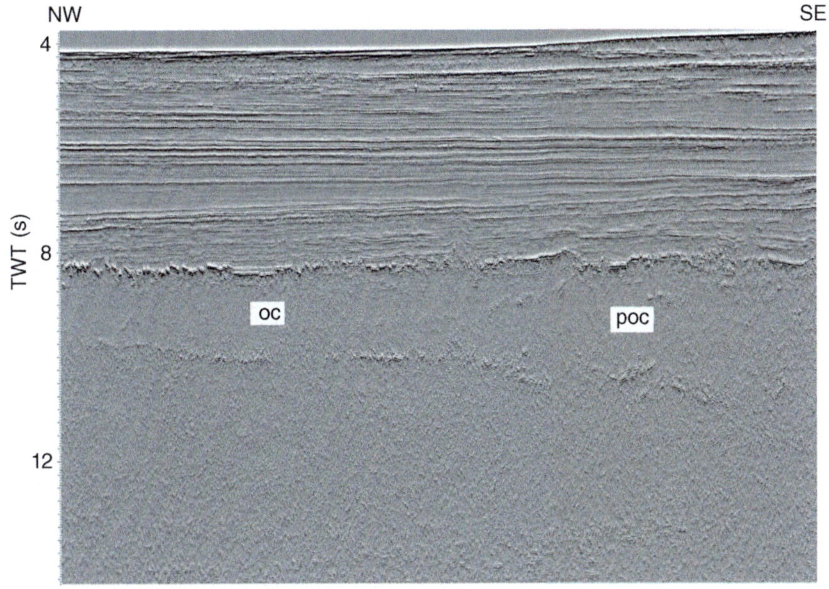

Figure 46.3a Uninterpreted dip-oriented reflection seismic profile through the transition between oceanic crust and proto-oceanic crust (exhumed continental lithospheric mantle) in the East Indian offshore.

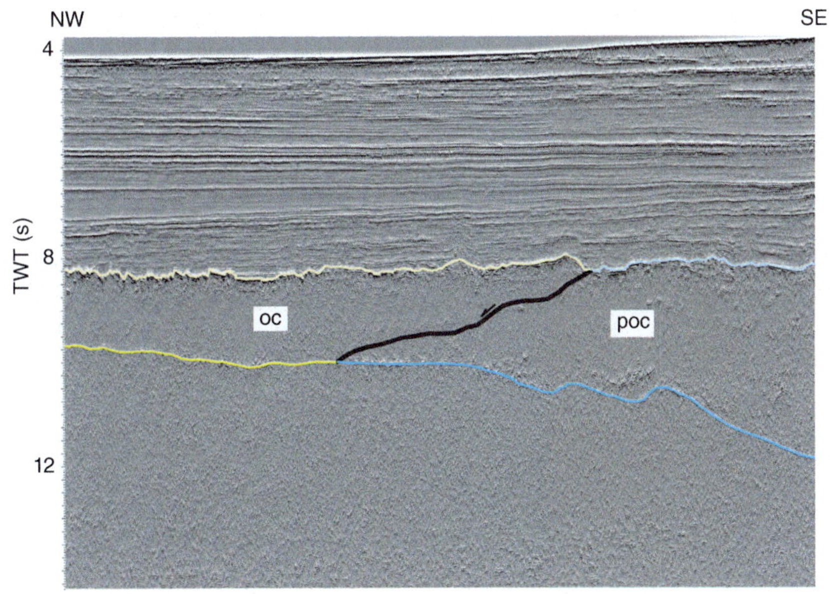

Figure 46.3b Interpretation of seismic section shown in Figure 46.3a. (Adapted from Nemčok *et al.*, 2013. Seismic data courtesy Directorate General of Hydrocarbons (DGH), India.) The breakup process was delivered in the crust first-mantle second (magma-poor) scenario. Abbreviations: oc – oceanic crust, poc – proto-oceanic crust. Horizons: light blue – top poc, dark blue – poc Moho, light yellow – top oc, dark yellow – oc Moho. Interpretation indicates that the poc breakup was accommodated by the upward-convex fault with normal fault displacement sense. Profile location is not given for confidentiality reasons. ION Geophysical is gratefully acknowledged for allowing the seismic image publishing in the original article.

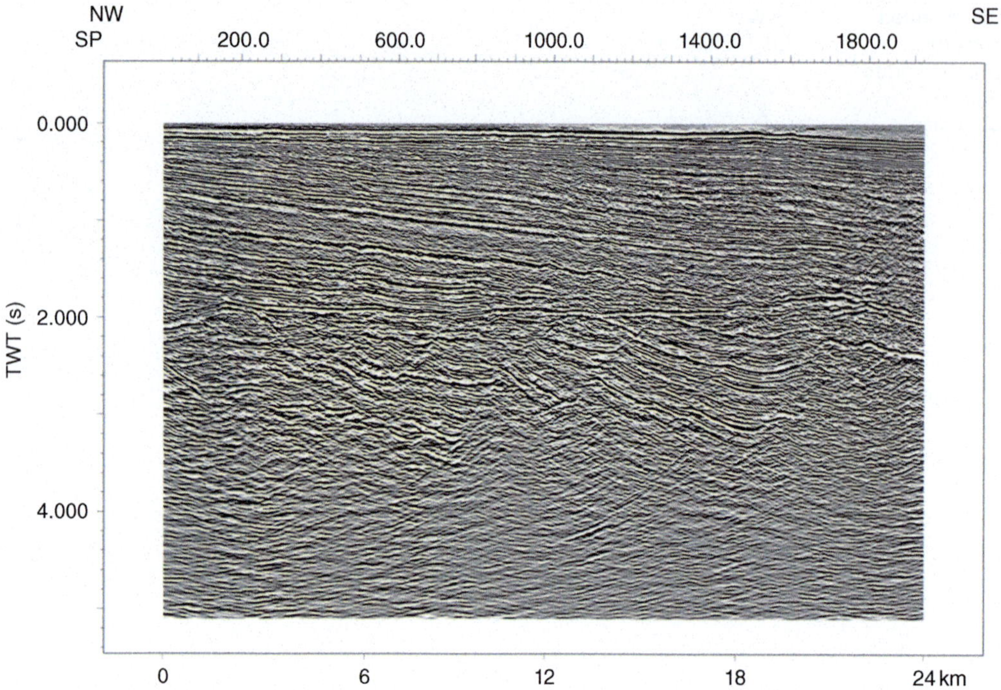

Figure 46.4a Uninterpreted dip-oriented reflection seismic profile through the proximal continental margin (stretching zone), offshore East India.

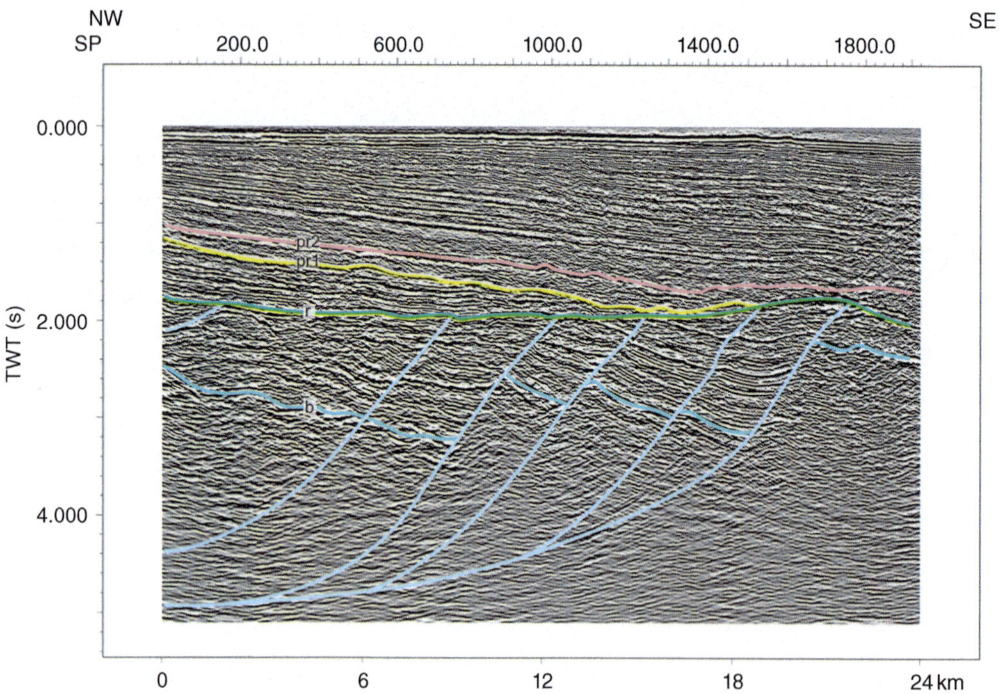

Figure 46.4b Interpretation of seismic section shown in Figure 46.4a. (Adapted from Nemčok *et al.*, 2013. Seismic data courtesy Reliance Industries Ltd.) The breakup process was delivered in the crust first-mantle second (magma-poor) scenario. Abbreviations: pr2 – top of the post-rift strata 2, pr1 – top of the post-rift strata 1, r – top of the syn-rift strata, b – top of the crystalline basement. Note a system of steeply-dipping listric normal faults controlling large sediment catchment areas. They accommodated relatively smaller extension. Profile location is not given for confidentiality reasons. Reliance Industries Ltd is gratefully acknowledged for allowing the seismic image publishing in the original article.

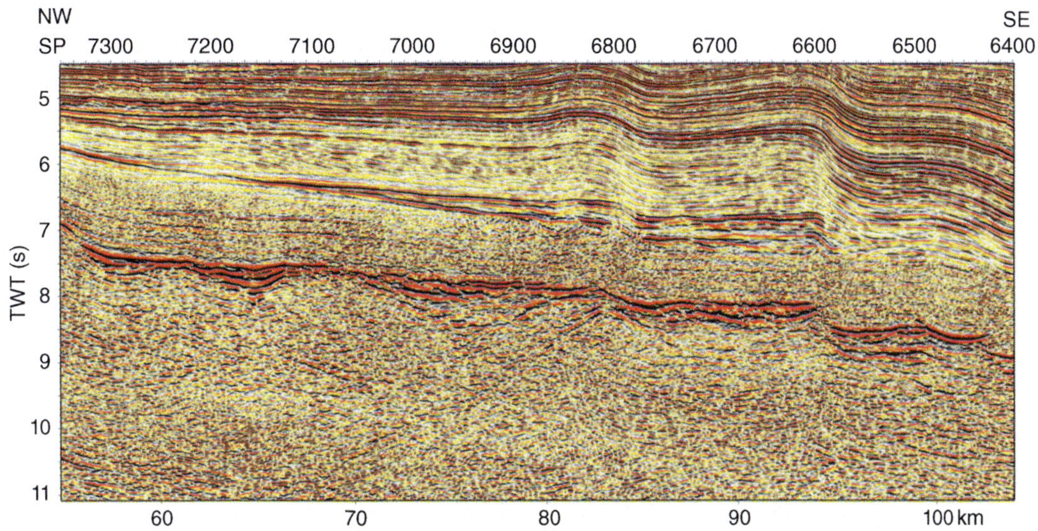

Figure 46.5a Uninterpreted dip-oriented reflection seismic profile through the distal continental margin (thinning zone), offshore East India.

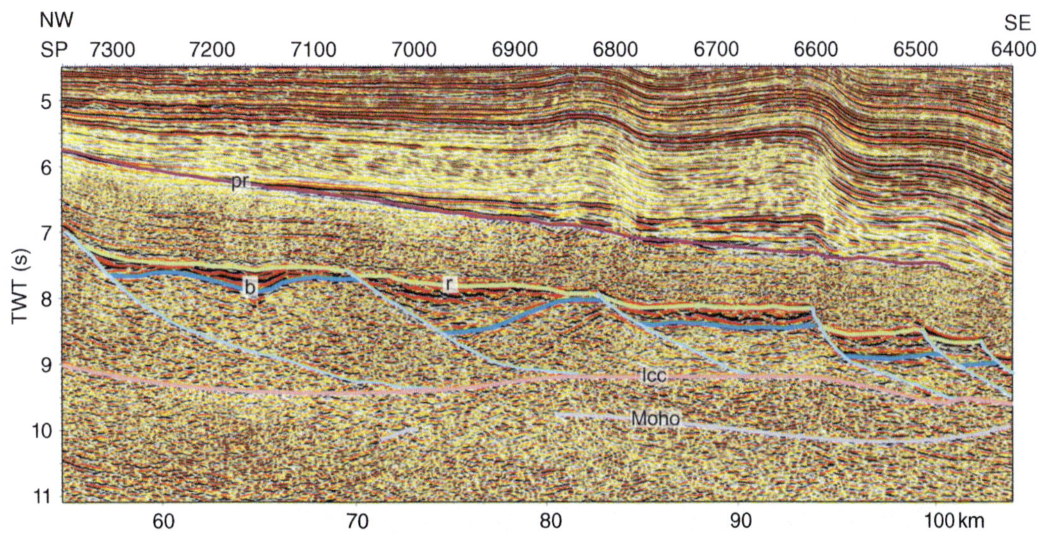

Figure 46.5b Interpretation of seismic section shown in Figure 46.5a. (Adapted from Nemčok *et al.*, 2013. Seismic data courtesy Reliance Industries Ltd.) The breakup process was delivered in the crust first-mantle second (magma-poor) scenario. Abbreviations: pr – top of the post-rift marker horizon, r – top of the syn-rift strata, b – top of the crystalline basement, lcc – top of the lower continental crust, Moho – top of the continental lithospheric mantle. Note a system of shallowly-dipping listric normal faults controlling small sediment catchment areas. Most of their fill is dominated by volcanic rocks. Normal faults accommodated relatively larger extension. Profile location is not given for confidentiality reasons. Reliance Industries Ltd is gratefully acknowledged for allowing the seismic image publishing in the original article.

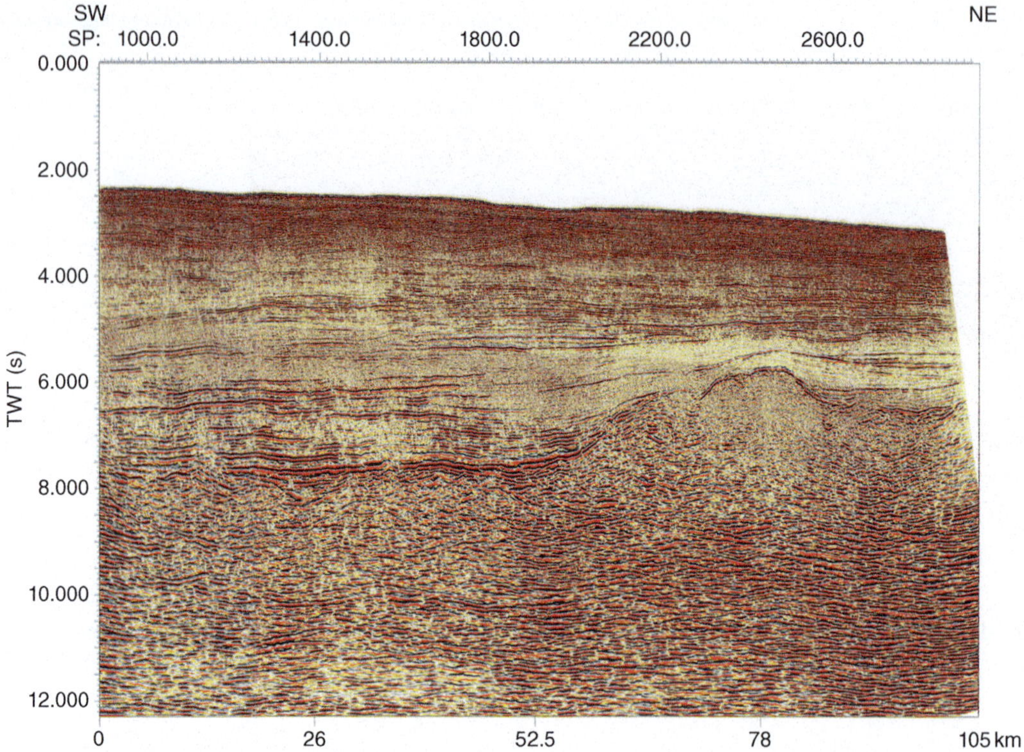

Figure 46.6a Uninterpreted strike-oriented reflection seismic profile through the distal continental margin (thinning zone), offshore East India.

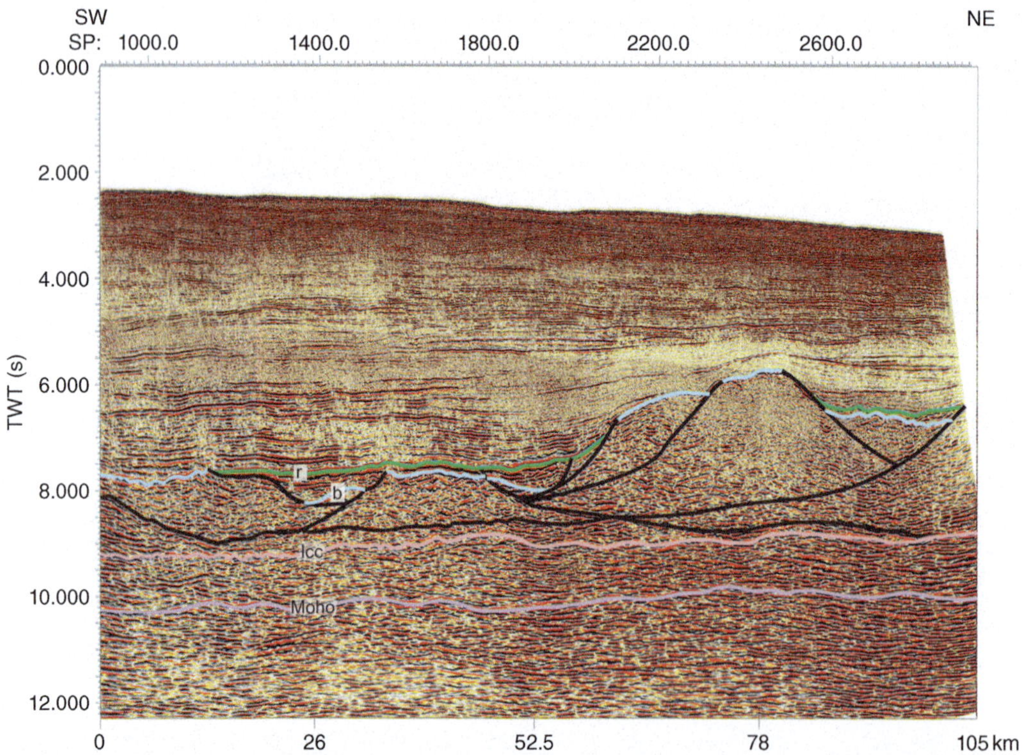

Figure 46.6b Interpretation of seismic section shown in Figure 46.6a. (Adapted from Nemčok *et al.*, 2013. Seismic data courtesy Reliance Industries Ltd.) The breakup process was delivered in the crust first-mantle second (magma-poor) scenario. Abbreviations: r – top of the rift fill, b – top of the basement, lcc – top of the lower continental crust, Moho – base of the continental crust. Black lines represent faults detached at the top of the ductile lower crust. Note that the boundary between the upper continental crust and the lower continental crust mapped (lcc) is not sharp – that is, following a single reflector – but rather is transitional. While the seismic image of the lower continental crust is characterized by subparallel and subhorizontal highly reflective set of reflectors, the upper crust is rather transparent. Profile location is not given for confidentiality reasons. Reliance Industries Ltd is gratefully acknowledged for allowing the seismic image publishing in the original article.

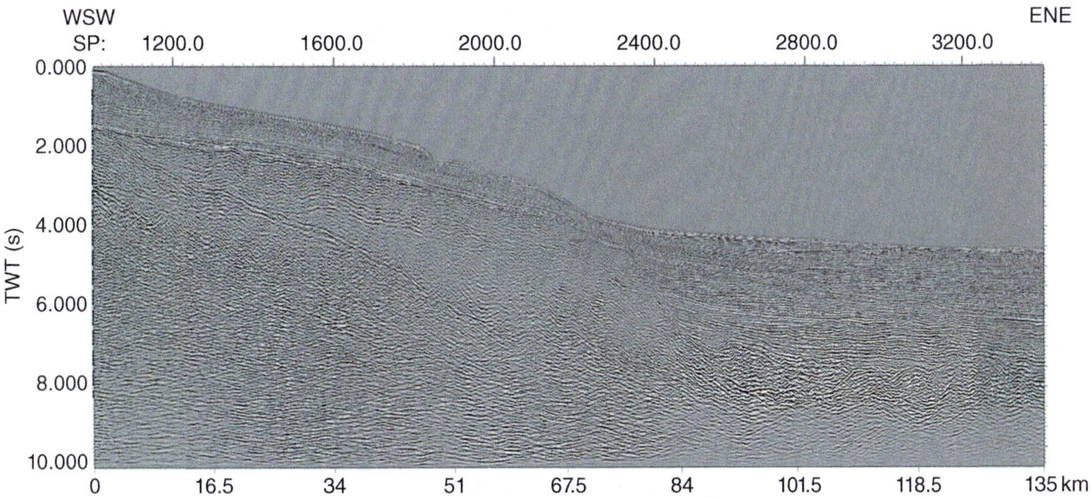

Figure 46.7a Uninterpreted dip-oriented reflection seismic profile through the rifted continental margin of East India.

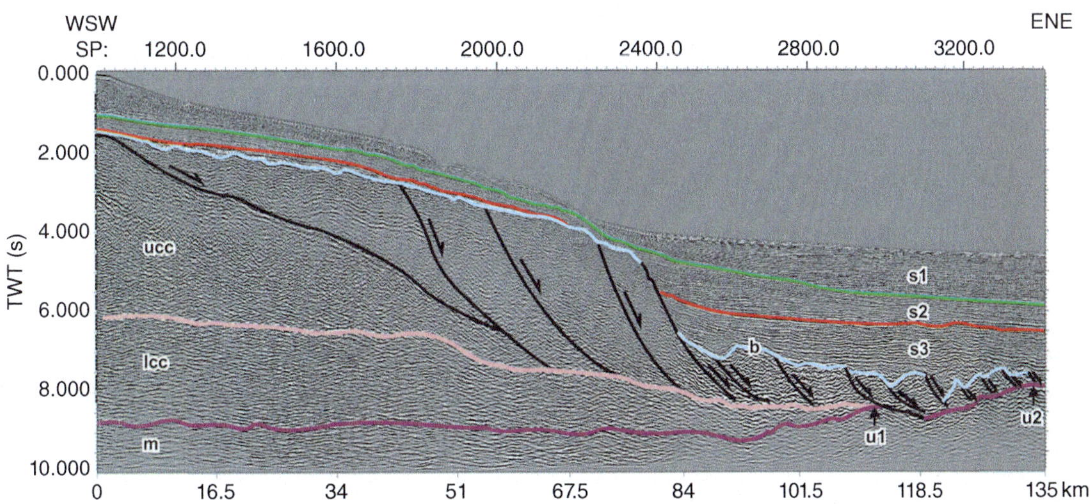

Figure 46.7b Interpretation of seismic section shown in Figure 46.7a. (Adapted from Nemčok *et al*., 2013. Seismic data courtesy Reliance Industries Ltd.) The breakup process was delivered in the crust first-mantle second (magma-poor) scenario. Abbreviations: ucc – upper continental crust, lcc – lower continental crust, m – continental upper lithospheric mantle, b – top of the continental basement, s1–s3 – sedimentary layers. u1 shows the place where the continental lithospheric mantle became unroofed first from underneath the lower crust. u2 shows the place where it eventually became unroofed from underneath a system of fault blocks formed by highly extended upper continental crust (i.e., extensional allochthon) and formed a proto-oceanic crust directly in contact with postrift sediment deposition. Note a fault that accommodates the differential movement between uplifting buoyant continental margin and subsiding dense proto-oceanic crust with portion of extensional allochthon during the early post-breakup history. It is the seventh fault from the toe of the continental margin. Profile location is not given for confidentiality reasons. Reliance Industries Ltd is gratefully acknowledged for allowing the seismic image publishing in the original article.

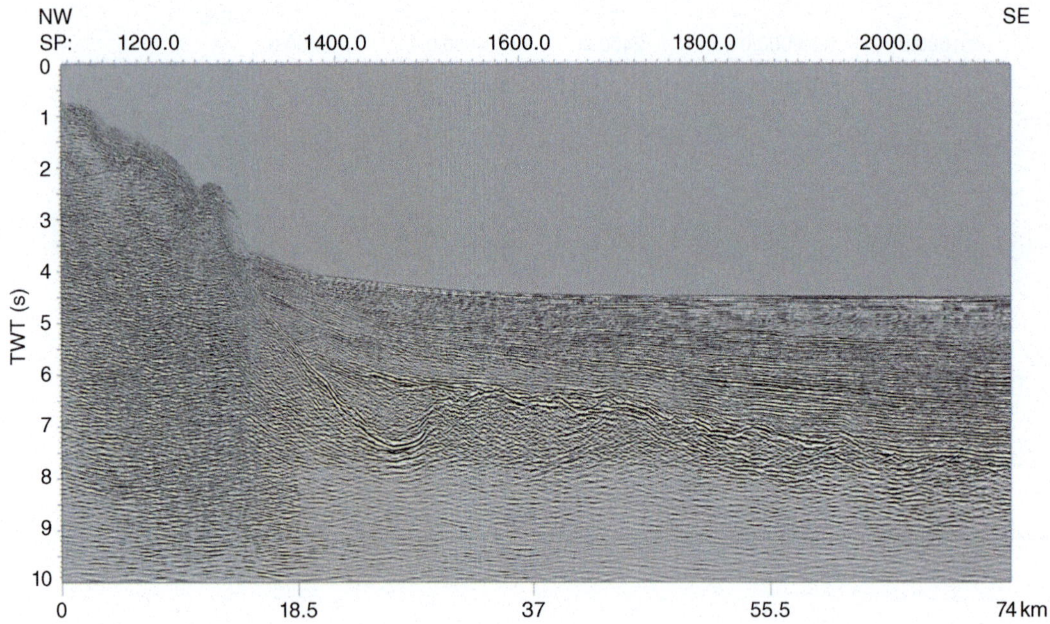

Figure 46.8a Uninterpreted dip-oriented reflection seismic profile through the transform continental margin of East India.

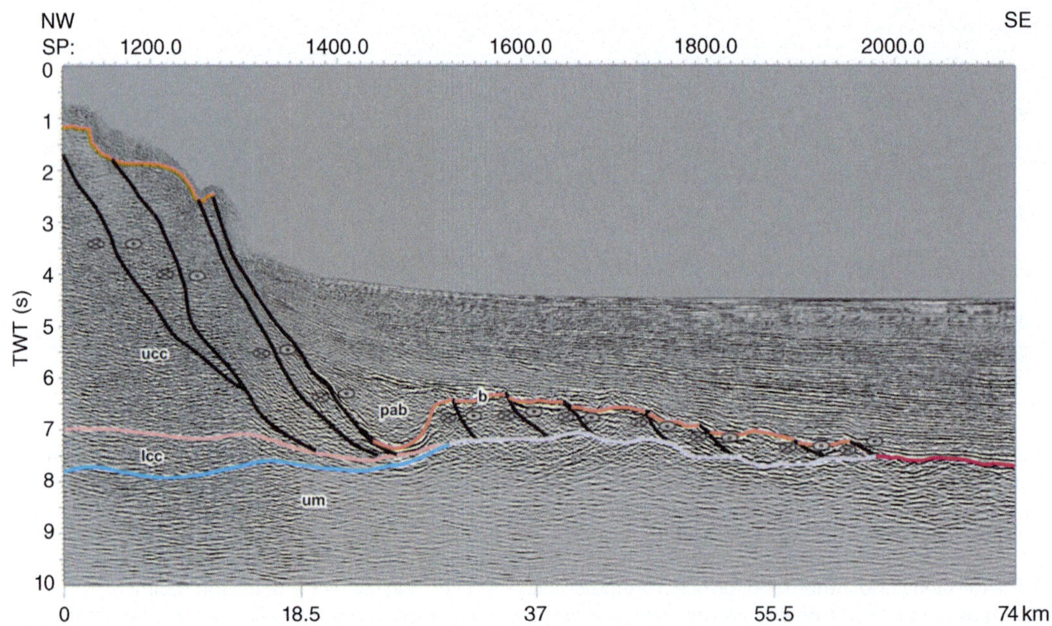

Figure 46.8b Interpretation of seismic section shown in Figure 46.8a. (Adapted from Nemčok *et al.*, 2013. Seismic data courtesy Reliance Industries Ltd.) The breakup process was delivered in the crust first-mantle second (magma-poor) scenario. Abbreviations: ucc – upper continental crust, lcc – lower continental crust, um – continental upper lithospheric mantle, b – top of the continental basement, pab – pull-apart basin. Note the extensional allochthon, formed by the upper-crustal fault blocks on top of the continental mantle, located oceanwards of the pull-apart basin (pab), indicating transtensional character of the transform fault. Strike-slip faults landwards of the pull-apart basin seem to be detached at the upper levels of the lower crust. Profile location is not given for confidentiality reasons. Reliance Industries Ltd is gratefully acknowledged for allowing the seismic image publishing in the original article.

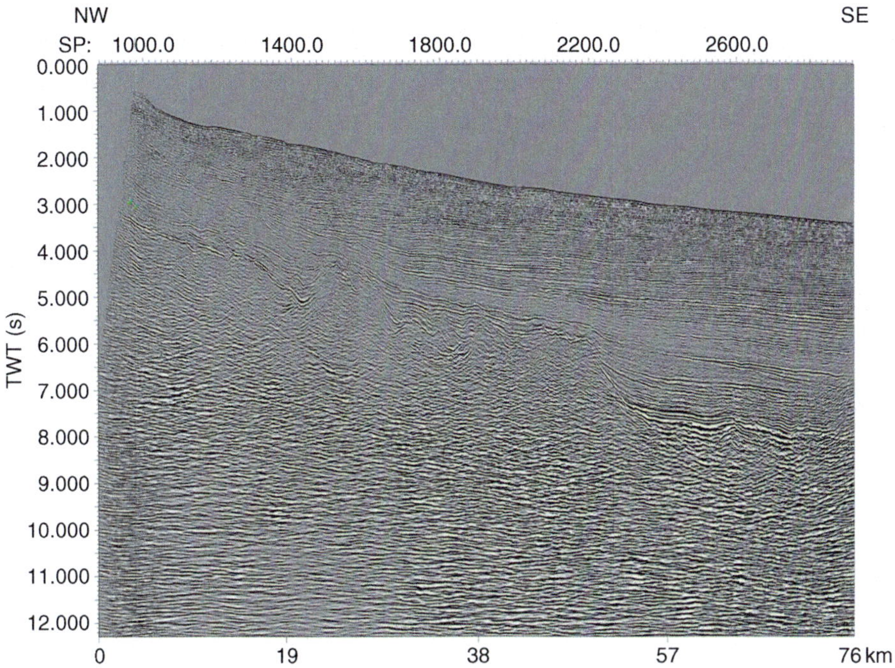

Figure 46.9a Uninterpreted dip-oriented reflection seismic profile through the transition between proximal margin (stretching zone) and distal margin (thinning zone) in the East Indian offshore.

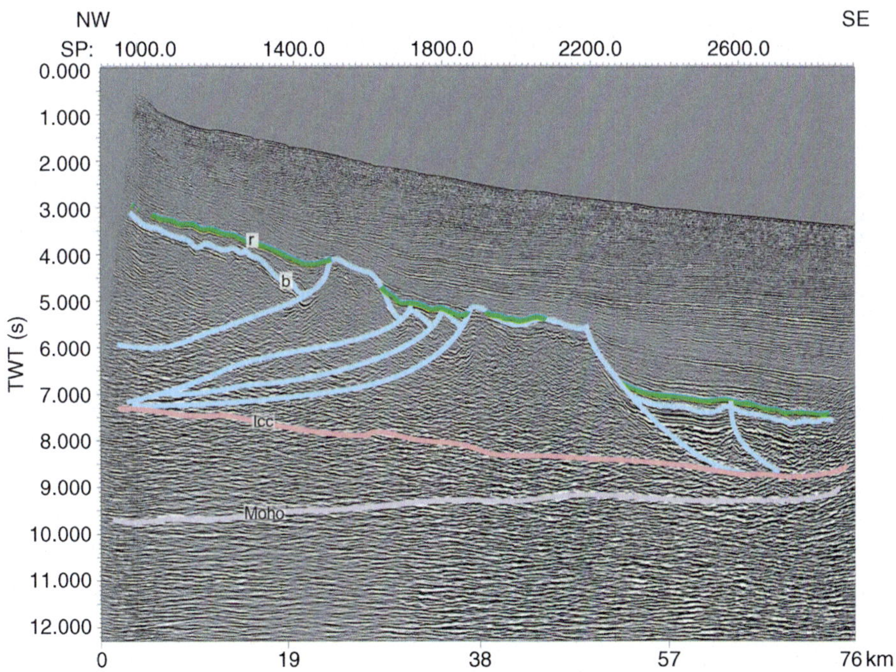

Figure 46.9b Interpretation of seismic section shown in Figure 46.9a. (Adapted from Nemčok *et al*., 2013. Seismic data courtesy Reliance Industries Ltd.) The breakup process was delivered in the crust first-mantle second (magma-poor) scenario. Abbreviations: r – top of the syn-rift strata, b – top of the crystalline basement, lcc – top of the lower continental crust, Moho – top of the continental lithospheric mantle. Profile location is not given for confidentiality reasons. Reliance Industries Ltd is gratefully acknowledged for allowing the seismic image publishing in the original article.

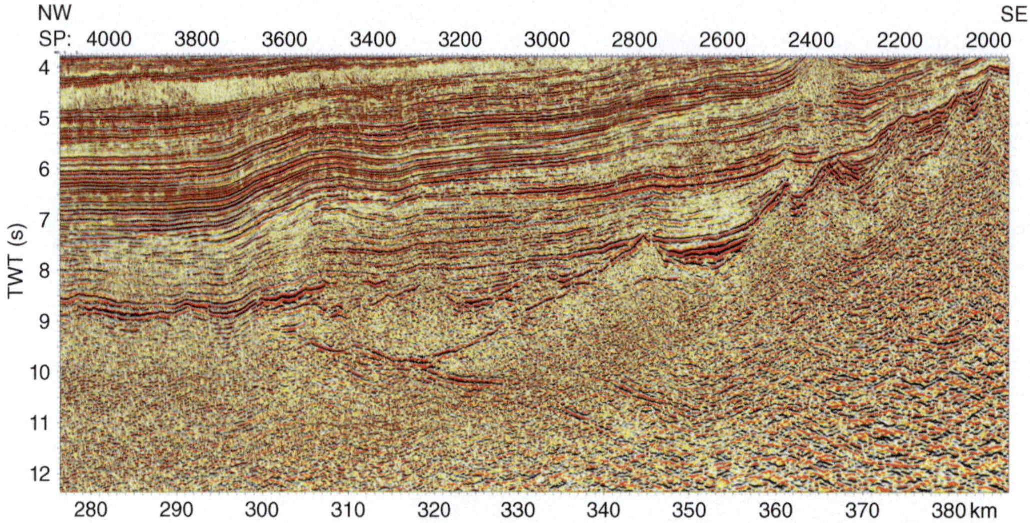

Figure 46.10a Uninterpreted dip-oriented reflection seismic profile through the rifted continental margin, offshore East India.

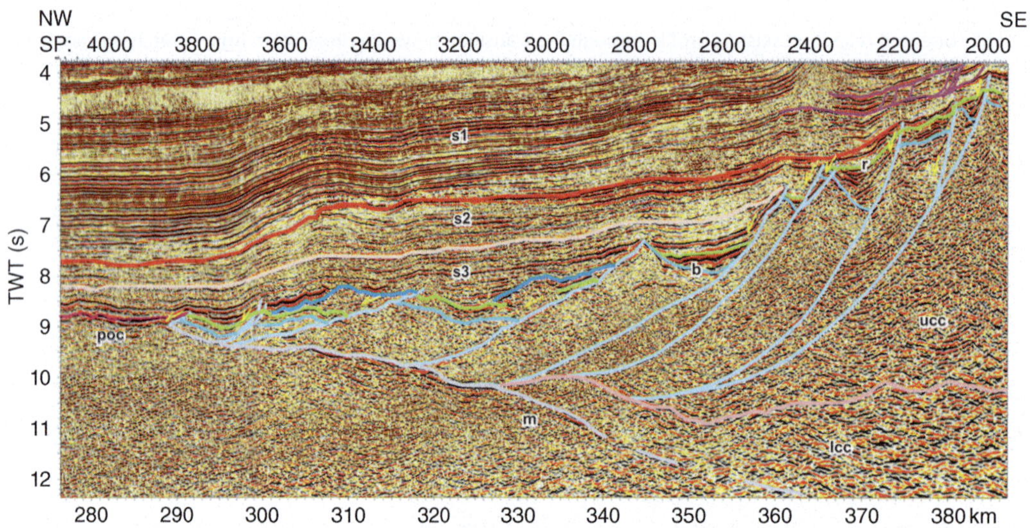

Figure 46.10b Interpretation of seismic section shown in Figure 46.10a. (Adapted from Nemčok *et al.*, 2013. Seismic data courtesy Reliance Industries Ltd.) The breakup process was delivered in the crust first-mantle second (magma-poor) scenario. Abbreviations: s1, s2, s3 – postrift sedimentary layers, r – top of the rift fill, b – top of the crystalline basement, ucc – upper continental crust, lcc – lower continental crust, poc – proto-oceanic crust, m – lithospheric mantle. The first four normal faults from SE to NW are detached at the top lower-crustal level and then at the top of the lithospheric mantle in the distal margin (thinning zone), which does not contain the lower crust. Note that the ductile lower crust tends to swell in the foot of the large footwall blocks and vacate the neighbouring hangingwalls or, at least, reduce its thickness there. Profile location is not given for confidentiality reasons. Reliance Industries Ltd is gratefully acknowledged for allowing the seismic image publishing in the original article.

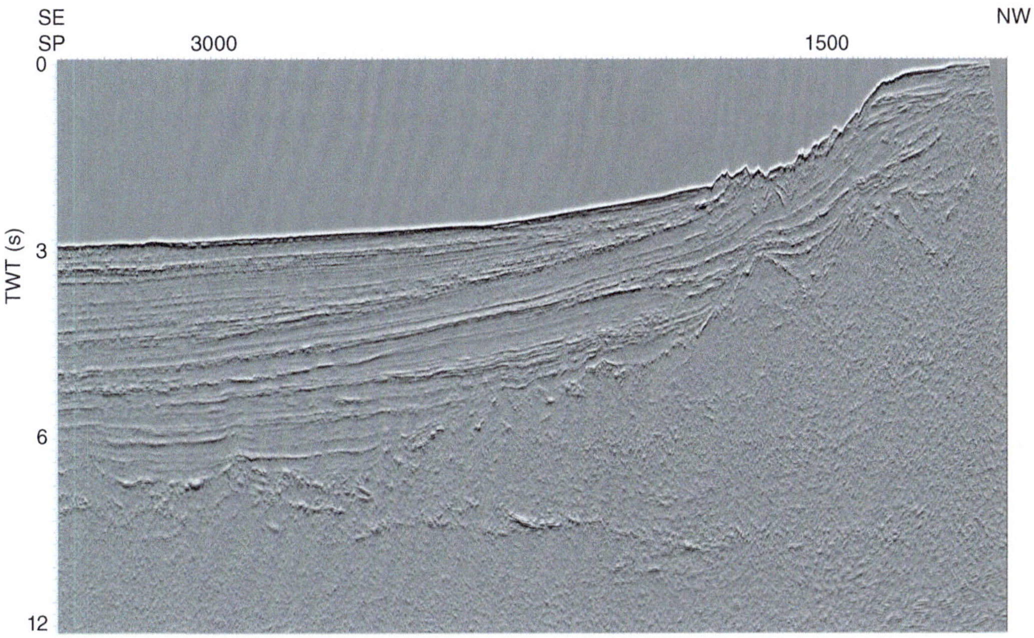

Figure 46.11a Uninterpreted dip-oriented reflection seismic profile through the transition between proto-oceanic crust (exhumed continental lithospheric mantle) and continental crust in the East Indian offshore.

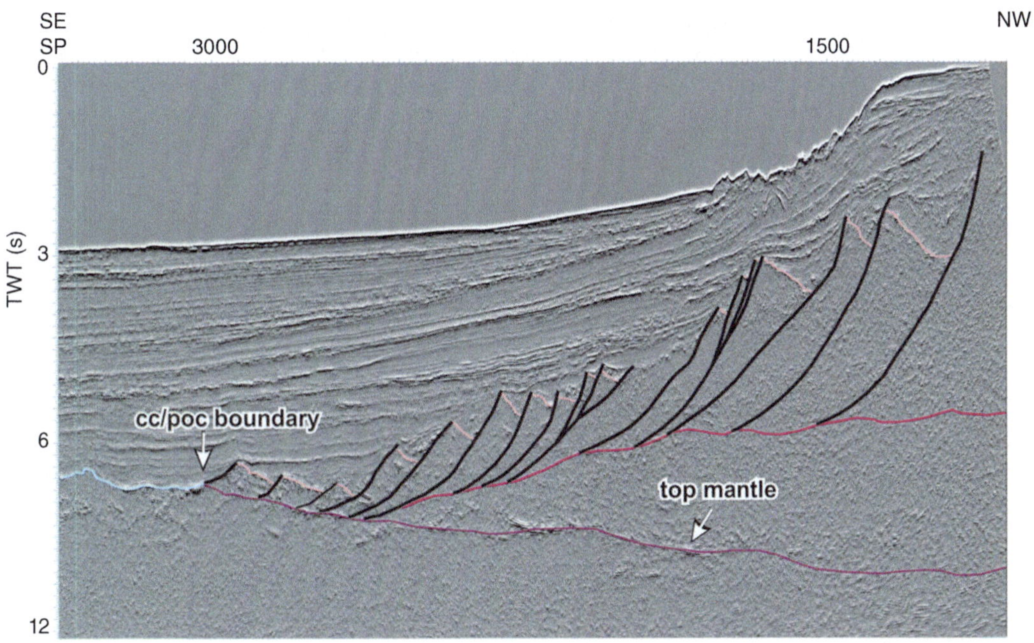

Figure 46.11b Interpretation of seismic section shown in Figure 46.11a. (Adapted from Nemčok *et al.*, 2013. Seismic data courtesy Directorate General of Hydrocarbons (DGH), India.) The breakup process was delivered in the crust first-mantle second (magma-poor) scenario. Abbreviations: cc – continental crust, poc – proto-oceanic crust. Horizons: light blue – top poc, pink – top continental crust, red – top lower continental crust, purple – cc Moho. Profile location is not given for confidentiality reasons. ION Geophysical is gratefully acknowledged for allowing the seismic image publishing in the original article.

Reference

Nemčok, M., Stuart, C.J., Rosendahl, B.R. *et al.* 2013. Continental break-up mechanism; lessons from intermediate- and fast-extension settings. In: Mohriak, W.U., Danforth, A.L.; Post, P.J. *et al.* (eds) *Conjugate Divergent Margins*, Special Publications vol. 369, The Geological Society, London, pp. 373–401. doi: 10.1144/SP369.14.

47

Distal Margin Examples from the Continental Passive Margin of Gabon

M. Nemčok[1,2], C. Welker[1], S. Sinha[3], M. Choudhuri[3], S. Sharma[3], S. Venkatraman[4], and N. Sinha[3]*

[1] EGI, University of Utah, Salt Lake City, UT, USA
[2] EGI Laboratory at SAV, Bratislava, Slovakia
[3] Reliance Industries Ltd, Mumbai, India
[4] ION Geophysical, Houston, TX, USA
** mnemcok@egi.utah.edu*

The south Atlantic passive margin formed during the Late Jurassic – Aptian due to rifting of South America and West Africa. The passive margin in offshore Gabon depicts excellent examples of structures in magma-poor passive margins. Details of the structures are provided in the captions to the figures. Further information can be found elsewhere (Nemčok *et al.*, 2013).

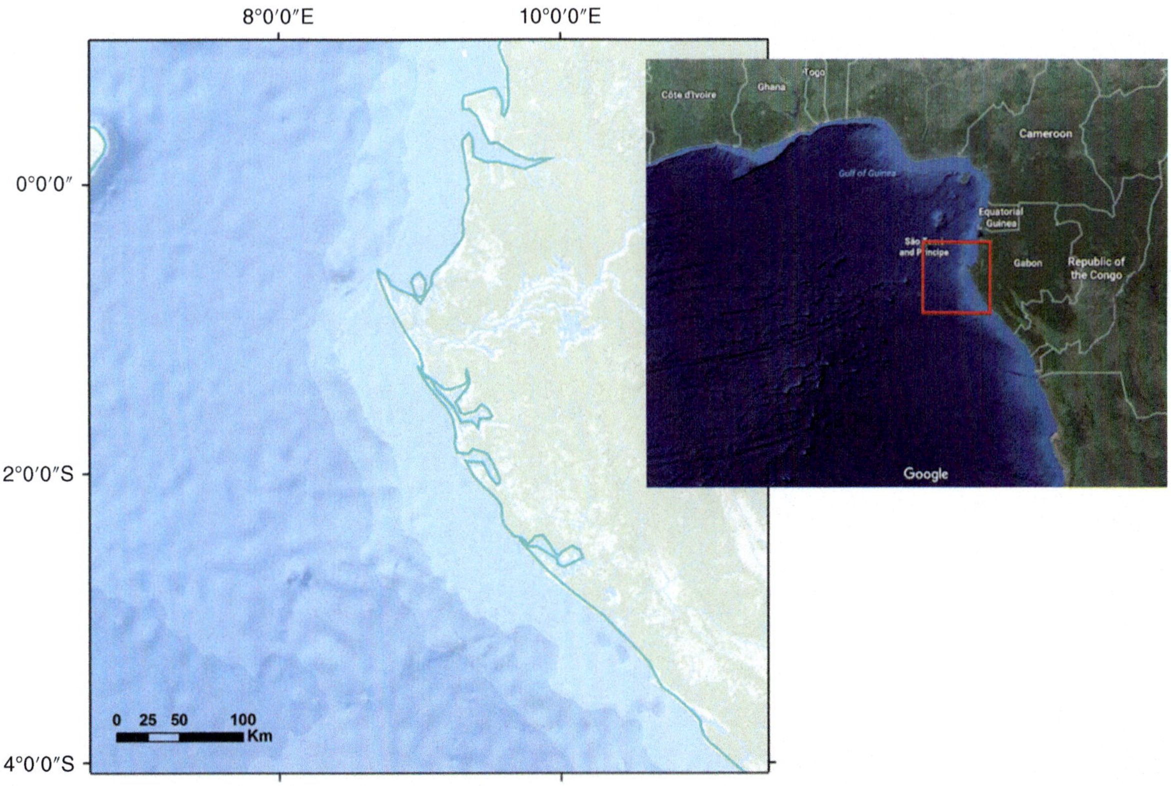

Figure 47.1 Map of West African passive margin near Gabon. Inset: Location of the study area.

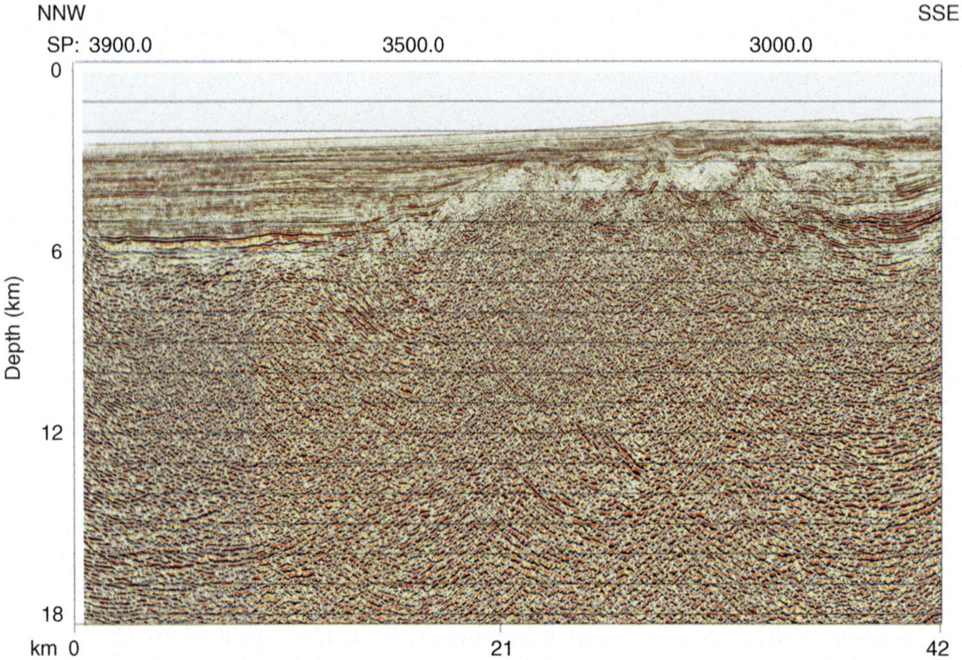

Figure 47.2a Uninterpreted dip-oriented reflection seismic profile through the rifted continental margin of Gabon.

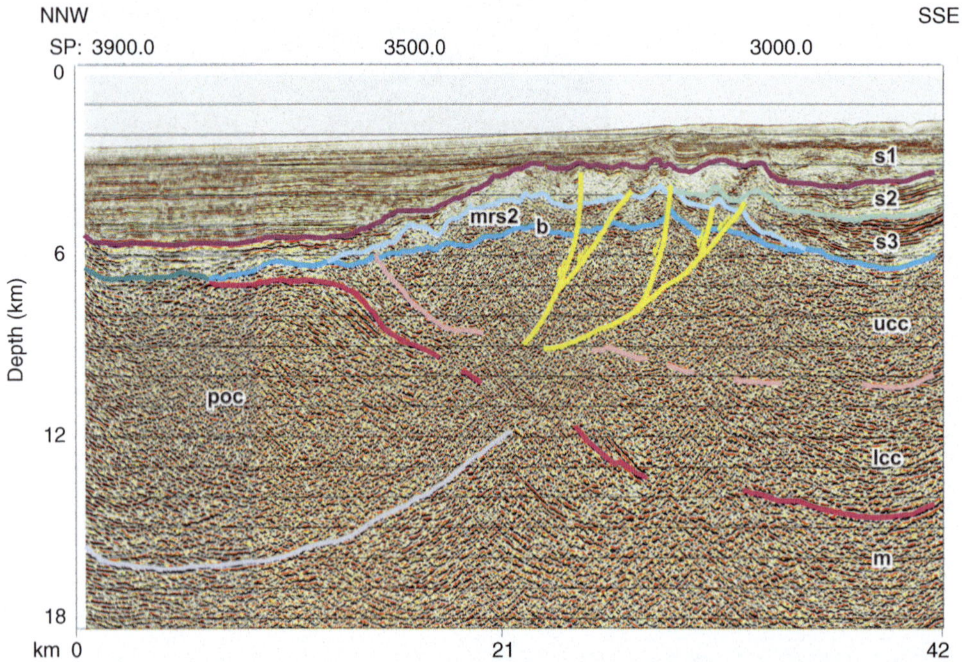

Figure 47.2b Interpretation of seismic section shown in Figure 47.2a. (Adapted from Nemčok et al., 2013. Seismic data courtesy ION Geophysical). The breakup process was delivered in the crust first-mantle second (magma-poor) scenario. Abbreviations: ucc – upper continental crust, lcc – lower continental crust, m – continental upper lithospheric mantle, b – top of the continental basement, s1–s3 – sedimentary layers, poc – proto-oceanic crust, mrs2 – magmatic rock-dominated rock suite 2. Profile location is not given for confidentiality reasons. ION Geophysical is gratefully acknowledged for allowing the seismic image publishing in the original article.

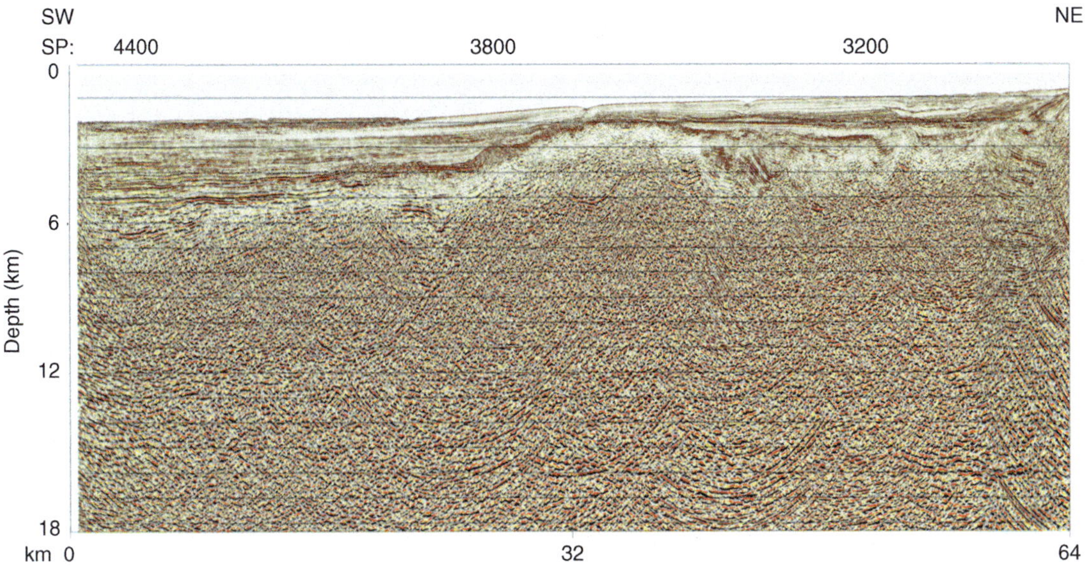

Figure 47.3a Uninterpreted dip-oriented reflection seismic profile through the rifted continental margin of Gabon.

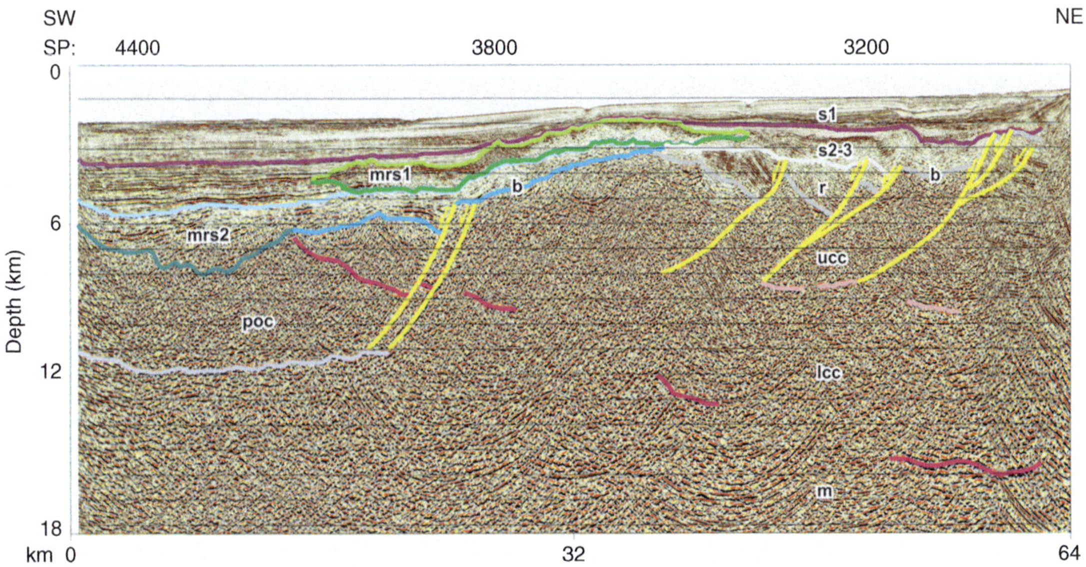

Figure 47.3b Interpretation of seismic section shown in Figure 47.3a. (Adapted from Nemčok et al., 2013. Seismic data courtesy ION Geophysical). The breakup process was delivered in the crust first-mantle second (magma-poor) scenario. Abbreviations: ucc – upper continental crust, lcc – lower continental crust, m – continental upper lithospheric mantle, b – top of the continental basement, s1–s3 – sedimentary layers, poc – proto-oceanic crust, mrs1 and mrs2 – magmatic rock-dominated rock suites 1 and 2, r – syn-rift strata. Note two normal faults that accommodate the differential movement between the uplifting continental margin and the subsiding proto-oceanic crust. Profile location is not given for confidentiality reasons. ION Geophysical is gratefully acknowledged for allowing the seismic image publishing in the original article.

Reference

Nemčok, M., Stuart, C.J., Rosendahl, B.R. *et al.* 2013. Continental break-up mechanism; lessons from intermediate- and fast-extension settings. In: Mohriak, W.U., Danforth, A.L.; Post, P.J. *et al.* (eds) *Conjugate Divergent Margins*, Special Publications vol. 369, The Geological Society, London, pp. 373–401. doi: 10.1144/SP369.14.

48

Pitfalls in Interpretation

Achyuta Ayan Misra and Ashok Yadav

Reliance Industries Ltd, Mumbai, India
**Achyuta.Misra@ril.com; achyutaayan@gmail.com*

Let us compare a couple of interpretations in this atlas. Compare (i) Figures 46.2 and 47.8 (Cauvery Basin, eastern Indian passive margin), and (ii) Figures 42.2 and 47.11 (Krishna Godavari Basin, eastern Indian passive margin). They are the same sections in different display properties. Figure 47.11 used a different geophysical attribute, which has come up quite recently. Are the interpretations exacly similar? They are not! Is one correct and the other incorrect? No! The overall tectonic interpretation from (i) is that it is a strike-slip set up and from (ii) is that is an orthogonally rifted margin. The major structures, grabens and crustal domains also match. However, at a small scale, when comparing every fault, interpretations vary. This is where the pitfalls in seismic interpretation need to be considered. The disparity between individual interpretations should be minimized. Problems or 'pitfalls' in seismic analysis may arise from glitches during acquisition, processing and interpretation (Nanda, 2016). In simple geological areas (e.g. Figure 46.2e) the seismic images correctly depict the surface geological setting. However, in geologically complex areas such as fold and thrust belts (e.g. Figure 35.2e), it is difficult for seismic images to correctly replicate the subsurface geological setting. This also leads to pitfalls in seismic data and interpretation.

Usually, the geophysical, seismic attribute, stratigraphic and workstation aspects are emphasized in discussing pitfalls in seismic interpretations (Tucker and Yorston, 1973; Herron, 2000; Brown, 2005; Marfurt and Alves, 2014). Such pitfalls can be grouped in four way (Tucker and Yorston, 1973; Brown, 2005; Torvela and Bond, 2011; Marfurt and Alves, 2014; Bond, 2015).

- Velocity accociated pitfalls: these occur because seismic data are presented in two-way-travel time in place of depth.
- Geometry associated pitfalls: these occur because reflections from 3D space are always plotted and visualized in 2D sections.
- Acquisition and processing associated pitfalls: these appear because all recorded events may not be geologic. Improper processing can mask geology.
- Human related pitfalls: these pitfalls occur because the interpreter prefers (i) specific interpretation due to their prior experience, (ii) visualization colours, vertical exaggeration and so on, (iii) certain seismic attributes, (iv) absolute dependence on the workstation and many more.

Only on the final point is discussed here, that is, human errors in interpretation during seismic structural analysis. For the other pitfalls, other works can be consulted (Tucker and Yorston, 1973; Nanda, 2016, Chapter 12). An example seismic section from Lebanon (Figure 48.1) will be used to point out how an interpretation can be bettered. This section shows a series of reflections that appear as high and low amplitude bands. Nearby data show that the high amplitude reflections correspond to calcareous sediments and the low amplitude ones are shale sections. Some reflections that are neither high nor low amplitude are marl sediments. The lower part, comprising chaotic reflections, is a pure shale section. The lithological information has to be kept in mind while starting the structural interpretation. Careless or biased interpretation of seismic data in geologically complex areas bring errors due to pitfalls mentioned above.To avoid human bias in seismic interpretation, the interpreter must consider:

- Making an interpretation that is geologically viable: for example, faults are seldom perfectly straight or planar (Mukherjee and Agarwal, submitted) like two mouse button clicks on the work station, one on the top and

Atlas of Structural Geological Interpretation from Seismic Images, First Edition. Edited by Achyuta Ayan Misra and Soumyajit Mukherjee.

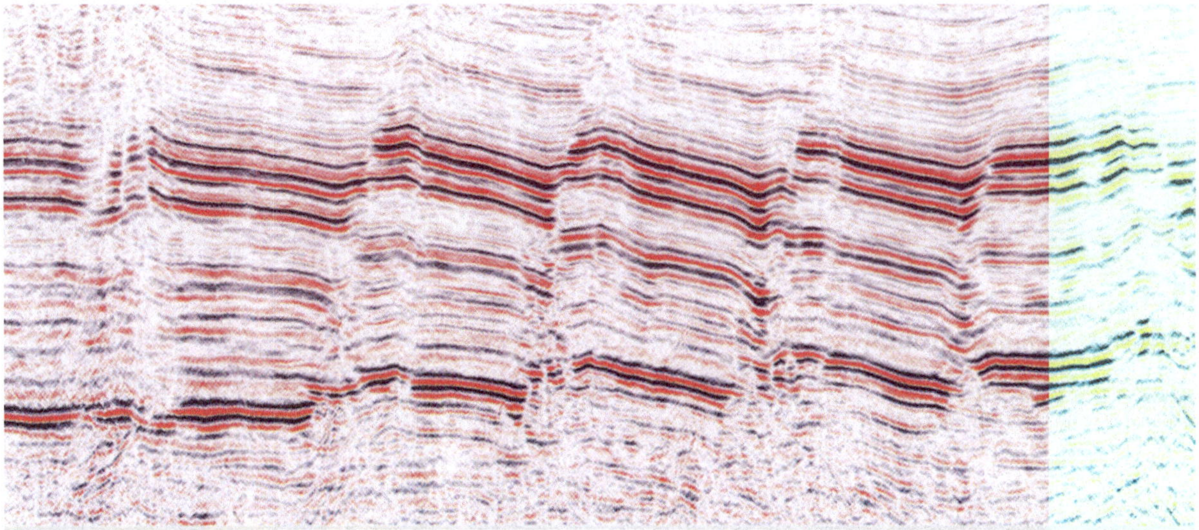

Figure 48.1 A reflection seismic section showing planar faults from Lebanon. *Source:* TGS-NOPEC, 2003. AAPG ©2002 whose permission is required for further use.

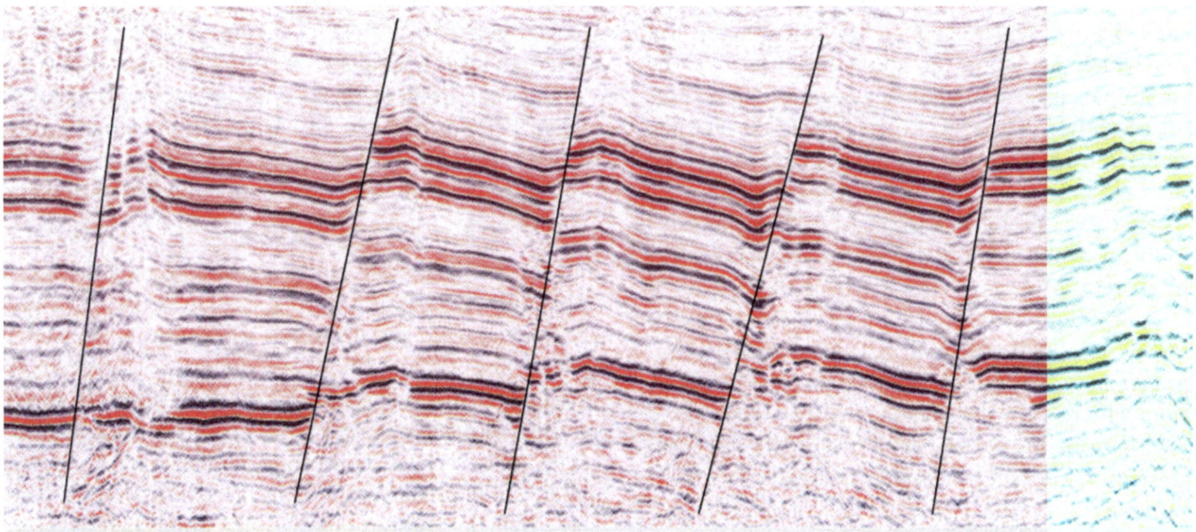

Figure 48.2 One possible interpretation for seismic section in Figure 48.1. Though the faults are identified correctly, the interpretation is very simple. Faults are not as planar as here. This can be categorized as an erroneous interpretation. Data from TGS-NOPEC, 2003.

the other at the bottom (Figure 48.2). Natural surfaces are curvi-planar in 3D or curvilinear in 2D, more like those shown in Figure 48.3. It is necessary to think beyond the straight/planar image of faults in conventional structural geological textbooks and accept the fact that faults are not so in nature (Alcalde *et al.*, 2017). Again, folds usually follow certain geometries (i.e., Mukherjee *et al.*, 2015) understood by classification schemes, for example those of Ramsay (Ramsay, 1967) or Huddelston (Huddleston, 1986), so the interpretation may follow one of those. However, in deep-water settings, folds may follow neither Ramsay's nor Hudleston's schemes (Torvela and Bond, 2011). Therefore, interpretation should utilize geological understanding and a model that fits the particular terrain. For example, in an extensional tectonic setting normal faults rather than reverse faults would be more expected, and vice versa (Mukherjee and Koyi, 2010a, 2010b). In other words, the geoscientist should be adept at pure structural geology and tectonics (Mukherjee, 2013, 2014a, 2015). Another important point is that fault throws *usually* increase with depth. More information on the variation of fault throws along depth is available elsewhere (Cartwright *et al.*,

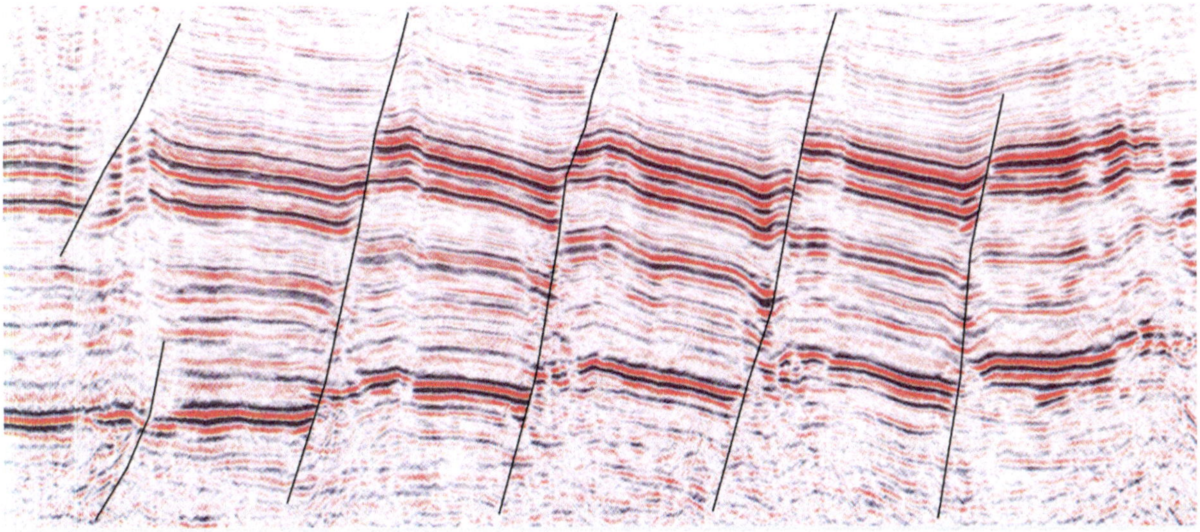

Figure 48.3 A second possible interpretation for the seismic section in Figure 48.1. This interpretation is better than the one in Figure 48.2. The faults are curvi-planar. However, the interpretation is simplistic. The faults cannot be defined as single lines. They rather appear as a zone on the seismic section. This can be categorized as underinterpretation. Data from TGS-NOPEC, 2003.

1998; Mansfield and Cartwright, 1996; Hongxing and Anderson, 2007; Baudon and Cartwright, 2008a, 2008b, 2008c).

- Understanding that geological interpretations are valid for a data set and additional data may revise the interpretation. Seismic interpretation is inherently subjective and should also be iterative.
- Using all available data: this comes as a rider to the previous point. Data are the prime strength of the interpreter. One biostratigraphic, lithologic, geophysical, petrophysical data or an extra well (Lark *et al.*, 2014) in an area can revise the interpretation.
- Using the workstation to its full capability: using all existing visualization capability of modern workstations is fruitful (Bacon *et al.*, 2007). Automated interpretation may work with horizons, but they are difficult to work with in the case of faults. Vertical exaggerations, too, are good techniques to make the interpretation viable. Colour palettes or scales are very important here. The color scale should be changed according to the type of data, namely sequential/discrete/continuous or the information the interpreter is trying to decipher. In seismic structural analysis, small changes in elevation or stratal terminations and alterations in amplitude or discontinuities and elevation in sedimentological analyses are looked for. A very commonly used colour scale is the grey scale because it supresses noise. The most abused color palette is the ubiquitous "Rainbow"! It just creates confusion in the guise of being colourful. The proper use of colour scales is discussed elsewhere (Matteo, 2013; Simmon, 2013; Hall, 2013).
- Understanding the data: knowing the data is of prime importance. For instance, without knowing whether the data are two-way-travel time or depth, migrated or unmigrated, the different attributes in the display and so on,very little can be achieved by interpretation.
- Over/underinterpreting: being too excited and interpreting faults even where there are none, or not picking them up where faults are prominent, is common with novices. See Chapter 2.3 for evidence to look for when picking faults in seismic data.

So, now let us look again at the seismic section in Figure 48.1 and find a suitable interpretation (Figure 48.4). First, observe the seismic sections and identify the fitness of the data. The extreme left and right of the seismic section have high noise content. Interpretation will be very subjective in those zones. However, we will try to interpret logically as much as possible from the data. Now, when we look at the faults, they are neither planar nor continuous. So, we honour the data and interpret the faults as curvilinear discontinuous 'sticks' on this section. Chapter 2.3 reviews how to identify faults. There are many more faults in this seismic section that cannot be seen as they are below seismic resolution. We must not try to identify those, as this might be an over-interpretation and, erroneous. Note that the faults stop suddenly at the bottom of the seismic section. Remember then that this is a pure shale section, which can accommodate stresses without being faulted. This is an instance where we use our knowlwdge of the lithology in the area.

Let us interpret two conspicuous horizons (blue and green lines in Figure 48.4). The horions give another

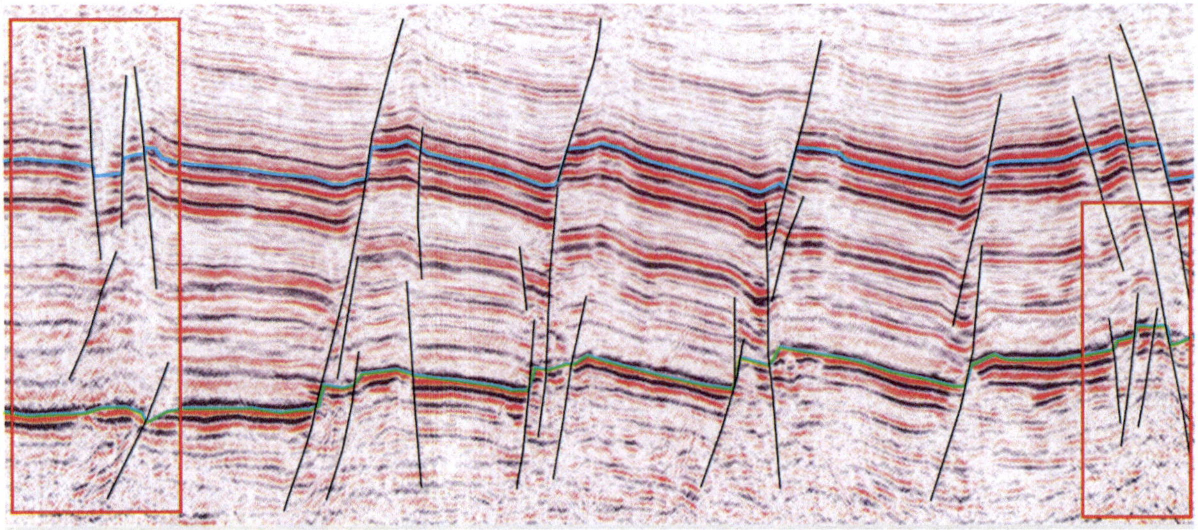

Figure 48.4 The most probable interpretation for the seismic section in Figure 48.1. Note high noise zones in red rectangles. Interpreted horizons in blue and green. Data from TGS-NOPEC, 2003.

angle to interpreting this seismic section. Few more minor faults can be picked from sudden changes in dip or small offsets of the horizons (see Chapter 2.3). Also note that the overall thickness of the section increases towards the left of the seismic image. Note the throws on the individual faults. Since the faults are mostly discontinuous and frequently branch, gradual decrease of throw towards younger stratigraphy is difficult to correlate. Also notice that the faults die out both at the top and bottom of the seismic section. So, the maximum throw will be, as expected (Mukherjee 2014b), somewhere in the middle of the fault length and decreases at its two ends, i.e., both upwards and downwards. This can be noted in the second and third fault zone from the left. These fault zones have at least one fault plane that offsets both the high amplitude carbonate horizons and continues for significant distance above and below them. Both the fault planes are continuous i.e. unsegmented or unbroken. Now, these are not the final or the only or even the best interpretations. There may be more faults picked if we have more information.

The coeditors and authors have tried throughout to provide examples of varied structural interpretations on seismic data. But this is not exhaustive. See Jaiswal *et al.* (2017). The reader can look at the Virtual Seismic Atlas (http://www.seismicatlas.org/) for more examples. Also, there could be many structures that have not so far been imaged in seismic sections! Experience, imagination (within logical limit), and practice are crucial in seismic interpretation.

References

Alcalde, J., Bond, C.E., Johnson, G. *et al.* 2017. The importance of structural model availability on seismic interpretation. *Journal of Structural Geology*, **97**, pp.161–171.

Bacon, M., Simm, R. and Redshaw, T. 2007. *3-D Seismic Interpretation*. Cambridge University Press, Cambridge.

Baudon, C. and Cartwright, J. 2008a. 3D seismic characterisation of an array of blind normal faults in the Levant Basin, Eastern Mediterranean. *Journal of Structural Geology*, **30**, 746–760.

Baudon, C. and Cartwright, J. 2008b. Early stage evolution of growth faults: 3D seismic insights from the Levant Basin, Eastern Mediterranean. *Journal of Structural Geology*, **30**, 888–898.

Baudon, C. and Cartwright, J. 2008c. The kinematics of reactivation of normal faults using high resolution throw mapping. *Journal of Structural Geology*, **30**, 1072–1084.

Bond, C.E. 2015. Uncertainty in structural interpretation: Lessons to be learnt. *Journal of Structural Geology*, **74**, 185–200.

Brown, A.R. 2005. Pitfalls in 3D seismic interpretation: Keynote presentation at the 11th Annual 3-D Seismic Symposium, Denver. *The Leading Edge*, **24**, 716–717.

Cartwright, J., Bouroullec, R., James, D. and Johnson, H. 1998. Polycyclic motion history of some Gulf Coast growth faults from high-resolution displacement analysis. *Geology*, **26**, 819–822.

Hall, M. 2013. Colouring maps. https://www.agilegeoscience.com/blog/2013/8/29/colouring-maps.html; last accessed 28 September 2017.

Herron, D.A. 2000. Pitfalls in seismic interpretation: Depth migration artifacts. *The Leading Edge*, **19**, 1016–1017.

Hongxing, G. and Anderson, J.K. 2007. Fault throw profile and kinematics of normal fault-conceptual models and geologic examples. *Geological Journal of China Universities* **13**, 75–88.

Hudleston, P.J. 1986. Extracting information from folds in rocks. *Journal of Geological Education*, **34**, 237–245.

Jaiswal, P., Holman, R., and Grammer, M. 2017. Pitfalls in three-dimensional seismic interpretation: Footprints of an irregular source-receiver layout. *AAPG Bulletin*, **101**, 1747–1751.

Lark, R.M., Thorpe, S., Kessler, H. and Mathers, S.J. 2014. Interpretative modelling of a geological cross section from boreholes: sources of uncertainty and their quantification. *Solid Earth*, **5**, 1189.

Mansfield, C.S. and Cartwright, J.A. 1996. High resolution fault displacement mapping from three-dimensional seismic data: evidence for dip linkage during fault growth. *Journal of Structural Geology*, **18**(2), 249–263.

Marfurt, K.J. and Alves, T.M. 2014. Pitfalls and limitations in seismic attribute interpretation of tectonic features. *Interpretation*, **3**, SB5–SB15.

Matteo. 2013. An introduction to color palettes for seismic amplitude – teaser. https://mycarta.wordpress.com/2013/04/20/an-introduction-to-color-palettes-for-seismic-amplitude-teaser/; last accessed 28 September 2017.

Mukherjee, S. 2013. *Deformation Microstructures in Rocks*. Springer.

Mukherjee, S. 2014a. *Atlas of Shear Zone Structures in Meso-scale*. Springer.

Mukherjee, S. 2014. Review of flanking structures in meso- and micro-scales. *Geological Magazine*, **151**, 957–974.

Mukherjee, S. 2015. *Atlas of Structural Geology*, Elsevier. Amsterdam.

Mukherjee, S., Agarwal, I. Submitted. Shear heat model for gouge free dip-slip listric normal faults. *Solid Earth Sciences*.

Mukherjee, S. and Koyi, H.A. 2010a. Higher Himalayan shear zone, Sutlej section – structural geology & extrusion mechanism by various combinations of simple shear, pure shear & channel flow in shifting modes. *International Journal of Earth Sciences*, **99**, 1267–1303.

Mukherjee, S. and Koyi, H.A. 2010b. Higher Himalayan shear zone, Zanskar section – microstructural studies & extrusion mechanism by a combination of simple shear & channel flow. *International Journal of Earth Sciences*, **99**, 1083–1110.

Mukherjee, S., Punekar, J., Mahadani, T. and Mukherjee R. (2015) A review on intrafolial folds and their morphologies from the detachments of the western Indian Higher Himalaya. In: Mukherjee, S., Mulchrone, K.F. (Eds) Ductile Shear Zones: From Micro- to Macro-scales. Wiley Blackwell pp. 182–205.

Nanda, N.C. 2016. *Seismic Data Interpretation and Evaluation for Hydrocarbon Exploration and Production: A Practitioner's Guide*. Springer.

Ramsay, J.G. 1967. *Folding and Fracturing of Rocks*. McGraw-Hill, New York.

Simmon, R. 2013. Subtleties of Color (Part 2 of 6). http://earthobservatory.nasa.gov/blogs/elegantfigures/2013/08/06/subtleties-of-color-part-2-of-6/; last accessed 28 September 2017.

TGS-NOPEC 2003. Problems in exploration – Part 1. Need to find fault? *AAPG Explorer*, **11**.

Torvela, T. and Bond, C.E. 2011. Do experts use idealised structural models? Insights from a deepwater fold–thrust belt. *Journal of Structural Geology*, **33**, 51–58.

Tucker, P.M. and Yorston, H.J. 1973. *Pitfalls in Seismic Interpretation* (No. 2). SEG Books, Society of Exploration Geophysicists, Tulsa, OK, pp. 1–50.

Index

Atlas of Structural Geological Interpretation from Seismic Images, First Edition. Edited by Achyuta Ayan Misra and Soumyajit Mukherjee.

Wiley The manufacturer's authorized representative according to the EU General Product Safety Regulation is Wiley-VCH GmbH, Boschstr. 12, 69469 Weinheim, Germany, e-mail: Product_Safety@wiley.com.

Printed and bound by CPI Group (UK) Ltd, Croydon, CR0 4YY
15/09/2026
02217280-0001